Springer Series in Optical Sciences

Volume 174

For further volumes:
http://www.springer.com/series/624

Springer Series in Optical Sciences

The Springer Series in Optical Sciences, under the leadership of Editor-in-Chief William T. Rhodes, Georgia Institute of Technology, USA, provides an expanding selection of research monographs in all major areas of optics: lasers and quantum optics, ultrafast phenomena, optical spectroscopy techniques, optoelectronics, quantum information, information optics, applied laser technology, industrial applications, and other topics of contemporary interest.
With this broad coverage of topics, the series is of use to all research scientists and engineers who need up-to-date reference books.
The editors encourage prospective authors to correspond with them in advance of submitting a manuscript. Submission of manuscripts should be made to the Editor-in-Chief or one of the Editors. See also www.springer.com/series/624

Rashid A. Ganeev

Nonlinear Optical Properties of Materials

Springer

Rashid A. Ganeev
Quantum Optics and Laser
Science Group
Department of Physics
Imperial College London
London
UK

ISSN 0342-4111 ISSN 1556-1534 (electronic)
ISBN 978-94-007-6021-9 ISBN 978-94-007-6022-6 (eBook)
DOI 10.1007/978-94-007-6022-6
Springer Dordrecht Heidelberg New York London

Library of Congress Control Number: 2012954019

Printed on acid-free paper

Springer is part of Springer Science+Business Media (www.springer.com)

Rashid A. Ganeev

Nonlinear Optical Properties of Materials

Springer

Rashid A. Ganeev
Quantum Optics and Laser
Science Group
Department of Physics
Imperial College London
London
UK

ISSN 0342-4111 ISSN 1556-1534 (electronic)
ISBN 978-94-007-6021-9 ISBN 978-94-007-6022-6 (eBook)
DOI 10.1007/978-94-007-6022-6
Springer Dordrecht Heidelberg New York London

Library of Congress Control Number: 2012954019

Printed on acid-free paper

Springer is part of Springer Science+Business Media (www.springer.com)

To my parents, wife, son, and daughter

Preface

The interest in nonlinear optics was initially mostly related to practical applications of those findings, such as efficient frequency conversion of laser radiation in "nonlinear" crystals toward the visible and ultraviolet ranges. Later it became a useful method of study of the nonlinear optical response of the media during interaction with ultrafast laser pulses. Currently, it is a well-developed branch of optics where various nonlinear optical processes are studied in many laboratories worldwide.

The topics presented in this book are mostly concerned with experimental research of the nonlinear optical characteristics (such as nonlinear refraction and absorption) of various media, low- and high-order harmonic generation in different materials, and formation and nonlinear optical characterization of clusters. These topics became the area of interest from the very beginning of my career. The motivation in writing this book was based on my willingness to show the inter-connection between these areas of nonlinear optics and demonstrate the experimental achievements in this field.

I carried out these studies in collaboration with many researchers from Japan, India, Uzbekistan, France, Canada, Germany, Ukraine, United Kingdom, Russia, Spain, and other countries. The studies presented in this book are based on my collaboration with those researchers and could not be realized without their generous efforts. I would like to thank all of them for long lasting scientific ties and discussions of various aspects of nonlinear optics.

The book is generally based on studies carried out in the Institute of Electronics (Tashkent, Uzbekistan), Institute for Solid State Physics, University of Tokyo (Kashiwa, Japan), Raja Ramanna Centre for Advanced Technology (Indore, India), Institut National de la Recherché Scientifique (Montreal, Canada), and Imperial College London (London, United Kingdom). I thank T. Usmanov, H. Kuroda, P. D. Gupta, P. A. Naik, T. Ozaki, J. P. Marangos, and J. W. G. Tisch for the fruitful discussions and support during our joint collaborative studies.

I would like to emphasize the role of my colleagues from Institute of Electronics, Uzbekistan, Tashkent, where I started and initiated some of these studies. I appreciate the collaboration with I. A. Kulagin, V. V. Gorbushin, S. T. Khudaiberganov, S. R. Kamalov, V. I. Redkorechev, A. A. Gulamov, G. S. Boltaev, N. K. Satlikov, R. I. Tugushev, I. A. Begishev, I. A. Khusainov, and my other colleagues from Laser-Matter Interaction Laboratory who stayed at the beginning of this research and currently continue the nonlinear optical studies of various media.

An important component of this research is the collaboration with many researchers involved in these studies outside the above-mentioned laser centers. I enjoyed the discussions with A. L. Stepanov (Physics-Technical Institute, Kazan, Russia), M. Danailov (ELETTRA, Trieste, Italy), H. Zacharias (Westfäliche Wilhelms-Universität, Münster, Germany), M. K. Kodirov (Samarkand State University, Samarkand, Uzbekistan), N. V. Kamanina (Vavilov State Optical Institute, Saint Petersburg, Russia), M. Castillejo (Instituto de Química Física Rocasolano, Madrid, Spain), D. B. Milošević (University of Sarajevo, Sarajevo, Bosnia and Herzegovina), A. A. Ishchenko (Institute of Organic Chemistry, Kiev, Ukraine), M. Lein (Leibniz Universität Hannover, Hannover, Germany), E. Fiordilino (University of Palermo, Palermo, Italy), and V. Strelkov (General Physics Institute, Moscow, Russia) regarding the past, present, and future joint experimental and theoretical studies of the nonlinear optical properties of materials.

I am also indebted to the members of those and other teams for the fruitful collaboration. Especially, I would like to emphasize the contribution of A. I. Ryasnyansky, P. V. Redkin, M. Suzuki, M. Baba, L. B. Elouga Bom, H. Singhal, J. A. Chakera, M. Tudorovskaya, and C. Hutchison during these nonlinear optical studies.

Finally, I would like to thank my family (Lida, Timur, and Dina) for their patience during my long trips and support at all stages of my life.

London, October 2012 Rashid A. Ganeev

Contents

Introduction

The fundamental properties of materials play an extremely important role in developing new technologies in various areas of our life. Among the important properties of various surrounding materials, the nonlinear optical features of transparent samples are of great importance, since their applications can considerably improve the characteristics of different optoelectronics devices.

The consideration of the fundamentals of nonlinear optics caught the attention from the very beginning of laser-matter studies, when pioneering work by Franken et al. [1] had opened the doors to the search for new nonlinear optical responses of matter during propagation of strong electromagnetic wave. There are plenty of textbooks, which mostly focus on the theoretical aspects of nonlinear optics (for example [2–8]), with few exceptions [5, 6] where the experimental studies were thoroughly analyzed. During the last few years many publications have appeared on different aspects of the nonlinear optical properties of various materials. In particular, there were many book series containing separate chapters dedicated to the experimental studies of the nonlinear optical characteristics of media (see in particular [9–12]). It seems timely to show the current status of the experimental work carried out in this field of optics during the last decade and discuss the main trends and advances of some newly developed branches of experimental nonlinear optics.

Nonlinear optical properties of media can be applied in various areas of science and technology. An example of such applications is optical limiting. Optical limiting effect, which allows a stabilization of propagation of intense light through the medium, can be extremely useful for protection of the eyes and sensitive photonic registrars against overexposure. This effect was observed in a few materials (in particular, fullerenes) and it was admitted that the application of some of them could be useful for this newly developed technique. To define suitable materials for these applications, one has to carefully analyze the nonlinear optical characteristics of various media, such as the nonlinear refractive indices, coefficients of nonlinear absorption, saturation absorption intensities, etc.

Knowing the nonlinear optical parameters of materials is also important for describing the propagation effects, self-interaction of intense laser pulses, and optimization of various nonlinear optical processes. Among those processes one can admit the importance of the studies on the frequency conversion of coherent laser sources. Currently, the wavelengths of these sources are fixed due to the specific properties of those lasers. In the meantime, the tuning of the wavelength of coherent radiation allows the use of coherent radiation at different wavelengths for various applications in medicine, biology, chemistry, physics, etc. Change in the wavelength of well-established sources of laser radiation, which currently occupy restricted spectral ranges (such as 1064 and 1054 nm in the case of Nd:glass and Nd:YAG lasers and 800 nm in the case of Ti:sapphire lasers), using nonlinear optical methods offers many new interesting effects. This area of research has a long history starting from the first observation of frequency conversion of laser light reported in 1961 [1]. The enormous achievements in this area include the creation of efficient crystalline converters of laser radiation toward visible and ultraviolet ranges, parametric up-conversion toward the near-infrared range (1–5 μm), generation of high-order harmonics of intense laser pulses through frequency conversion in gases and plasmas allowing the creation of coherent sources in vacuum ultraviolet and extreme ultraviolet ranges, etc.

The area of interest for nonlinear optical characterization of materials is also closely related to the new field of nanostructures formation during laser-matter interaction. The formation of nanoparticles, nanowires, and other exotic nanostructures during laser ablation of different materials, formation of plasma plumes containing clusters, characterization of their structural, morphological, optical, and nonlinear optical properties, use of nanostructures for various applications, such as, for example, high-order harmonic generation, and many other features of newly developed and commercially available nanoparticles open the doors for their use in photonics, electronic industry, medicine, and other scientific and industrial applications.

This book describes the achievements in all the above-mentioned areas (nonlinear optical characterization of media, low- and high-order harmonic generation of laser radiation, and formation, characterization of nanostructures and their applications for frequency conversion). We will discuss the results of the studies carried out in different spectral and temporal ranges, as well as show their applications and "self-interaction" of these areas of research carried out in different branches of nonlinear optics. Recent achievements, such as new approaches in single-shot characterization of nonlinear refractivities of materials, nanoparticle-contained plasma formation, and efficient conversion of laser light into laser plumes towards the shorter wavelength range, are among many topics, which will be analyzed from the point of view of their further applications. We also discuss the comparative studies and future developments in these fields.

The novelty of the approach developed in this book is related to the interconnection of the studies with each other. We show how the nonlinear optical analysis of materials leads to improvement in their high-order nonlinear optical response during the interaction with strong laser fields. Ablation-induced nanoparticles formation is correlated with their applications as efficient sources of coherent short-wavelength photons. From the other side, recent achievements in harmonic generation in plasmas are closely related with the knowledge of the properties of materials in the laser plumes. Finally, all these studies are interconnected with the low-order nonlinear optical features of various materials. The studies presented in this book are combined into three branches: low- and high-order harmonic generation of laser radiation in various media (with emphasis in the latter case of the advanced properties of plasma harmonic experiments carried out during the last few years), low-order nonlinear optical characterization of materials (colloids, transparent materials, dyes, semiconductors, liquids, fullerenes, plasmas, etc), and nanoparticles studies (laser-induced, ion bombardment, and chemical formation, characterization of low-order nonlinearities, application for optical limiting and high-order harmonic generation). Overall, this book shows the interconnection between different fields of material studies using nonlinear optical methods.

The book is organized as follows. In Chap. 1, the low-order harmonic generation is analyzed using various media (nonlinear crystals, fullerene- and metal-doped organic films, colloidal metal suspensions, organic dyes, plasma and ionic media, gases and surfaces). High-order harmonic generation from the plasma plumes produced during ablation of metal and organic targets is discussed in Chap. 2. In Chap. 3, studies on nonlinear refraction and nonlinear absorption of media are presented. Laser-ablation-induced cluster formation is analyzed in Chap. 4. Low-order nonlinear optical properties of clusters are discussed in Chap. 5. Finally, in Chap. 6, we describe the applications of nanoparticle-contained plasmas for high-order harmonic generation of ultrashort laser pulses.

References

1. P.A. Franken, A.E. Hill, C.W. Peters, G. Weinreich, Phys. Rev. Lett. **7**, 118 (1961)
2. N. Bloembergen, *Nonlinear Optics* (Benjamin, New York, 1964)
3. Y.R. Shen, *The Principles of Nonlinear Optics* (Wiley-Interscience, New York, 2003)
4. J.F. Reintjes, *Nonlinear Optical Processes in Liquids and Gases* (Academic Press, Orlando, 1984)
5. F. Zernike, J.E. Midwinter, *Applied Nonlinear Optics* (Wiley, New York, 1973)
6. M.G. Kuzyk, C.W. Dirk (eds.), *Characterization Techniques and Tabulations for Organic Nonlinear Materials* (Marcel Dekker Inc., New York, 1998)
7. R.W. Boyd, *Nonlinear Optics* (Academic Press, New York, 2008)
8. P.P. Banerjee, *Nonlinear Optics Theory, Numerical Modelling, and Applications* (CRC Press, New York, 2004)
9. J.M. Hales, J.W. Perry, *in Introduction to Organic Electronic and Optoelectronics Materials and Devices*, ed. by S.A. Sun, L.P. Dalton (CRC, Berlin, 2008), p. 513
10. M. Lewenstein, A. L'Huillier, *in Strong Field Laser Physics, Springer Series in Optical Sciences*, vol. 134, ed. by T. Brabec, (Springer, Berlin, 2008), p. 147
11. S.E. Chin, *in Femtosecond Laser Filamentation. Springer Series on Atomic Optical and Plasma Physics*, vol. 55 (Springer, Berlin, 2010), p. 77
12. I.N. Mihailescu, J. Hermann, *in Laser Processing of Materials: Fundamentals, Applications and Developments. Springer Series in Materials Science*, vol. 139, ed. by P. Schaaf (Springer, Berlin, 2010), p. 49

Chapter 1
Low-Order Harmonic Generation of Laser Radiation in Various Media

One of the main objectives of optics and quantum electronics is a search of new sources of coherent radiation in different spectral ranges. A commonly used technique is the frequency conversion (FC) of polarized radiation of well-developed laser sources via nonlinear optical processes. Presently, along with widely used nonlinear crystalline media, the gases, plasma plumes, nanoparticles, and reflection from surface have been successfully demonstrated as attractive nonlinear media for harmonic generation (HG) of laser radiation. Gaseous sources are the subject of great interest in last three decades due to their wide applications as nonlinear optical media. They permit to convert the laser frequency into ultraviolet (UV), extreme ultraviolet (XUV), and infrared (IR) ranges using different parametric processes. Conversion efficiencies attained in such a way are favour to create the coherent XUV sources with intensities higher than that of synchrotron radiation. Another important application of HG in gases is the generation of attosecond pulses.

At the same time, a search of new nonlinear media allows expanding the FC in different spectral ranges. An interest in FC capabilities of various media also connected with the studies of their nonlinear optical properties that can be used in different applications (optical switching, optical limiting, etc.). The studies of surface enhanced nonlinear optical properties of various materials using FC technique took an attention during last two decades. An analysis of laser plasma characteristics through the harmonic generation studies is another application of FC process.

Various HG processes were recently examined in different media. In order to enhance optical third-harmonic generation (THG) of laser radiation in polymer films, linear and nonlinear optical properties of polymer films containing ethyl-violet with varying thickness were examined in [1]. The THG intensity increased monotonically with thickness, as opposed to the theory for non-phase-matched THG in films

We discuss the studies of low-order harmonic generation in various media (crystals, plasma, ion beams, fullerenes, gases, colloidal metals, dye vapours, solid surfaces, and metal-doped polymers). We examine various processes of frequency conversion optimisation and conversion efficiency limitation in these media.

R. A. Ganeev, *Nonlinear Optical Properties of Materials*,
Springer Series in Optical Sciences 174, DOI: 10.1007/978-94-007-6022-6_1,

thicker than the coherence length. By measuring the refractive indices, it was found that the difference between the indices of the fundamental and the third-harmonic was smaller in thicker films. Thus, the coherence lengths for thicker films were evaluated to be larger than those for the thin films, explaining the monotonic increase of the THG signal with thickness. THG of femtosecond laser pulses in sputtered nanocrystalline TiO_2 thin films was investigated in [2]. Using layers of graded thickness, the dependence of THG on the film parameters is studied. The maximum THG signal is observed at a thickness of 180 nm. The corresponding conversion efficiency is 26 times larger compared to THG at the air-glass interface. The orientation dependency of nonlinear optical effects, including the second harmonic generation (SHG) and THG, as well as two-photon luminescence, in semiconductor nanowires were reported in [3]. The mismatch of dielectric constants between nanostructures and their environment governs the rise of optical nonlinearities causing SHG even in materials with a high symmetry crystal lattice that would not generate second harmonic in the bulk state. Due to the depolarisation effects, the intensity of the optical electric field inside illuminated nanowires depends dramatically on their orientation related to the exciting light polarization.

Noncollinear generation of second and third harmonics in three-dimensional photonic crystals of artificial opals is experimentally realized in [4] and interpreted in terms of nonlinear diffraction. A factor-of-two enhancement of second- and third-harmonic generation intensity in diffraction directions is observed due to effective phase matching provided by photonic crystal periodicity. The enhanced THG was experimentally achieved from hole-array in gold film [5]. The surface plasmon polarition excitations, considered as the major contribution to this THG enhancement, were studied by a spectrum-resolved femtosecond laser system. The momentum matching conditions between the fundamental and harmonic waves are well established, giving results of a series of clear THG beam spots with respect to different excitation wavelengths. This efficient THG beaming from the pure metallic structure would considerably broaden the research in the field of nonlinear plasmonics. The second and third harmonics in tellurite microstructured fibers pumped by a 1557 nm femtosecond fiber laser were demonstrated in [6]. The intensities of the second and third harmonics are enhanced by increasing the nonlinear coefficient of the tellurite microstructured fiber.

Generation of low-order harmonics (third and fifth) of the fundamental radiation of a Q-switched Nd:YAG laser (1064 nm, pulse 15 ns) was observed in a CaF_2 laser ablation plume [7]. The ablation process is triggered by a second Q-switched Nd: YAG laser operating at 532 or 266 nm. In the scheme employed, the fundamental laser beam propagates parallel to the target surface at controllable distance and temporal delay, allowing to the probing of different regions of the freely expanding plume. The intensity of the harmonics is shown to decrease rapidly as the distance to the target is increased, and for each distance, an optimum time delay between the ablating laser pulse and the fundamental beam is found. In situ diagnosis of the plume by optical emission spectroscopy and laser-induced fluorescence serves to correlate the observed harmonic behaviour with the temporally and spatially resolved composition and velocity of flight of species in the plume. It is concluded that harmonics are selectively generated by CaF species through a two-photon resonantly enhanced sum-mixing process. Polar molecules have been shown to be the dominating species for harmonic generation in an ablation plume. Implications of these results for the generation of high harmonics in strongly polar molecules, which can be aligned in the ablation plasma are discussed.

In this Chapter, we discuss the studies of low-order harmonic generation in various media (crystals, plasma, ion beams, fullerenes, gases, colloidal metals, dye vapours, solid surfaces, and metal-doped polymers). We examine various processes of FC optimisation and conversion efficiency limitation in these media.

1.1 Nonlinear Crystals

The crystals are the media where the nonlinear optics has started from in 1961 [8]. Currently, new nonlinear crystals grow and characterize frequently, while the "old" crystals continue to be a subject of studies and applications. In particular, a novel multipass scheme for enhancement of second harmonic generation in barium borate crystal reported in [9] allows the 2.5 enhancement of second harmonic yield. THG of high-intensity, sub-100 fs idler pulses from a Ti:sapphire-laser-pumped optical parametric amplifier is demonstrated in [10] by using a single nonlinear crystal of BiB_3O_6. Maximum internal energy conversion as high as 11 % from the fundamental to the third harmonic is achieved by phase- and group-velocity matching for the direct cubic nonlinear process together with the velocity-mismatched cascading quadratic processes. High-efficient THG at 355 nm based on $La_2CaB_{10}O_{19}$ (LCB) was measured in [11]. For type I LCB, a 355 nm UV light output of 5.0 mW corresponding to the conversion efficiency of 28.3 % was generated under a picosecond Nd:YAG laser, and 16 W with the efficiency of 17.5 % was generated under a nanosecond 1064 nm pumping source. The third-order nonlinear optical properties of the lithium potassium borate ($LiKB_4O_7$) single crystal have been investigated by means of the rotational Maker fringe technique using Nd:YAG laser at 1064 nm working in picosecond regime [12]. The value of the third-order nonlinear optical susceptibility was found to be about $1.4 \times 10^{-21}\,m^2\,V^{-2}$ that is one order higher than that of fused silica. Large third-order susceptibility and efficient THG was also reported in centrosymmetric Cu_2O crystals [13]. High-efficient ultraviolet laser generation was analysed in a nonlinear optical crystal $BaAlBO_3F_2$ (BABF) [14]. 355 nm UV light was generated by using a type I BABF crystal as a sum-frequency mixing of the fundamental light and the second harmonic from a picosecond Nd:YAG laser operated at a repetition rate of 10 Hz with a pulse duration of 20 ps. The conversion efficiency was 26.4 %. All these recent studies have demonstrated the interest in further developments of new nonlinear crystals. In the meantime, "old" crystals (KDP, ADP, BBO, LBO, etc) for a long time attracted the attention due to their various interesting features.

In this section we analyse the studies of HG conversion efficiencies in KDP and ADP nonlinear crystals, pulse compression in such processes, and tuneable IR and UV radiation generation via parametric waves generation.

1.1.1 Efficient Fifth Harmonic Generation

Cascade generation of laser radiation harmonics (up to the fifth order) was achieved for the first time using KDP crystals more than four decades ago [15]. The frequency of laser radiation was doubled in the first two crystals, while in the third one the fourth harmonic and the fundamental waves were mixed. This method of FC of the radiation of Nd:glass and Nd:YAG lasers toward the far UV (~200 nm) is believed to be most efficient. It should be noted that phase-matched generation of the fifth harmonic of Nd:glass laser radiation was achieved at negative Celsius temperatures in KDP and ADP crystals. However, relatively low-energy lasers with small sizes of output beams were used in those studies, and the conversion efficiency (η) of fifth HG (5HG) has been achieved at the level of a few percents.

In this subsection we describe the studies of cascade harmonic generation of a wide-aperture Nd:glass laser in KDP and ADP crystals [16]. Cascade second, fourth and fifth harmonics generation in KDP and ADP crystals can be achieved using two frequency conversion schemes: (1) $o_1o_1 \rightarrow e_2, o_2o_2 \rightarrow e_4, o_1o_4 \rightarrow e_5$ and (2) $o_1e_1 \rightarrow e_2, o_2o_2 \rightarrow e_4, o_1o_4 \rightarrow e_5$. Here o and e mean the ordinary and extraordinary polarization waves, respectively, and the subscripts refer to the order of harmonic. Since 5HG process implies the mixing of waves according to Mainley-Rowe relationship, the energies of the mixed waves have to be maintained within the ratio of 4:1 in case of plane waves of fourth harmonic and fundamental radiation. This relationship underlines the importance of high-efficient fourth HG (4HG) for the following wave mixing.

Experiments on cascade SHG, 4HG, and 5HG were carried out using a wide-aperture (45 mm) Nd:glass laser. The laser output characteristics were as follows: $I = 5 \times 10^9\,\mathrm{W\,cm^{-2}}$, $t = 0.5\,\mathrm{ps}$, $\lambda = 1064\,\mathrm{nm}$, $\theta = 55\,\mu\mathrm{rad}$ (I, t, λ, and θ are the intensity, pulse duration, wavelength, and radiation divergence, respectively). The spatial distribution of the intensity was fitted well by a hyper-Gaussian function of sixth power ($N = 6$, $M = 2$ in intensity distribution of $I(\mathrm{r,t}) = I_o \exp[-(r/a)^N - (t/\tau)^M]$, where a is the beam radius, τ is the half of the pulse width at 1/e level, N and M are the parameters of modulation). Such spatial distribution was proven to be optimal for the efficient FC in crystals [17]. The output radiation was directed onto three nonlinear crystals positioned one by one along the axis of beam propagation. In order to achieve 90° phase matching conditions for 5HG, the third nonlinear crystal was placed in a cryostat.

A 17.5 mm long KDP crystal with an aperture of 50 × 50 mm was cut to achieve type-II synchronism for SHG. η in that crystal was 87×10^{-2}. A 50 × 50 × 19 mm type-I KDP crystal was used for 4HG, and the $\eta = 75 \times 10^{-2}$ of this process was achieved. The 4HG conversion efficiency was improved by substantial adjusting of the length of first crystal (10 mm). The highest 4HG conversion efficiency ($\eta = 92 \times 10^{-2}$) was achieved in that case at second harmonic radiation intensity of $1.5 \times 10^9\,\mathrm{W\,cm^{-2}}$.

The 5HG experiments were carried out using the second scheme ($o_1e_1 \rightarrow e_2$, $o_2o_2 \rightarrow e_4, o_1o_4 \rightarrow e_5$) in which the SHG crystal was rotated on angle of 50.8° relative to polarization of pump radiation in order to achieve the optimal ratio of fundamental and fourth harmonic intensities in the crystal-mixer. ADP was chosen as a mixer due to smaller nonlinear absorption coefficient at fourth harmonic frequency (0.5×10^{-9} cm W^{-1} and 0.35×10^{-9} cm W^{-1} in the cases of KDP and ADP, respectively) and taking into account that the 5HG phase matching temperature in ADP (−70 °C) was higher than that in KDP (−140 °C).

A cooling system, in which the heat exchange with the crystal took place via gaseous medium, was found to be the best. An ADP crystal was mounted in an additional optical cell with quartz windows filled with nitrogen at a pressure of 8×10^5 Pa. This cell was placed into a cryostat with two quartz windows. Using such cooling system, a suitable temperature could be maintained within 0.1°C accuracy over the cross-section of 50 × 50 mm crystal. The maximal 5HG ($\lambda = 211$ nm), efficiency (19×10^{-2}) was achieved at fundamental intensity of 2×10^9 W cm^{-2} and phase-matching temperature of −67.5 °C. The nonlinear absorption of ultraviolet radiation in ADP was proven to be a main factor of limitation of 5HG conversion efficiency at higher intensities of laser radiation.

1.1.2 Laser Pulse Compression During Second Harmonic Generation in KDP

The efficiency of frequency doubling of the pulses of the duration ranging from hundreds femtoseconds to a few picoseconds is hindered by the influence of group velocity dispersion of the interacting pulses in nonlinear crystal. This process restricts the efficiency of FC of sub-picosecond pulses. The method proposed in [18] for delaying of the *e* wave relative to the *o* wave of the fundamental radiation when it reaches a KDP crystal (where the type-II interaction takes place) has made it possible to increase the energy conversion efficiency and also to compress the converted pulses. The method is as follows. If *e* pulses are delayed relative to *o* pulses by τ (where τ is the pulse duration), then at some radiation intensities the length of interaction in the negative nonlinear crystal is doubled and this peculiarity increases the efficiency of nonlinear optical conversion. Below we discuss the application of this method for both the increase of SHG conversion efficiency in type-II KDP crystal and for the compression of pulses ($\tau = 5$ ps) of Nd:glass laser [19].

Linearly polarized radiation ($\lambda = 1054$ nm, $E = 10$ mJ) was passed through the quarter-wave plate and was divided by a KDP polarizing prism into two beams with orthogonal polarization. The beams were then reflected by movable mirrors and combined in the same prism so that they passed through a nonlinear crystal (KDP, type-II, 20 × 20 × 40 mm) along the same direction. The second harmonic radiation was filtered from fundamental radiation and directed in single-shot autocorrelator for the measurement of pulse duration.

An optimal ratio between the intensities of interacted *o* and *e* waves $k = I_o/I_e$ in the crystal should be close to the ratio of the group lengths (L_o/L_e), where $L_o = \tau/[1/(\upsilon_{e,f}) - 1/(\upsilon_{e,2\omega})]$ and $L_e = \tau/[1/(\upsilon_{o,f}) - 1/(\upsilon_{e,2\omega})]$ [20]. In present studies, the ratio L_o/L_e was ~1.4. The ratio of intensities of *o* and *e* waves was kept close to this value in the experiments of pulse compression. The dependence of second harmonic pulse duration on the delay between *o* and *e* waves was measured. Since the group velocity of *e* wave ($\lambda = 1054$ nm) in the KDP crystal exceeded the group velocity of *o* wave ($\upsilon_{o,f} = 1.966 \times 10^{10}$ cm s^{-1}, $\upsilon_{e,f} = 2.019 \times 10^{10}$ cm s^{-1}), the delay of *o* wave relative to *e* wave simply resulted in a steep fall of the conversion efficiency because the waves did not overlap in the crystal. An opposite situation (the delay of *e* wave relative to *o* wave) caused the compression of converted radiation pulses. There was an optimal delay between the pulses with orthogonal polarization. The duration of second harmonic pulses was varied from 5 ps (in the case when *e* and *o* waves coincided at the entry of the nonlinear crystal) to 2 ps for $\Delta\delta/2 = 1$ mm ($\Delta\delta$ is the difference of the optical paths of *e* and *o* waves).

Figure 1.1 shows the dependence of second harmonic pulse duration on the intensity ($I = I_o + I_e$) of fundamental wave (at $k = 1.4$) measured at a fixed delay between the pulses (4 ps). The pulse duration decreased up to 2 ps with the increase of intensity. An optimisation of the delay between *e* and *o* waves also led to the growth of conversion efficiency from 33×10^{-2} (at zero delay between the orthogonal polarizations) to 50×10^{-2} (at 4 ps delay, for $k = 1$).

1.1.3 Parametric Waves Generation

The sources of continuously tuneable (over a range of hundreds reciprocal centimetres) coherent UV radiation can be useful in spectroscopy of atoms and molecules, biology, laser isotope separation, nonlinear spectroscopy, laser chemistry, etc. The generation of tuneable UV coherent radiation by non-resonant harmonics and sum

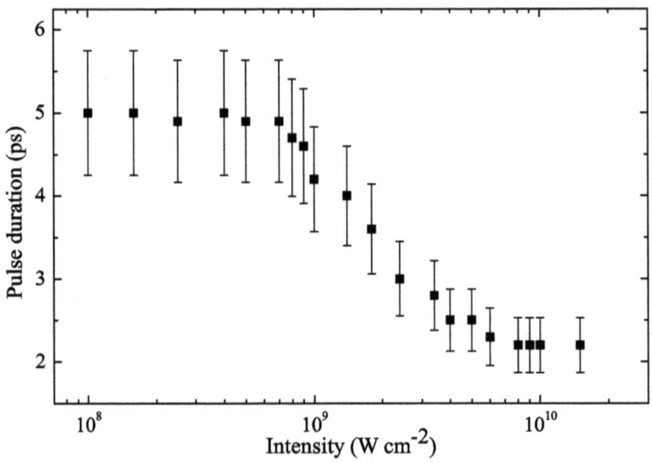

Fig. 1.1 The dependence of the duration of converted pulses on the intensity of pump radiation. Adapted from [19] with permission from Springer Science + Business Media

frequencies generation, which will be discussed in this subsection, is based on crystalline optical parametric oscillators (OPO) using as the nonlinear media [21]

Output parameters of laser radiation in these studies were as follows: $\lambda = 1054.8$ nm, radiation bandwidth 0.3 nm, pulse duration 7 ps, pulse energy 80 mJ. Laser radiation was converted to the second harmonic in type-I KDP crystal ($20 \times 20 \times 10$ mm). Conversion efficiency ($\eta = 44 \times 10^{-2}$) was saturated at $I \sim 10^{10}\,\mathrm{W\,cm^{-2}}$. Second harmonic radiation ($\lambda = 527.4$ nm) excited a double-crystalline (KDP, type-II, $20 \times 20 \times 40$ mm and $20 \times 20 \times 60$ mm) travelling-wave OPO + optical parametric amplifier.

Generation and amplification of parametric waves in double-crystalline scheme was carried out using parallel beams. The distance between crystals (70 cm) was chosen both from the point of view of the effective amplification of parametric waves, the limitation of bandwidth, and further optimal mixing of parametric IR radiation with second harmonic waves in a crystal-mixer. The bandwidth and the divergence of parametric waves were 2 nm and 3 mrad, respectively. Maximal η of parametric waves generation was 30×10^{-2} at $I = 1.4 \times 10^{10}\,\mathrm{W\,cm^{-2}}$. IR radiation was tuned from 870 to 1220 nm. This radiation than was mixed with second harmonic radiation in the crystal-mixer (KDP, type-II, $20 \times 20 \times 20$ mm). Generated UV radiation was tuned in the range of 338–366 nm.

One can note the importance of the influence of picosecond parametric IR radiation spectral bandwidth on the process of efficient mixing with second harmonic radiation. The spectral bandwidth of phase-matching conditions in the KDP crystal-mixer is 0.4 nm. At the same time the bandwidth of parametric waves was 2 nm. The narrowing of the bandwidth of parametric waves by increasing the distance between crystals in double-crystalline scheme led to a decrease of the η of tuned IR radiation. Maximal η in UV range was measured to be 12×10^{-2} (for $\lambda = 349$ nm).

1.2 Fullerene-Doped Polyimide Films, Fe- and Zn-Doped Polyvinylpyrrolidone, Colloidal Metals Solutions and Organic Dyes

HG in various media opens the doors for their characterization through nonlinear optical methods [22]. Below we show the developments in this field in the case of fullerene-doped polyimide films, Fe- and Zn-doped polyvinylpyrrolidone, colloidal metal suspensions and organic dyes. A few studies were devoted to FC of laser radiation in fullerenes [23, 24]. The spectral dependences of $\chi^{(3)}(3\omega)$ have been investigated in [24]. $\chi^{(3)}(3\omega)$ of fullerenes was found to be 8.7×10^{-11} esu. A larger value of $\chi^{(3)}(3\omega)$ (2×10^{-10} esu) has beenreported in [23]. Metal injection into the polymer structures leads to the variations of their both optical and nonlinear optical characteristics. Note the interest on metal-doped polymers with conjugated stabilizers, in particular, polyphenilenevinylene and polyvinylpyrrolidone (PVP). PVP has an advantage of accepting of high concentrations of various guest molecules and atoms without the loss of its good optical properties. An interest to colloidal metal suspensions is connected with the perspectives of their applications in optical limiting, optoelectronics, etc. Particles aggregation of such samples leads to the growth of nonlinear optical response due to the increase of local field amplitude in growing fractals.

In this section, we discuss the studies of third harmonic (TH, $\lambda = 354.7$ nm) generation using the Nd:YAG laser radiation in fullerene-doped polyimide films [25], Fe- and Zn-doped PVP solutions [26], and colloidal metals [25]. Laser radiation ($t = 35$ ps, $\lambda = 1064$ nm, pulse energy $E = 2$ mJ) was focused into the investigated media using a 25 cm focal length lens. The converted radiation ($\lambda = 354.7$ nm) was spectrally extracted from fundamental one and registered by a photomultiplier tube (PMT). C_{70}-doped 1 μm thick polyimide 6B films were coated on the surface of quartz substrates. C_{70} concentrations in these films were 0.2 wt. % and 0.5 wt. %. The sizes of metal clusters (Pt and Cu) in prepared colloidal suspensions were in the range of 10–60 nm, depending on the aggregation rate. The aqueous PVP solutions doped with iron (4.25 wt. %) and zinc (0.85 wt. %) were also used in those studies. The investigated suspensions were kept in 2 mm thick silica glass cells.

The dependence of TH intensity on fundamental radiation intensity $I_{3\omega}(I)$ for polyimide film containing 0.5 % C_{70} was analysed and the slope of this dependence was close to the cubic one till $I = 5 \times 10^{10}$ W cm^{-2}. At further growth of intensity the deviation from the cubic power dependence was observed. In the case of films containing 0.2 % C_{70}, the slope of $I_{3\omega}(I)$ dependence was found to be close to the cubic and remained constant up to maximal intensities used. The maximum η of TH for the films with 0.2 % and 0.5 % C_{70} were measured to be 10^{-6} and 6×10^{-6}, respectively. The $\chi^{(3)}(3\omega)$ of 0.5 % C_{70}-doped film was calculated to be 9×10^{-13} esu. No TH was detected in the case of pure polyimide films.

The phase-matching conditions ($\Delta k = k_3 - 3k_1 = 0$, where k_1 and k_3 are the wave numbers of fundamental and harmonic radiation) that are determined by fundamental radiation focusing parameters should be fulfilled for efficient FC in parametrical nonlinear optical processes. In these experiments a confocal parameter of focusing radiation was $b = 6$ mm, i. e. the conversion process took place at the conditions of weak focusing (in case of 1 μm thick C_{60}-doped films).

The phase mismatching and linear absorption of C_{70}-doped films did not significantly influence the FC process. At these conditions, the dependence of the TH intensity on the nonlinear medium concentration should be close to quadratic that was observed in these experiments. The dependence of the TH intensity on the fundamental radiation at small intensities should have a cubic slope. The reason of observed deviation from the cubic slope at high intensities is the Kerr–induced nonlinearity that leads to the variations of phase correlation between the fundamental and generated waves. The analysis of relations between the TH phase mismatch and conversion efficiency shows that, with the growth of fundamental intensity, the $I_{3\omega}(I)$ dependence declines from cubic one due to Kerr-induced nonlinearities responsible for the changes of the refractive indices at fundamental and TH frequencies.

The procedure of HG studies in colloidal metals was the same as the one for fullerene-doped films. The slope of $I_{3\omega}(I)$ dependence for colloidal platinum was found to be (2.8 ± 0.3) and remained approximately constant up to the maximal intensities used ($I = 1 \times 10^{11}\,\mathrm{W\,cm^{-2}}$). The maximum THG efficiencies of for colloidal platinum and colloidal copper were measured to be 7×10^{-7} and 3×10^{-7}, respectively.

Both colloidal platinum and colloidal copper possess normal dispersion at $\lambda = 354.7$ nm. This conclusion was drawn on the basis of the optical absorption spectra studies. Variations of surface plasmon frequency with the growth of aggregation determine the optical characteristics of these colloidal metals. $\chi^{(3)}(3\omega)$ of colloidal platinum and colloidal copper were measured to be 1.5×10^{-14} esu and 1×10^{-14} esu, respectively.

The slope of $I_{3\omega}(I)$ dependence for iron-doped PVP suspension (PVP:Fe) was close to 3 and remained constant in the whole intensity range. The same features were observed in the case of zinc-doped PVP suspension (PVP:Zn). The maximal η for PVP:Fe and PVP:Zn were measured to be 8×10^{-7} and 5×10^{-7}, respectively. $\chi^{(3)}(3\omega)$ of PVP:Fe and PVP:Zn were calculated to be 5×10^{-13} esu and 3×10^{-13} esu, respectively. No THG was observed in pure PVP and other metal-doped PVP suspensions (PVP:Mn, PVP:Co).

The nonlinear optical properties of vapours of organic molecules were for the first time studied in the benzene (C_6H_6) vapour [27], where radiation was converted to the vacuum UV region, and also in the acetylene (C_2H_2) vapour, where THG in the UV range was performed using a dye laser [28]. The nonlinear optical parameters of organic dye vapours were studied in [29], where various schemes for radiation frequency conversion were proposed. The FC of picosecond IR pulses in organic dye vapours was analysed in [30]. The dipole moments and nonlinear susceptibilities of organic dye vapours were calculated using the free electron model. Although this model is too simplified, it allowed one to calculate quite accurately the optical and nonlinear optical spectral parameters of the media possessing the conjugated double bonds [31].

In [30], a Nd:YAG laser emitting trains of picosecond pulses was used as a radiation source. The pulse duration was 35 ps and the energy of the pulse train was 5 mJ. Laser radiation was focused to the centre of a cell containing naphthalene vapour ($C_{10}H_8$). The cell was heated up to 220 °C to obtain the naphthalene concentration of $N \sim 10^{18}\,\mathrm{cm^{-3}}$. The pump and third harmonic radiation at 354.7 nm was coupled out through the output quartz window and directed to a spectrograph. The naphthalene was chosen because its vapours have a high enough concentration already at relatively low temperatures (~200 °C), which is required to achieve the synchronous interaction between the pump and harmonic waves. By using the energy levels, frequencies, and transition dipole moments calculated for naphthalene, the third-order nonlinear susceptibility $\chi^{(3)}(3\omega)$ was calculated for a single molecule to be $\chi^{(3)}(3\omega) = 0.42 \times 10^{-34}$ esu (for the THG of the Nd:YAG laser radiation, $\omega + \omega + \omega = 3\omega$). Third harmonic generation in isotropic media can be performed not only in the scheme $\omega + \omega + \omega \Rightarrow 3\omega$, but also in the scheme $\omega + \omega + \omega + \omega - \omega \Rightarrow 3\omega$. The fifth-order nonlinear susceptibility $\chi^{(5)}(3\omega)$ calculated according to the latter scheme was 1.3×10^{-46} esu.

Figure 1.2 shows the dependence of the TH intensity on the naphthalene vapour temperature. The maximum intensity of converted radiation was achieved at 170 °C.

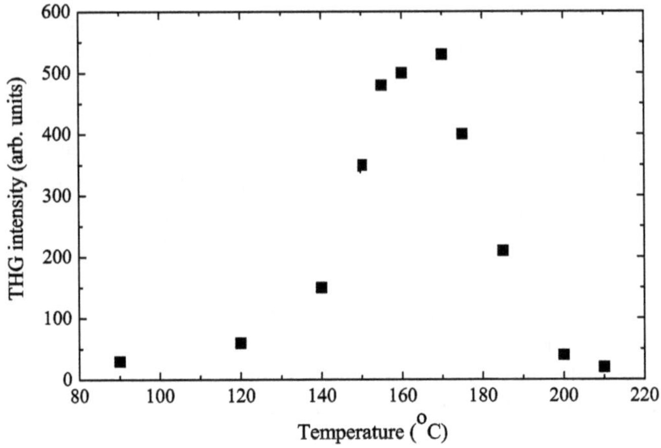

Fig. 1.2 Dependence of the TH intensity of the naphthalene vapour temperature. Adapted from [30] with permission from Elsevier

The maximum THG efficiency in naphthalene vapour was 10^{-8}. Such a dependence is typical for generation of odd harmonics upon tight focusing ($b \ll L_c$, where $b = 8$ mm is the confocal diameter of focused radiation, and $L_c = 180$ mm is the length of the nonlinear medium) in an isotropic nonlinear medium with the anomalous dispersion and also for generation of the difference frequency by neglecting the sign of the medium dispersion.

As mentioned above, THG in naphthalene vapour can occur through two channels: the direct four-photon ($\omega + \omega + \omega \Rightarrow 3\omega$) process and six-photon generation of the difference frequency ($\omega + \omega + \omega + \omega - \omega \Rightarrow 3\omega$). The necessary condition for synchronous THG to proceed through the first channel upon tight focusing of the incident radiation to the nonlinear medium is the falling of the third harmonic wavelength to the anomalous dispersion region (i.e. the condition $k_3 - 3k_1 < 0$ should be fulfilled, where k_3 and k_1 are the wave numbers of the harmonic and pump radiation, respectively). Only in that case the additional phase shift appearing after the propagation of radiation through the focal region is compensated by the dispersion of the nonlinear medium, which provides the conditions for the synchronous interaction of the pump and harmonic waves.

The analysis of single-photon absorption in naphthalene vapour shows that no absorption bands exist near the TH wavelength (354.7 nm) [32]. Naphthalene has the normal dispersion in this spectral range. For this reason, the condition of synchronous interaction of the waves in the channel $\omega + \omega + \omega \Rightarrow 3\omega$ for naphthalene vapour is not fulfilled for moderate output powers of the Nd:YAG laser, when the influence of the optical Kerr effect is not sufficient for changing the sign of dispersion.

For THG via the second channel, the situation is different. The presence of the negative dispersion region is no longer required for the $\omega + \omega + \omega + \omega - \omega \Rightarrow 3\omega$ process to occur, because the difference frequency can be generated in the synchronous regime in the media with different signs of dispersion [33]. Therefore, in that case, the most probable mechanism for THG is the fifth-order FC. A similar process of the difference-frequency generation in naphthalene vapour at the second harmonic wavelength of a Nd:YAG laser was studied earlier upon four-photon interaction ($2\omega + 2\omega - \omega \Rightarrow 3\omega$) [29].

1.3 Plasma and Ionic Media

1.3.1 Laser-Ablated Targets

The lowest-order (i.e. third) harmonic generation in laser-produced plasma plumes [34] is a rarely studied process compared with conventional THG in other media. An analysis of low-order harmonic generation in such plasmas produced by a laser heating pulse on the surface of solid target allowed formulating several recommendations as regards further advancement toward shorter wavelengths (see the following chapter). The use of femtosecond pulses enabled producing lowest-order harmonics from laser ablation with a relatively high conversion efficiency (of the order of 10^{-3}), which allows developing the recommendations for generation of shorter wavelength coherent radiation through the same nonlinear optical process. Thus the analysis of THG in laser plumes can be considered as a versatile method allowing for choosing the appropriate plasma media for efficient conversion of the wavelength of conventional laser sources (Ti:sapphire, Nd:glass, Nd:YAG, etc) toward the XUV range through the high-order harmonic generation (HHG). In this section, we present the studies of the THG in various plasmas using two lasers (Ti:sapphire and Nd:YAG) generating femtosecond and picosecond pulses [35].

The interest to the THG in some plasma plumes (in particular B and Mn) has been drawn by the early observations of strong high-order harmonics producing from the above plasmas in the spectral range of below 90 nm [36, 37]. One can expect that low-order harmonics producing in these nonlinear media could also possess the high conversion efficiency. In that case, the highly efficient sources of coherent radiation in the 100–200 nm range can be elaborated and used for various applications. The first step in these studies is an analysis of the lowest (third) harmonic generating in plasma media. These studies allow identifying the most efficient plasma media, which can be used for harmonic generation in the longer-wavelength spectral range. Note that generation of lowest order harmonic does not require vacuum equipment, since the wavelength of TH of commonly used laser sources (i.e. Ti:sapphire, Nd:glass, and Nd:YAG lasers) lies in the near ultraviolet range (265 nm and 355 nm respectively).

A Ti:sapphire laser ($\lambda = 795$ nm, $t = 110$ fs, $E = 8$ mJ) at a 10 Hz pulse repetition rate was one of the laser sources used in the studies described below. The laser radiation was focused inside a vacuum chamber by a 100 mm focal-length lens. The beam waist radius of the focused radiation was 16 μm. The spectral characteristics of TH radiation (265 nm) were analysed by a spectrometer and registered by a PMT. To create plasma plume, a heating pulse was split from the amplified laser beam by a beam splitter before a compression of the fundamental pulse. The pump pulse duration was 210 ps. A spherical lens focused this beam on a target placed in a vacuum chamber to generate a laser ablation plume (Fig. 1.3a). The area of ablation was adjusted to be approximately 0.6 mm. The heating picosecond pulse intensity at the target surface was varied in the range of $I_{pp} = (0.5-5) \times 10^{10}\,\mathrm{W\,cm^{-2}}$. A compressed fundamental pulse at a centre wavelength of $\lambda = 795$ nm had the energy of 12 mJ with a pulse duration of 110 fs after the propagation through the compressor stage. After the proper delay with regard to the pump pulse this radiation was focused by a 100 mm focal length lens on the ablation plume from the orthogonal direction. The lowest harmonic generating in metal plasmas was analysed using the fiber spectrometer (Ocean Optics: USB2000). The application of nonlinear crystals (β-barium borate, KDP) in front of the focusing lens allowed for the absolute calibration of THG efficiency.

A passive mode locking Nd:YAG laser was generated a train of pulses ($\lambda = 1064$ nm) of 50 ps duration. A single pulse was separated from the central part of the

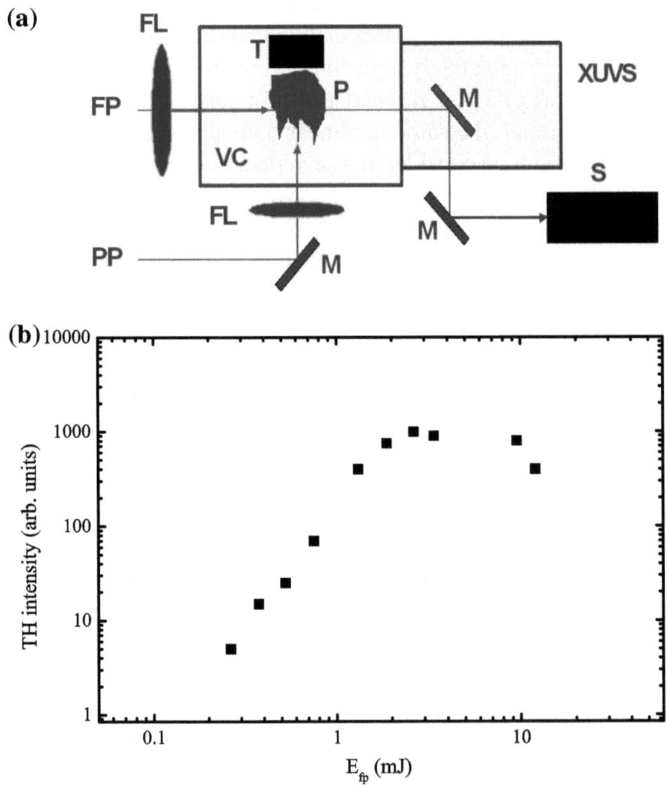

Fig. 1.3 **a** Experimental setup for the low-order HG studies in laser plasma. *FP* fundamental pulse; *PP* heating picosecond pulse; *FL* focusing lenses; *T* target; *VC* vacuum chamber; *P* plasma; *XUVS* XUV spectrometer; *S* USB2000 spectrometer. **b** TH intensity as a function of fundamental pulse energy measured in the Mo plasma at a delay between pulses of 15 ns. Adapted from [35] with permission from IOP Publishing

pulse train and a few first pulses before the separated pulse were used for plasma formation on the target surfaces placed in the vacuum chamber. These picosecond pulses were focused by a 100 mm focal length lens onto the different targets. After some delay with regard to the last heating pulse of train, the fundamental single pulse was focused by a 100 mm focal length lens in the area of plasma plume, and the intensity of this single pulse at the focus was varied up to 3×10^{13} W cm^{-2}. The TH of fundamental laser radiation generated in the plasma was directed to a fiber spectrometer. In these experiments, the In, C, Mg, Be, Cr, Mn, W, Mo, and B were used as the bulk targets.

In the case of 110 fs pulses, the lowest harmonic generating in Mo and B plasmas was studied at different experimental conditions. The influence of third harmonic intensity on the laser pulse energy showed a monotonic increase up to 3 mJ with further saturation and gradual decrease of harmonic generation efficiency (Fig. 1.3b, molybdenum plasma). The slope of $I_{3\omega}(I_{1\omega})$ dependence at small energies of fundamental pulse (>2 mJ) was equal to 2.6. The maximum intensity of femtosecond laser pulse at the focal spot could reach 8×10^{16} W cm^{-2} at the maximum pulse energy available from the laser. As this intensity considerably exceeds the barrier suppression intensity for singly charged Mo and B ions, the position of the laser focus was adjusted to be either before or after the laser plume to optimise the TH yield. The intensity of the femtosecond laser pulse at the position of the preformed plasmas was varied between 2×10^{14} and 5×10^{15} W cm^{-2}.

The harmonic generation efficiency was optimised by choosing the optimal position of the focus of the fundamental laser beam relative to the plasma plume. It was observed that, for very intense laser pulses (>10^{16} W cm^{-2}) focused in the centre of the plume, the harmonic radiation was considerably decreased. However, the harmonic intensity showed a maximum when the fundamental laser beam was focused either before or after the plume, depending on the experimental conditions. The 1×10^{-3} and 4×10^{-4} conversion efficiencies of the TH radiation were achieved in the boron and molybdenum plasmas respectively at the optimal position of laser focus. The analysis of TH spectrum showed the unsaturated profile, which repeated the spectral distribution of fundamental radiation.

Figure 1.4a shows that initially, with increase of delay between the heating and fundamental pulses, the intensity of harmonic radiation increases. Further increase of delay leads to saturation and gradual decrease of TH yield. The optimal delay depends on the target material. It was shown that lighter targets demonstrated higher conversion efficiency (Fig. 1.4b). These measurements were carried out at a 25 ns delay. One can expect that this delay becomes optimal for lighter neutrals and ions, since, for heavier particles, the time of flight from the target toward the interaction area becomes longer (due to smaller velocities compared with lighter particles).

In the case of picosecond pulses, the harmonic generation in the laser plasma produced on the surfaces of different targets was carried out at two experimental conditions. The harmonic efficiency was analysed at ambient conditions, when the air pressure above the target surface was maintained at normal (atmospheric) conditions. The harmonic radiation was also analysed when target was ablated at vacuum conditions. The THG experiments demonstrated that the plasma produced at the presence

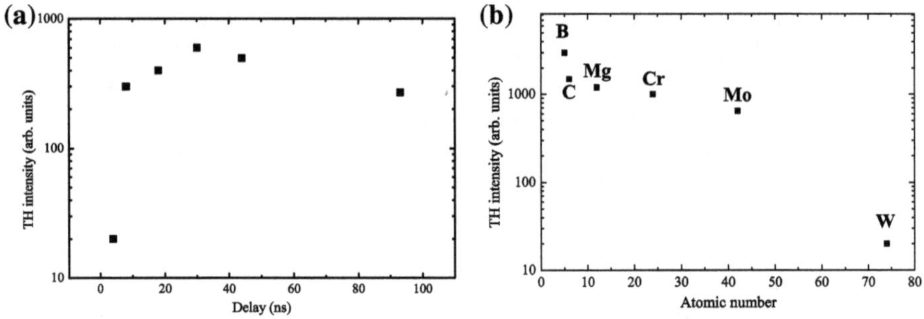

Fig. 1.4 Dependence of the TH intensity on **a** the delay between heating and fundamental pulses in the case of boron plasma and, **b** atomic number. Reproduced from [35] with permission from IOP Publishing

of air could be considered as a suitable medium for efficient conversion of coherent infrared radiation toward the UV range and limited by the atmospheric transmission edge (200 nm). Note that THG in Mn, B, and In plasmas at the conditions when plasma formation was accomplished at vacuum conditions did not demonstrate the enhancement of conversion efficiency compared with the case of plasma formation in air conditions. The maximum THG efficiency at ambient conditions of plasma formation ($\sim 10^{-4}$) has been achieved during the conversion of infrared picosecond radiation to the TH using the manganese plasma.

The distance between the target and optical axis of propagation of the fundamental radiation was varied in the course of experiment by a manipulator, which controlled the position of the target relative to the waist of the fundamental radiation. Figure 1.5 demonstrates a considerable influence of this parameter on the THG, which was

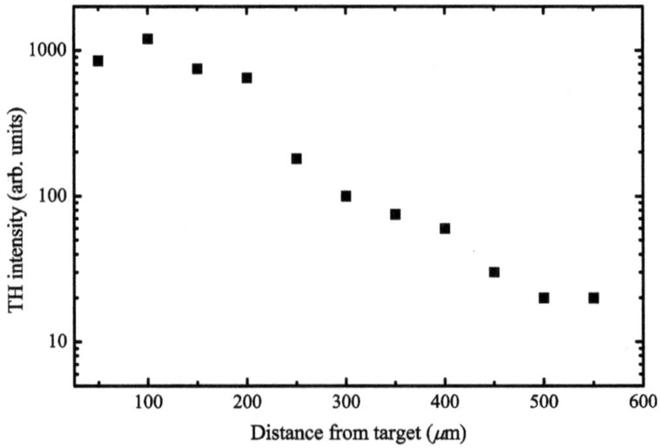

Fig. 1.5 Dependence of the TH intensity on the distance between the fundamental beam and target surface. Reproduced from [35] with permission from IOP Publishing

associated with optimisation of the plasma characteristics (plasma density and degree of excitation of higher states of the target plasma and atmospheric molecules) in the interaction zone. The optimal distance of propagation of the fundamental beam above the target surface was 100 μm.

The formation of target plasma at ambient atmospheric conditions differs compared with plasma formation in vacuum chamber. In the former case, the atmospheric gases act as the medium preventing fast spreading of the plasma plume. The spectral and temporal measurements of the plasma plume produced by the train of pulses on the surface of various targets at atmospheric conditions have shown the extension of excitation of laser plume for a longer period compared with vacuum plasma. In the meantime, the harmonic generation in laser plasma at atmospheric ambient conditions has approximately the same conversion efficiency of TH compared with THG in the plasma produced at vacuum conditions. It means that the presence of atmospheric air did not change drastically the conditions of THG in laser-produced plasma on the surfaces of different targets.

These results show that optimisation of the lowest harmonic allows defining the conditions of plasma formation for efficient higher-order harmonic generation [38]. While the efficiency of THG using picosecond pulses still remains at the moderate level (due to small intensity of the laser beam and lower thresholds of multiphoton ionisation of particles in the plasma), the efficient frequency conversion using femtosecond pulses ($\sim 10^{-3}$ in the boron plasma) demonstrates the applicability of proposed technique for generation of effective harmonics in the shorter wavelength range (100–200 nm) restricted by the transmission of optical materials. The indirect confirmation of this assumption was presented in [39], where the first observation of efficient HHG in laser plasma (up to the 61st order in the boron plume) was demonstrated. The same can be said about the manganese plasma. Earlier studies of this medium have demonstrated its suitability for highest harmonic cut-off observed in laser ablation (101st harmonic, [36]).

1.3.2 Gallium Ion Beams

Theoretical studies have shown that HG above 400 eV can be achieved using various ions [40]. These media possess (a) high thresholds of optical breakdown, (b) broad ranges of transparency, and (c) high nonlinear susceptibilities as compare with atomic media due to possibility of FC in quasi-resonant conditions. The ion beams can be shaped by the methods of electronic optics thus achieving optimal geometrical and density parameters. In this subsection, THG studies in gallium ion beams using liquid metal ion sources (LMIS) are discussed [22]. The nonlinear optical parameters of ionic medium responsible for conversion processes were determined for the analysis of THG using the Nd:glass laser radiation in single-charged gallium ions (Ga II). Third-order nonlinear susceptibilities responsible for the THG ($\chi^{(3)}(3\omega)$) were calculated for ground state and excited states of Ga II.

Output characteristics of Nd:glass laser used for HG in Ga ions were as follows: $E = 15\,\text{mJ}$, $t = 8\,\text{ps}$, $\lambda = 1054$ nm. Laser radiation was focused into the vacuum chamber containing LMIS. The violet emission at the tip of LMIS needle was observed indicating the presence of excited ions in ionic beam. An analysis of components of ion beam using time-of-flight spectrometer showed that 95 % of the whole beam consisted of singly charged Ga II whereas the others were the doubly charged ions and clusters. FC was carried out in the intersection zone of laser and ion beams at the distance of 1 mm beneath the extracted electrode. Converted radiation passed through the output quartz window of the chamber, spectrally extracted from fundamental radiation and measured by a PMT. The intensity of laser radiation in the intersection zone was $2 \times 10^{13}\,\text{W cm}^{-2}$. The dependences of THG ($\lambda = 351.7\,\text{nm}$) conversion efficiency as the function of laser intensity and ionic current were investigated. Following procedure was adopted to determine the THG conversion efficiency. Fundamental radiation was converted to the TH using KDP crystals, and its energy was measured using a standard energy meter. After decreasing TH energy using calibrated filters, this radiation propagated through the same detection channel as the TH generated in ion beam. Thus a calibration factor between the PMT output signal and the TH energy was established.

The η(I) dependence (Fig. 1.6) had a slope of 2. Maximal THG efficiency in Ga II beam was measured to be 10^{-7} at ion concentration of $10^{10}\,\text{cm}^{-3}$ [41]. This efficiency considerably exceeded the one that could be obtained taking into account the calculated susceptibilities at ground state and the used concentration of Ga ions. An increase of η can be attributed to the presence of excited ions. In accordance with calculations, the nonlinearities of excited gallium ions should considerably exceed the nonlinearities of unexcited ions. The presence of small concentration of excited ions in the beam should improve the conversion efficiency in this medium.

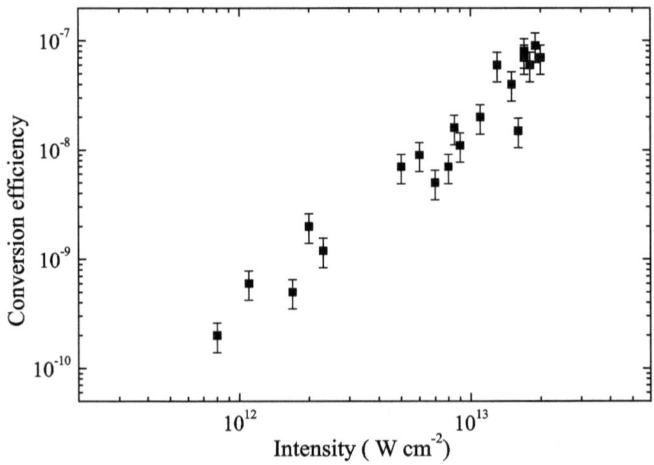

Fig. 1.6 Dependence of the THG conversion efficiency in Ga II ion beams on the incident radiation intensity. Reproduced from [22] with permission from IOP Publishing

Third-order susceptibility per excited ion was calculated from these experimental data to be 10^{-29}–10^{-30} esu.

1.3.3 Generation of Backscattered 2ω and 3ω/2 Harmonics of Femtosecond Radiation from the Plasmas with Different Atomic Number

Second harmonic (2ω) and three-halved harmonic ($3\omega/2$) emission from the laser plasma produced on different surfaces has been experimentally studied by many authors since the first observation by Bobin et al. [42]. These phenomena itself are interesting from the viewpoint of nonlinear physics. The other reason is that the emitted $3\omega/2$ is visible when a 800 nm laser radiation is used and quite suitable for optical measurements. Its measurement is a very useful diagnostic of laser-plasma interaction, in particularly in the case of plasma waves generation. The emission of radiation near $3\omega/2$ frequency has long been used as an indicator of two-plasmon decay (TPD) [43]. Some observations of red-shifted backscattered 2ω and $3\omega/2$ harmonic component were reported in [44, 45], whereas the calculations predict dominantly blue-shifted component. In experiments using nanosecond laser pulses, these harmonics showed red-shifted components in the longer-wavelength side of the main peak at the laser intensities above 5×10^{13} W cm^{-2}. These red-shifted waves have been predominantly attributed to the parametric decay instability of an incident laser radiation around the critical density. In the experiments using 20 ps pulses, the broadening of generated harmonic spectra at the intensities above 10^{15} W cm^{-2} has been attributed to the oscillating two-stream instability.

Most of those studies were carried out using the nanosecond and picosecond radiation. The investigation of 2ω and $3\omega/2$ femtosecond harmonics generated in specular direction was analysed in [46]. At underdense plasma conditions created by ultrashort pulses ($n_e \leq n_c/4$, n_e and n_c are the electron density and critical density, respectively) the product of growth rate and pulse duration is much smaller compared with the case of longer pulses. At these conditions no hydrodynamical motion occurs during the interaction time. Veisz et al. [46] observed no split of $3\omega/2$ harmonic spectrum on two parts that was the characteristic of previous observation of this radiation [45, 47]. They found the dependences of HG efficiencies on pulse duration. These peculiarities were explained on the basis of TPD growth rate in heterogeneous plasma in the linear regime.

In this subsection, we discuss the studies of backscattered 2ω and $3\omega/2$ harmonics generating from the plasmas using femtosecond radiation operated at 10 Hz pulse repetition rate [48]. The femtosecond radiation from Ti:sapphire regenerative amplifier ($\lambda = 795$ nm, $E = 10$ mJ, $t = 150$ fs) was used in these studies. The p-polarized radiation was focused by a 10 cm focal length lens on the surface of targets placed in the vacuum chamber. The radiation was interacted with target surfaces at the incidence angle of 45°. The characteristics of generated radiation were analysed at specular (90°) and backscattered (180°) directions using fiber optics spectrometer and monochromator. The targets with different atomic numbers (Z) were used at these studies (boron, graphite, polyethylene, chromium, molybdenum, and tungsten). The maximal intensity of pump radiation at the target surfaces was measured to be $1 \times 10^{16}\,\mathrm{W\,cm^{-2}}$.

The backscattered 2ω radiation was generated from the initial pulses interacted with the target. After 1–5 s from the beginning of interaction the backscattered $3\omega/2$ harmonic propagating through 45°, 99.5 % reflectance ($\lambda = 795\,\mathrm{nm}$) mirror was observed in the case of low-Z targets. This mirror also served as a filter of fundamental backscattered radiation. Depending on experimental conditions (target, pulse duration, focusing lens position, etc.), this process lasted during 10–20 s and then disappeared believed to be due to changes of the optimal conditions of HG and modification of target. After that the shutter was closed and the target was shifted so that the next set of laser radiation interaction with the target was performed at the fresh place of material. No $3\omega/2$ harmonic generation was observed in the case of single shots, as well as in the case of variations of target position from shot to shot at used pulse repetition rate conditions.

In these experiments, the low-divergence, collimated, backscattered harmonics were observed. The ratio of 2ω and $3\omega/2$ harmonic intensities was varied from shot to shot. Figure 1.7a shows the harmonic and plasma spectra obtained from low-Z targets (B, C, CH). These spectra were also observed from the moderate (Cr) and heavy (Mo, W) targets (Fig. 1.7b). Figure 1.7 presents the single shot spectra. The efficiencies of 2ω and $3\omega/2$ harmonic generation observed from low-Z targets were measured as 10^{-6} and 2×10^{-7}, respectively. The $3\omega/2$ radiation was considerably suppressed and blue-shifted in the case of Cr and fully disappeared when W and Mo were used as the targets.

At a laser energy of 3 mJ corresponding to an intensity of $3 \times 10^{15}\,\mathrm{W\,cm^{-2}}$, the emission of $3\omega/2$ radiation was appeared with a threshold character. This intensity was much higher than early reported thresholds of generation of such radiation using longer pulses. In nanosecond range the threshold of generation was estimated as $2 \times 10^{13}\,\mathrm{W\,cm^{-2}}$ [49], whereas in picosecond range it was $2 \times 10^{14}\,\mathrm{W\,cm^{-2}}$ [50]. The obtained results can be explained using the theory of the TPD instability. The pulse duration was varied while maintaining the same intensity and it was found that the threshold of $3\omega/2$ radiation generation decreases for longer pulses.

The point of interest here is whether some prepulse can change the plasma characteristics in multi-pulse regime inside the crater created by first pulses. All these measurements were performed at a prepulse intensity estimated to be $\sim 10^{11}\,\mathrm{W\,cm^{-2}}$ at maximal laser intensity and contrast ratio of 10^{-5}. The influence of the prepulse generating 10 ns before the main pulse on the harmonic efficiency at different prepulse/pulse ratio was analysed. The variations of this parameter were established by different time scale delays triggering Pockels cell of the regenerative amplifier of laser. A steady state decrease of harmonic conversion efficiency was observed with the growth of this ratio.

The spectrum of this emission was always characterized in previous studies by two peaks, asymmetrically displaced relative to $3\omega/2$ wavelength, with red-shifted

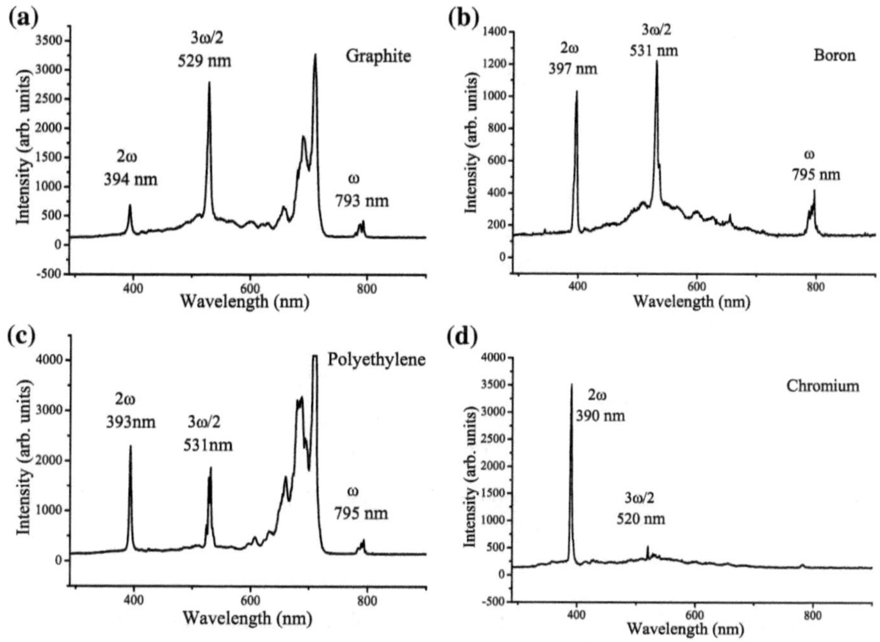

Fig. 1.7 Harmonic and plasma spectra from low-Z (**a** boron, **b** graphite, and **c** polyethylene) and high-Z (**d** chromium, **e** molybdenum, and **f** tungsten) targets. Adapted from [48] with permission from Springer Science + Business Media

component usually the stronger one. This peculiarity of emission spectrum was commonly attributed to Thomson, or Raman scattering of interaction beam photons off plasmons produced by the same beam through the TPD [51]. Various theories have since been proposed to precisely relate the plasmons generation with the observed spectral features. $3\omega/2$ harmonic radiation could be originated from the excitation in the plasma of a parametric instability of the type of decay of a light wave of frequency ω into two plasmons of frequency $\omega/2$ in the vicinity of plasma region of 1/4 of critical density. This in turn leads to the generation of frequency $3\omega/2$ due to various mechanisms. The first process is the coalescence of the parametrically excited plasma oscillation ($\omega/2$) with the incident light wave (ω), and the second one is a very unlikely process of coalescence of three plasma oscillations ($3 \times \omega/2$).

The analysis of backscattered harmonic spectra has shown that the $3\omega/2$ emission was slightly red-shifted (1–2 nm) in some of studies with light targets and was mostly centred at exact position of $3\omega/2$ harmonic of fundamental radiation. At the same time the $3\omega/2$ radiation from Cr target was strongly blue-shifted ($\Delta\lambda = 10$ nm). No modulation of registered spectra was detected in the case of light targets. The 2ω radiation was dominantly blue-shifted both for low- and high-Z materials. The spectral width of $3\omega/2$ harmonic was 4 nm (Fig. 1.8), whereas the spectral width of incident radiation was ~11 nm. The observed narrowing of $3\omega/2$ harmonic bandwidth

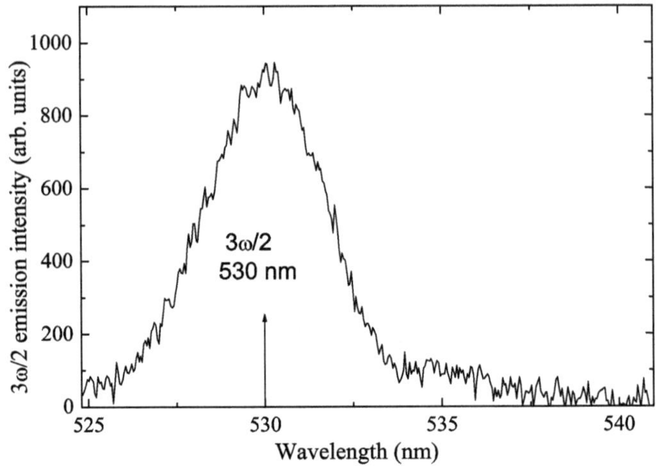

Fig. 1.8 $3\omega/2$ emission spectrum generated from boron target. Reproduced from [48] with permission from Springer Science + Business Media

can be attributed to the fulfilment of occasional phase-matched conditions only for central part of spectral distribution of incident radiation possessing higher intensity. The nanosecond laser experiments previously observed both red and blue wings, which were explained by simulations of TPD saturation and Langmuir wave decay instability. In the case of short pulses the influence of latter process is insignificant since this process develops on the time scale of ion-acoustic wave propagation. The fundamental radiation was frequency modulated and this modulation was considerably increased for backscattered fundamental wave in the case of high intensities and shortest pulses. The frequency modulation in that case was also observed in second harmonic spectrum. Backscattered fundamental radiation was red-shifted, whereas 2ω radiation was considerably blue-shifted (especially in the case of Mo target).

The harmonic yield was also measured as a function of pulse duration. Figure 1.9 shows the intensities of backscattered (1) fundamental, (2) 2ω, and (3) $3\omega/2$ radiation at different pulse durations at the conditions of constant fluence. The distance between gratings in Ti:sapphire laser compressor was varied that allowed changing the pulse duration in the range of 150–1600 fs. The measurements were carried out for both directions from the optimal compressor position. In these studies the harmonic generation and reflectivity of fundamental wave did not depended on the sign of the chirp of incident pulses. No variations in reflectivity of fundamental radiation were observed in that case. The efficiency of 2ω harmonic generation was steadily dropped for longer pulses, as was previously reported in [46, 52]. At the same time, the $3\omega/2$ yield was insignificantly decreased, without specific peculiarities reported in [46], where a maximum of $3\omega/2$ radiation conversion efficiency at 350 fs range was observed.

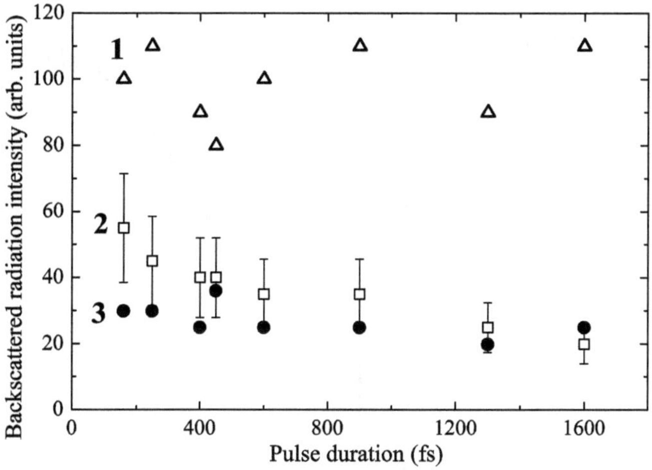

Fig. 1.9 Dependences of backscattered *1* fundamental radiation, *2* 2ω, and *3* 3ω/2 on the pulse duration of incident radiation in the case of boron target. Reproduced from [48] with permission from Springer Science + Business Media

1.4 Gases

Frequency multiplication of laser radiation in gases nowadays is one of the most efficient methods of generation of the short wavelength coherent radiation. An interest to such nonlinear medium was facilitated by three factors: their transparency in vacuum ultraviolet (VUV) range, the possibility of achieving the phase-matching conditions, and the high nonlinear susceptibilities due to resonant and quasi-resonant conditions. Below we analyse the studies of HG in noble gases at the conditions of negative and positive dispersion, as well as of generation of tuneable VUV radiation. We also discuss the studies of harmonic and parametric generation in air.

1.4.1 Third Harmonic Generation in Conditions of Negative and Positive Dispersion in Rare Gases and Tuneable Radiation Generation in 113.5–17.0 nm and 117.8–119.2 nm Ranges

In the first part of this subsection, the investigations of THG in Xe and Xe:Ar gas mixtures using TH pump radiation of picosecond Nd:YAG laser ($\lambda = 354.7$ nm) are presented. THG in xenon ($\lambda = 118.2$ nm) at these conditions is a quasi-resonant process. There are two transitions in the energy spectra of this gas ($5p^6\,{}^1S_o - 7s[I_2^1]^o$ and $5p^6\,{}^1S_o - 5d[I_2^1]^o$), with the oscillator strengths of 0.0968 and 0.395, respectively. The 354.7 nm radiation (TH of Nd:YAG laser radiation) was focused by a 25 cm focal length lens into a 18 cm long gas cell fixed in front of the input slit of

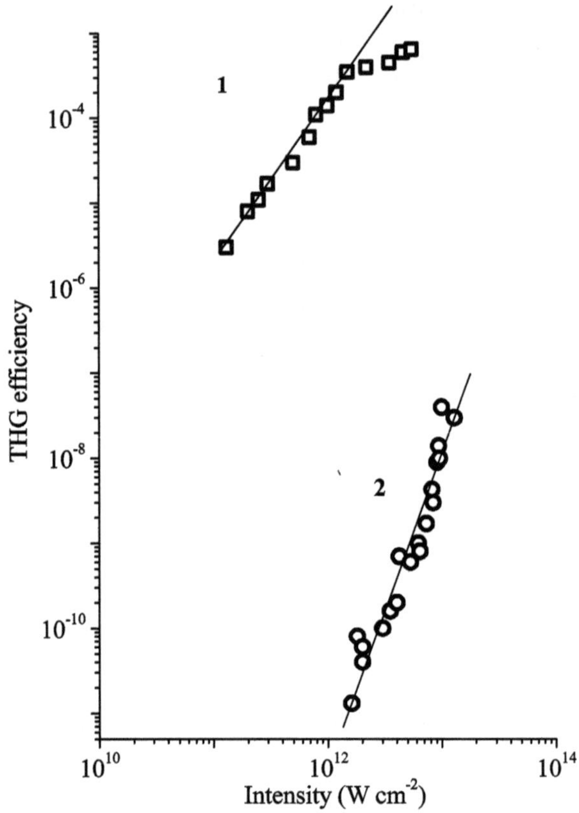

Fig. 1.10 TH conversion efficiency measurements at different input radiation intensities in the media with negative dispersion *1* and positive dispersion *2*. Adapted from [54] with permission from Springer Science + Business Media

vacuum monochromator. The intensity of TH of this radiation (118.2 nm) was measured by a calibrated channel electron amplifier and pyroelectric detector. Curve 1 of Fig. 1.10 presents the results of η measurements at different input radiation intensities. Maximum η was measured to be 8×10^{-4} [53]. The results of calculations neglecting Kerr effect influence are shown as a solid line. The $\chi^{(3)}(3\omega)$ per atom for pure xenon was calculated to be 5.0×10^{-35} esu. With the intensity growth ($I > 8 \times 10^{11}$ W cm^{-2}) the Kerr nonlinearities induced a decline from quadratic $\eta(I)$ slope. THG was also analysed in xenon-argon mixtures. The phase mismatch referred to one atom of argon was measured to be $C_{Ar} = 5.6 \times 10^{-18}$ cm^2. The maximum η was measured to be 3.1×10^{-3} at laser intensity of 2.3×10^{12} W cm^{-2}, Xe pressure of 1.2×10^3 Pa, and the ratio of argon and xenon pressures of $R = 11$.

The efficient (phase-matched) HG of focused radiation in isotropic medium is possible only in narrow spectral ranges of negative dispersion. Below we show that at high intensities the conditions of efficient HG caused by Kerr nonlinearities in the media with positive dispersion (MPD) can be achieved as well [54]. When the laser radiation is focused into the centre of MPD, a harmonic generated in the region in front of the Gaussian beam waist is completely compensated by a harmonic generated behind the beam waist due to additional phase shift caused by focusing. In other words, there is a reverse harmonic energy conversion into the incident beam energy behind the beam waist. However, when this phase shift is compensated by the phase shift caused by Kerr nonlinearities, the HG in MPD appeared to be possible.

The THG studies in MPD were carried out using the TH radiation of Nd:glass laser ($\lambda_l = 1055.1$ nm), as a pump source. This TH radiation ($\lambda = 351.7$ nm, $E = 0.5$ mJ, $t = 5$ ps) was focused in krypton and xenon possessing the positive dispersion in the spectral range of the TH of driving radiation (117.2 nm). The dependences of THG efficiency on incident radiation intensity were studied in Kr (Fig. 1.10, curve 2). Intensity dependence was found to be similar to $\eta \sim I^4$, in agreement with theoretical calculations presented below.

The efficiency of frequency conversion of Gaussian beams under the conditions of strong focusing is determined by the phase integral

$$I = \int_{-\infty}^{\infty} \frac{\exp\{i\,[ax + 2\,(a\rho - 1)\arctan x]\}}{1 + x^2}\,dx \tag{1.1}$$

here $\alpha = (3\pi\omega/c)bN\Delta\chi^{(1)}$; $\rho = (\pi\Delta\chi_k^{(3)}/cn_1n_3\Delta\chi^{(1)}) \times \Phi \times \exp(-r^2/r_o^2)$, b is the confocal parameter, ω is the frequency of fundamental radiation, $\Phi = (n_1c/8\pi)|E^2|$ is the intensity of the fundamental radiation, E is the amplitude of electromagnetic wave, N is the density of the medium, r_o is the beam waist radius, r is the transverse polar coordinate, n_1 and n_3 are the refraction indices at the fundamental and 3ω frequencies, $\Delta\chi^{(1)}$ is the difference of linear susceptibilities at the fundamental and 3ω frequencies, and $\Delta\chi^{(1)} = \chi^{(1)}(\omega) - \chi^{(1)}(3\omega)$; $\Delta\chi_k^{(3)} = n_3\chi^{(3)}(\omega;\ \omega,\ \omega,\ -\omega) - \chi^{(3)}(3\omega;\ 3\omega,\ \omega,\ -\omega)$ is the difference of the Kerr-induced third-order susceptibilities.

In the case of MPD ($\alpha < 0$) the conversion efficiency is given by

$$\eta = \left\{\frac{2\pi\gamma(\alpha\rho)^2}{\Gamma\,(1 + \alpha\rho)}\frac{e^{\alpha}}{1 - \alpha\rho}\left[\alpha\,(2 + \rho)\,\Psi\,(1 - \alpha\rho,\, 1;\, -2\alpha) + \Psi\,(-\alpha\rho,\, 1;\, -2\alpha)\right]\right\}^2 \tag{1.2}$$

here $\gamma = \chi^{(3)}(3\omega)/\Delta\chi_k^{(3)}$; $\chi^{(3)}(3\omega)$ is the nonlinear susceptibility responsible for THG; $\Gamma(x)$ is the gamma function; $\Psi(a,b;x)$ is the hypergeometric function.

In the case of a focused beam the influence of the phase shift caused by the optical Kerr effect is similar to the one of the phase shift caused by focusing. Kerr-induced variation of medium's optical properties changes the spatial parameters of the focused beam. This leads to the possibility of THG in the MPD. The conversion efficiency depends on the incident radiation intensity as $\eta \sim \Phi^4$, which is obtained from latter equation for $|\rho| \ll 1$.

The physical interpretation of such phenomenon is as follows. TH intensity is proportional to the third order of fundamental intensity and $\mathrm{sinc}^2(A\Delta k)$. A is the independent coefficient and Δk is the wavevector detuning. For small detuning, the $\mathrm{sinc}^2(A\Delta k)$ is equal to 1. In the case of high detuning, $\mathrm{sinc}^2(A\Delta k) \sim (\Delta \mathrm{k})^2 \sim (\Delta \mathrm{n})^2$.

The detuning of refractive indices in Kerr medium linearly depends on the intensity ($\Delta n = n_3 - n_1 = n_3^o + \gamma_3 I - n_1^o - \gamma_1 I$; γ_3 and γ_1 are the nonlinear refractive indices of the medium). So, $\mathrm{sinc}^2(A\Delta k) \sim I^2$, and overall intensity dependence looks like $I_{THG} \sim I^3 \times I^2 = I^5$. This simplified consideration shows that in the case of MPD the growth of intensity leads to the optimisation of phase-matching conditions, and the conversion efficiency depends on the fourth order of laser intensity.

No saturation or influence of other nonlinear processes was observed within the studied intensity range. Maximal η in Kr was 4×10^{-8} at the intensity of 9×10^{12} W cm^{-2}. Similar results were observed in xenon with the conversion efficiency five times less than in krypton.

There are two areas of negative dispersion of Xe (113.5–117.0 nm and 117.8–119.2 nm). Frequency tripling in these ranges provides the coherent VUV photons with the energy of ~10 eV. Here we briefly discuss the tuneable frequency conversion studies in these VUV ranges in xenon and gas mixtures [55, 56].

The parameters of pump radiation were as follows: tuning range 338–366 nm, pulse duration 7 ps. The influence of the xenon pressure and gas mixture on the THG conversion efficiency, as well as the influence of the pump radiation intensity on the THG efficiency at different wavelengths of converted radiation, was analyzed in those studies. Figure 1.11 shows the TH intensity generated in xenon in these ranges. The η in the 117.8–119.2 nm range was ten times higher compared with the 113.5–117.0 nm range. The dependence of the THG conversion efficiency on the driving UV radiation intensity was quadratic throughout the investigated tuning ranges. The maximum η (8×10^{-4}) in the 113.5–117.0 nm range was achieved at $\lambda = 116.6$ nm when the xenon pressure was 2×10^3 Pa, and the pressure ratio of argon and xenon was $R = 10$ [57].

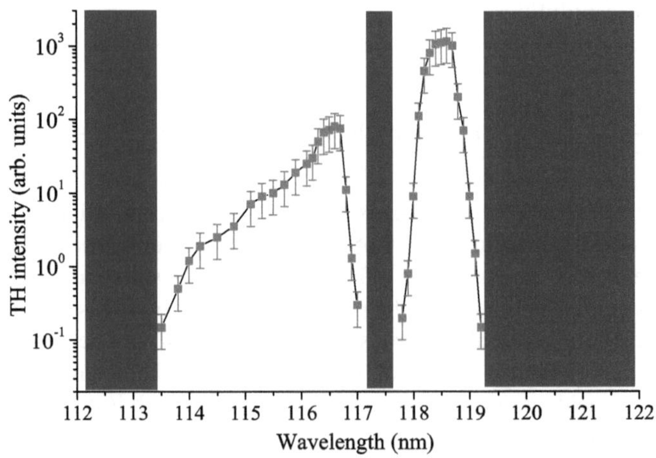

Fig. 1.11 Ranges of generation of VUV radiation using the THG in xenon. *Filled boxes* are the positive dispersion ranges of xenon. Reproduced from [56] with permission from Turpion

1.4.2 Third Harmonic Generation in Air Using Femtosecond Radiation at Tight Focusing Conditions

The THG in air was analysed since early stages of laser-gas interaction studies using femtosecond pulses. The self-guiding conditions of the femtosecond pulse propagating through the laser-induced plasma can enhance the effective length of interaction. Such an opportunity of TH emission enhancement was investigated by Yang et al. [58]. The effects of temporal and spatial self-action of light in atmospheric air were analysed by Fedotov et al. [59]. The influence of continuum generation on the THG in air using intense femtosecond IR radiation was reported in [60]. It was shown that, during laser pulse filamentation in air, an intense ultra-short TH pulse is generated forming a two-colour filament [61]. A preliminary excitation of atmospheric air using 532 nm, 15 ns pulses for the enhancement of the efficiency of TH emission from 150 fs, 800 nm pulses was studied in [62].

The conversion efficiencies as high as 10^{-3} were reported in those studies [58, 59, 63], although some calculations predict an efficiency of 10^{-2} [64]. It was found that this process is rather independent on the wavelength of input radiation [60]. The combined action of self-focusing and plasma generation in air plays an important role for on-axis and off-axis phase matching conditions along the filament resulting in high conversion efficiency [65]. The THG efficiency was also analysed in methane and noble gases at the conditions of self-action [66]. The measurements of spectral and propagation characteristics of TH pulses generated at sub-atmospheric pressures of noble gases were reported in [63, 66].

Most of these studies were performed in the conditions of the self-guided beam propagating through the gas medium when the self-focusing process induced by optical Kerr effect was compensated by the self-defocusing process caused by free electrons generated in the ionised channel. The majority of these THG experiments were carried out in a loosely focused geometry required for the phase matching in a positively dispersed medium. There are only few reports of THG studies in a tight focused geometry [65, 67, 68]. No detailed analysis was found in literature regarding the phase-matching conditions of THG in air at tight focused configurations. Such analysis might be important for further applications in the optimisation of harmonic generation in positively dispersed gases at different experimental conditions (spectral range, pulse duration, focusing conditions, etc). Below we discuss the studies of the THG in air using femtosecond pulses variable in the range of 110–1300 fs by analysing various aspects of this process from the point of view of phase matching conditions in tightly focused geometry [69].

The Ti:sapphire laser ($\lambda = 795$ nm, $t = 110$ fs, $E = 8$ mJ) at 10 Hz pulse repetition rate was used in these studies. The output pulse duration was varied from 110 fs to 1.3 ps for the analysis of the influence of the pulse characteristics on the THG in air. The laser radiation was focused inside the chamber by a 100 mm focal length lens. The beam waist radius of the focused radiation (16 μm) twice exceeded the one defined from diffraction-limited focusing. Air pressure inside the chamber was varied from 0.01 to 101 kPa. The spectral characteristics of TH radiation ($\lambda = 265$ nm) were analysed by a spectrometer and registered by both photomultiplier tube and charge coupled device. The CCD camera was also used for the analysis of the spatial characteristics of fundamental and harmonic radiation.

The theoretical analysis of THG was carried out on the basis of the equations describing the evolution of fundamental E_1 and harmonic E_3 field envelopes and density of electrons generated due to multi-photon ionisation [60].

The studies of the THG as a function of fundamental intensity were carried out using the radiation of different pulse duration. The intensity variations were performed using the calibrated filters. The experimental slope of $I_3 \sim (I_1)^l$ dependence for 230 fs laser pulses was close to $l = 4.9$ up to the fundamental intensity of $I_1 \approx 5 \times 10^{14}\,\mathrm{W\,cm^{-2}}$. Further growth of fundamental intensity led to a considerable decrease of a slope ($l \sim 1$). Similar dependence was observed using various pulse durations (up to 800 fs). In particular, at shorter pulse duration ($t = 140$ fs) this slope was found to be 5.5. The optical breakdown caused by tunnelling and/or multiphoton ionisation was observed at $I_1 = 4 \times 10^{14}\,\mathrm{W\,cm^{-2}}$ by the appearance of a spark in the focus, that was close to previously reported data ($2.9 \times 10^{14}\,\mathrm{W\,cm^{-2}}$ [70], and $2 \times 10^{14}\,\mathrm{W\ cm^{-2}}$ [71]).

The point of slope change corresponds to the saturation intensity (I_s). The saturation of ionisation is a typical effect, which occurs in multiphoton ionisation experiments when the probability of this process becomes close to unity leading to the depletion of neutral atoms. The interaction mechanism between the laser pulse and air experiences considerable changes when the laser intensity exceeds I_s. For $I_1 < I_s$, the fundamental light interacts mostly with neutral molecules, contrary to another situation ($I_1 > I_s$) when the radiation interacts with ionised plasma. The increase of ion and electron concentrations led to the considerable variations between the phases of fundamental and harmonic radiation due to free-electron-induced negative dispersion.

It was found that the small variations of laser intensity cause large fluctuations of TH emission in the intensity range of 9×10^{13}–$2 \times 10^{14}\,\mathrm{W\,cm^{-2}}$, whereas the fluctuations of THG efficiency at higher intensity become smaller. This fact underlines the strong intensity dependence of this process at relatively low intensities and the decrease of this dependence at higher intensity. Analogous variations of THG scale with laser intensity below and above I_s were reported in a number of studies [59, 68, 71, 72]. In particular in [71], the slope of $I_3(I_1)$ dependence was estimated to be between 3.5 and 5, whereas above the saturation intensity this parameter dropped to 3. One has to note that similar results were reported previously in various gases and using different sources of fundamental radiation (248, 800, and 1064 nm), which indicates that the processes involved do not depend much on the details of atomic and molecular spectra.

As it was already mentioned, in the case of focused beam the influence of the phase shift caused by Kerr effect is similar to the one of the phase shift caused by focusing. Kerr-induced variation of medium's optical properties changes the spatial parameters of focused beam. This leads to the possibility of THG in the medium with positive dispersion. The theoretical model describing THG at such conditions is presented in [73]. Notice that in this model the TH intensity depends on the fundamental radiation intensity as $I_1 \sim I_3^5$. Such a dependence of TH emission on the fundamental intensity in the case of the influence of Kerr effect was also predicted in the theoretical consideration of this process presented by Malcuit et al. [72].

The experimental dependences of the TH intensity versus fundamental radiation intensity showed a slope close to 5. Similar dependences are typical for the THG caused by both the direct four-wave process accompanied by the influence of intensity-induced change of refraction index and the six-wave mixing process ($\omega + \omega + \omega + \omega - \omega = 3\omega$). The origin of the THG mechanism can be revealed from the dependence of the TH yield on the nonlinear medium density.

The peculiarity of Kerr-induced intensity-dependent process is a $I_3(p)$ dependence (p is the gas pressure) that distinguishes from the well-known $I_3 \sim p^2$ dependence in gases predicted from the lowest-order perturbation theory. Such dependence was observed at small laser intensity in a broad range of pressure variations (Fig. 1.12). However, the above-presented approach based on transient intensity-dependent phase matching predicts that at small gas concentration and high intensities of fundamental radiation the $I_3 \sim p^d$ dependence should show the $d = 4$ slope [54]. In the meantime, when the THG is caused by six-wave mixing, the $I_3(p)$ should follow the second order

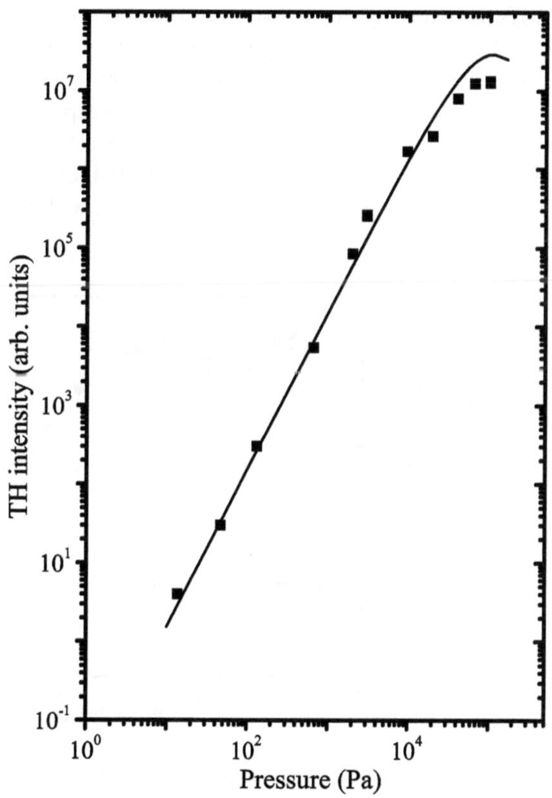

Fig. 1.12 $I_3(p)$ dependence measured using the 110 fs pulses. Reproduced from [69] with permission from Optical Society of America

power dependence. Thus the observed $I_3 \sim p^2$ dependence could indicate that the observed THG can be caused by a six-wave mixing process. However, it is unclear whether the susceptibility for the six-wave process is greater than that for the direct four-wave harmonic generation. The competition of these two processes can lead to the re-distribution of their influence through the intensity scale.

The theoretical analysis showed that the THG efficiency essentially depends on the relation between the Kerr nonlinearities responsible for the changes of the refraction indices of fundamental and harmonic radiation. At certain relations between Kerr nonlinearities, the density dependence of TH radiation generated due to third-order nonlinearity can also be close to the power dependence with a slope of 2. The influence of multi-photon ionisation can intensify the TH signal saturation. The calculated $I_3(p)$ dependence is shown in Fig. 1.12 by solid line. One can conclude from this that THG at tight focusing conditions may be caused by the influence of intensity-induced phase modulation rather than six-wave process.

The intensity- and pressure-dependent experiments did not reveal which of the two above processes was certainly responsible for the THG in air at the used experimental conditions. To define which of these—four-wave or six-wave—processes plays a dominant role, the THG at different pulse durations was carried out. The pulse duration of fundamental radiation was gradually decreased from 1300 fs to 120 fs by changing the distance between compressor gratings. Figure 1.13 shows the I_3 (t_p) dependence at the pulse energy $E = 0.34$ mJ. The "optimal" pulse durations were observed at different ranges of pulse intensities. The phase matching was achieved at $I_1 = 1 \times 10^{14}$ W cm^{-2} that corresponded to $t_p \approx 500$ fs. The TH output was then diminished with a decrease of pulse duration below 500 fs due to the phase mismatch caused by the influence of Kerr-induced variation of refractive index as well as initial stages of ionisation. However, with further decrease of pulse duration,

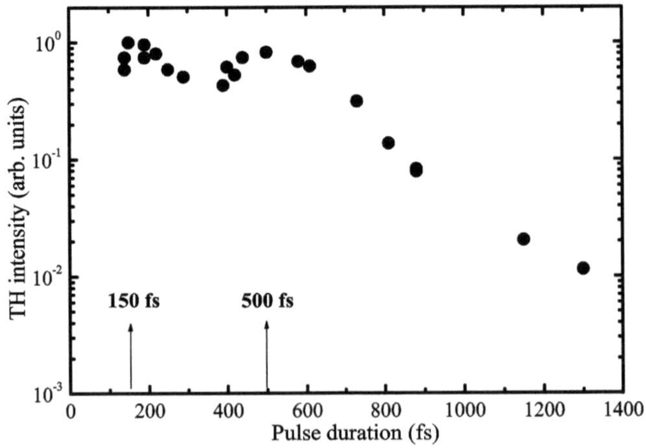

Fig. 1.13 Dependence of the TH emission on the pulse duration. Reproduced from [69] with permission from Optical Society of America

the optimal phase matching conditions were again achieved at $t \approx 150$ fs due to the growing influence of free-electrons-induced self-defocusing that compensated the self-focusing caused by Kerr nonlinearity.

These pulse duration dependent studies revealed that the observed THG process was originated from the Kerr-induced variations of phase relations between the fundamental and TH pulses, rather than from six-wave process. The $I_3(t_p)$ dependence presented in Fig. 1.13 can be considered as a confirmation of the approach that predicts the fulfilment of optimal conditions in the medium with positive dispersion through the transient variation of refractive index using the variations of pulse duration. The intensity-induced phase modulation can also lead to the variations of the spatial shape of fundamental and harmonic radiation. The observations of the shape of fundamental beam after tight focusing in air have shown the considerable deviations from the initial spatial distribution prior to achieving the multiphoton ionisation in air (Fig. 1.14). These measurements were carried out at different pulse durations and fixed pulse energy. "Hot rings" (Fig. 1.14b, $t = 210$ fs) and "hot spots" (Fig. 1.14c, $t = 130$ fs) were appeared with a decrease of pulse duration. Such patterns were observed even in the case of high-quality profile of laser beam. The spatial self-action of fundamental light was accompanied by a spectral broadening of laser radiation. A Gaussian-like shape of TH emission was observed at low fundamental intensity that was then changed to a ring-like structure at higher intensity.

The temporally and spatially inhomogeneous appearance of free electrons due to ionisation causes the self-phase modulation, defocusing, and blue shift of the light. The change in the refractive index of gaseous medium due to the generation of electrons is given by

$$\Delta n = \omega_p^2/2\omega^2, \tag{1.3}$$

where $\omega_p = (4\pi e^2 N_e/m)^{1/2}$ is the plasma frequency, and N_e is the electron density. The change of the refractive index at the TH frequency due to plasma formation is almost one order of magnitude less than Δn induced at the frequency of fundamental radiation (see Eq. 1.3). This peculiarity may account for the fact that the

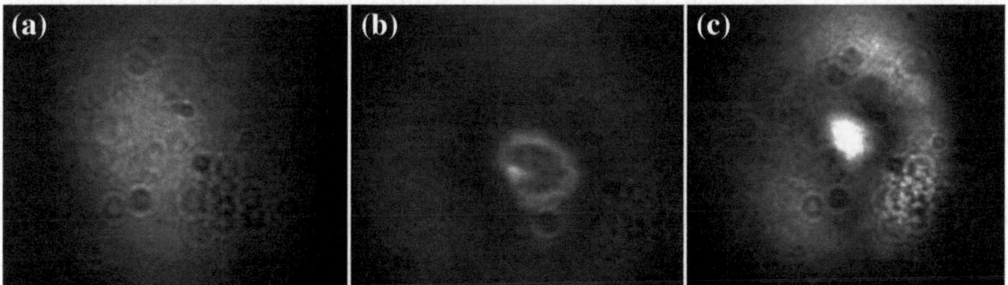

Fig. 1.14 Spatial distributions of the fundamental radiation propagated through the focal spot at the intensities below the optical breakdown **a** $I_0 = 9 \times 10^{13}$ W cm^{-2}, **b** $I_0 = 2 \times 10^{14}$ W cm^{-2}, and **c** $I_0 = 3 \times 10^{14}$ W cm^{-2}. Reproduced from [69] with permission from Optical Society of America

spatio-spectral structures caused by the self-action of TH pulse were less pronounced compared to those of fundamental radiation.

1.4.3 Fourth Harmonic Generation During Parametric Four-Wave Mixing in the Filaments in Ambient Air

Harmonic generation in air and other gases still attracts the attention and allows the observation of new features of frequency conversion. The comparative experimental investigations on third- and fifth-harmonic generation in atomic and molecular gases driven by mid-infrared ultrafast laser pulses at a wavelength of 1500 nm were reported in [74]. It was observed that the conversion efficiencies of both the THG and 5HG saturate at similar peak intensities ($\sim 1.5 \times 10^{14}\,\mathrm{W\,cm^{-2}}$) for argon, nitrogen, and air, whose ionisation potentials are close to each other. Near the saturation intensity, the ratio of yields of the 5HG and THG was found to be 10^{-1} for all the gases. These results show that high-order Kerr effect seems to exist; however, contribution from the fourth-order Kerr refractive index coefficient alone is insufficient to balance the Kerr self-focusing without the assistance of plasma generation. In the meantime in [75], THG and 5HG were studied by tightly focused femtosecond pulses at 2.2 μm wavelength in air. The measured ratio of yields of the third and fifth harmonics in their setup was found equal to 2×10^{-4}. This result contradicts the recent suggestion that the Kerr effect in air saturates and changes sign in ultra-intense optical fields.

THG was analysed in [76] by intense femtosecond laser pulse at a central wavelength of 800 nm superposed by its second harmonic in air. The TH radiation showed a periodic modulation with a period of 0.67 fs when the delay between fundamental and second-harmonic wave was continuously changed. The periodic modulation of THG can be attributed to the interference of third-harmonic signals generated from a direct THG channel ($3\omega = \omega + \omega + \omega$) and a four-wave mixing (FWM) channel ($3\omega = 2\omega + 2\omega - \omega$). It was shown in [77] that significant enhancement of THG during filamentation in air can be achieved by terminating a filament with either a pair of uncoated glass plates or a thin plasma string created by a second laser pulse. By comparing the results obtained under these two conditions, it appears likely that the inserted plasma string plays a role as a phase plate, by which the phase mismatch induced by the Gouy phase can be partially corrected. The TH generated during femtosecond filamentation in air was also studied in [78]. By establishing a gradient from atmospheric pressure to vacuum conditions, the filament was truncated abruptly at defined positions. The introduction of the pressure gradient led to an enhancement of the generated TH radiation by 3 orders of magnitude. This effect is attributed to an improved on-axis phase-matching condition.

The self-channelling (filamentation) in air using high-power 800 nm femtosecond laser pulses causes pulse self-compression, spatial self-filtering, intensity clamping with self-stabilized high intensity in the core, super-continuum generation, white light generation, and terahertz emission. At these conditions, one can expect the creation of optimal phase matching conditions for various parametric processes, in particular 4HG in the case of two-colour laser (i.e. beam containing fundamental as well as second harmonic radiation). One can assume that low-order nonlinearity of air allows for efficient generation of the fourth harmonic through the parametric processes of the wave mixing in filaments.

Parametric four-wave mixing (FWM) processes in atomic media have been studied for a long time starting from early laser-atom experiments. In particular, the $\omega + \omega + 2\omega \rightarrow 4\omega$ process was analysed in strontium vapours, and relatively high efficiencies (3×10^{-5}) were reported in the case of resonance-induced enhancement of wavelength conversion [79]. Four-wave mixing in gases was studied using different gases as well, and the vacuum ultraviolet radiation ($\lambda \approx 100-200$ nm) has been generated using different laser sources (see [33] and references therein). The interest in FWM has been growing as an alternative to the direct harmonic generation. Tuneable, visible, ultrashort pulses can be created using FWM by filamentation in a gas cell [80]. Few-cycle pulses in the deep UV and mid-IR have also been generated by this method [81, 82].

In this subsection, we discuss the generation of fourth harmonic through the parametric mixing of the fundamental wave and second harmonic of the Ti:sapphire laser ($\omega + \omega + 2\omega \rightarrow 4\omega$) in air filaments. We analyse the influence of various experimental parameters (laser intensity, polarization, chirp, and pulse duration) on the 4HG in air. Nonlinear frequency conversion demonstrated in this study, such as fourth harmonic (4H, $\lambda \approx 200$ nm) of infrared femtosecond pulse, is an efficient approach to obtain ultrashort laser pulses in the ultraviolet spectral region at the very edge of the transmittance of air [83].

In these studies, a Ti:sapphire femtosecond laser was used, which delivered 790 nm, 45 mJ pulses of 48 fs at a repetition rate of 10 Hz. The pulses were focused by a convex lens of focal length of 2 m in ambient air. The value of the Rayleigh length (before filamentation), with the lens used, was about 15 mm. In the focal area, 100–130 mm long filaments and white light generation were observed. At these conditions, broadening of the TH emission was also observed. A type-I phase matched beta-barium borate (BBO) crystal of 1 mm thickness was inserted in the beam between the lens and its geometrical focus. The conversion efficiency to the SH ($\lambda = 395$ nm) in BBO was measured to be 8 %. The residual fundamental radiation and SH were assumed to have orthogonal polarizations before entering the filaments. The intensity of fundamental radiation inside the second harmonic crystal was maintained at such a level that no self phase modulation (SPM), appreciable chirp, or white-light generation were introduced on the residual radiation by the BBO crystal. At this conversion efficiency, the filamentation induced by the second harmonic wave itself was observed, when the filter was inserted after the crystal to remove the unconverted fundamental radiation. The characteristics of the light filaments formed by SH femtosecond laser pulses have previously been experimentally investigated [84]. In particular, the critical power required for self-focusing of 395 nm radiation was found to be in the range of 0.3–0.6 GW, which is almost ten times less than the critical power in the case of 790 nm laser.

The influence of pulse characteristics on the 4HG in air was analysed by changing the duration of the output pulse between 48 fs and 2300 fs. The on-axis 4H radiation was studied after propagation through the 2 mm slit placed in front of the dispersive prism. The harmonic radiation propagating through the filaments was dispersed by a fused silica prism and analysed by a fiber spectrometer. The pre-dispersion through the prism was necessary to avoid saturation of the spectrometer by the unconverted fundamental radiation.

Generation of the off-axis TH component due to the phase matching between the fundamental and TH waves started at pump powers around the critical power related with filaments formation. In the far field, the off-axis component appeared as a ring whose diameter corresponded to a phase-matching cone. At pump powers above the critical power for self-focusing, most of the TH energy was concentrated in the ring pattern. It was a cone of diameter of 80 mm at the distance of 1600 mm from the filaments. The same can be said about the 4H radiation, which repeated the same properties. It may be noted that the maximum intensity of the observed 4H appearing as a result of the $\omega + \omega + 2\omega \rightarrow 4\omega$ sum mixing did not correlate with conditions of maximum TH yield. Also note that the maximum intensity of 4H was observed for a red-shifted wavelength ($\lambda \approx 201.5$ nm) with regard to the exact position of the 4H ($\lambda \approx 790/4$ nm $= 197.5$ nm). Efficient FWM was observed in the case of artificially chirped 790 nm pulses, which will be discussed below.

The combined action of self-focusing and plasma generation in air plays an important role in on-axis and off-axis phase matching along the filament, resulting in high conversion efficiency of the THG [65]. THG conversion efficiency in air up to 10^{-3} was achieved, which is in the range of previously reported data [85]. This high conversion efficiency is induced by phase locking between the generated TH wave and the fundamental wave inside the filament, which maintains the phase-matching condition for THG over a long distance.

In the case of two-colour pump, at small fundamental beam energy, only TH radiation on the coaxial direction was observed, while the increase of fundamental beam energy led to the appearance of ring-like structure of the TH and 4H radiation. The transformation of the TH beam shape following the increasing pump energy implies that one or more nonlinear optical effects are playing important role in the THG during the filamentation.

The phase modulation of the laser radiation can easily change the spectrum and intensity of the harmonics. It may be noted that the effective spectral broadening and tuning of the harmonics in the range of few nanometers using the chirp variation can be achieved only in the case of broadband radiation. SPM of the laser pulse is widely used to generate additional frequencies. In particular, insertion of optical medium (like a glass slab) in the path of laser pulse leads to moderate spectral broadening of the laser pulse by SPM. One can expect that propagation of focused intense laser pulse through the air can also cause significant SPM-induced broadening of the driving laser pulse and remarkably change the harmonic spectrum during the formation of filaments.

It was observed that, during filamentation, the propagating laser pulse experiences a modification of its spectral and temporal structures. The SPM due to the refractive index change by plasma formation and by the Kerr effect induces spectral broadening. The SPM in air induces spectral broadening towards the shorter wavelength in the leading edge of the pulse, while the Kerr effect causes the spectral broadening towards the longer wavelength in the leading edge and shorter wavelength in the trailing edge. The spatial and spectral characteristics of the 790 nm radiation creating air filaments were analysed in those studies. The initial spatial shape of the compressed driving beam was relatively smooth, without any significant variations of laser intensity along the beam. The spatial shape of the laser beam creating the filaments and propagating through the air, even after some smoothening during long pass, showed the hot spots created due to the filamentation of the separated parts of beam. Strong SPM and white light generation in the area of filament formation also caused a considerable change of the spectral shape of driving laser beam. The spectra of the driving pulse before and after propagation through the air filaments, considerably differ from each other. While the spectral width of the laser pulse before focusing always remained in the range of 20 nm, the observation of a dramatic increase of spectral width of the focused laser radiation propagating through the air filaments can be attributed to above-mentioned nonlinear optical processes. At the same time, the bandwidth and spectral shape of the 4H remained mostly unchanged in a broad range of variable parameters of experiments, which indicates the nonsaturated conditions of 4HG. The 4HG was observed at the conditions when long filaments were produced in air.

A comparison of the fourth and third harmonic spectra and fundamental spectra was made at different intensities. While the 3H radiation, in most of cases, showed a blue shift, the 4H radiation shifted towards red. The spectral width of 4H emission remained constant ($\sim$1.4 nm) compared to the $\sim$20 nm bandwidth of the propagated fundamental radiation. A significant variation in the fundamental spectrum was observed at intensities close to the ionisation threshold, whereas the 4H spectrum showed a smooth envelope even at considerably higher intensity. This feature can be explained by the dynamics of the sum frequency generation during the initial stage of laser–air interaction. The intense ultrashort pulse produces a temporal variation of the electron density gradients, which makes the spatial and spectral shapes of the self-modulated pulse propagating through the focal spot complicated. The intensity-induced phase modulation could also lead to variations in the spatial shape of fundamental and harmonic radiation [69]. However, in the case of 4H emission, no significant spectral modulation was observed.

The influence of the chirp variation of fundamental radiation on the 4H efficiency was studied. The chirp of fundamental radiation was varied by changing the separation between the gratings in the compressor chamber. Reducing the separation from the chirp-free conditions resulted in positively chirped pulses, and an increase of the distance between the gratings provided negatively chirped pulses. At negative chirp, the pulse contains short wavelength components of the laser spectrum in the leading part of the pulse and vice versa. The chirp contained in the laser pulse also affects the temporal structure of the laser pulse.

The influence of pulse duration and chirp variation on the 4H emission was studied at fixed pulse energy. The pulse duration of the fundamental radiation was gradually increased from 48 to 2300 fs by changing the distance between the compressor

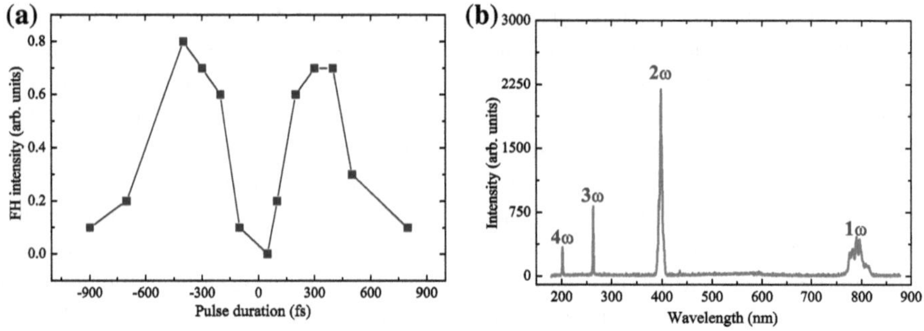

Fig. 1.15 **a** 4H output as a function of pulse duration for positively and negatively chirped laser radiation. Positive and negative values of pulse duration correspond to positively and negatively chirped pulses. **b** Harmonic spectra at optimal conditions for the 4HG. Reproduced from [83] with permission from American Physical Society

gratings. Figure 1.15a shows the $I_{4\omega}(t_p)$ dependence on pulse duration, at a fundamental pulse energy of $E = 44$ mJ. The optimal pulse duration (300–400 fs) was defined, which allowed generation of the maximum 4H intensity. No 4H was obtained at zero chirp.

Another behaviour was observed when the intensity of fundamental radiation was varied by changing the pulse energy. The reason for this could be some change in the divergence of the fundamental beam, as well as different roles of the radiation in the generation of free electrons at different pulse durations. The partial separation between fundamental and second harmonic pulses in the filaments cannot fully exclude the FWM. At the same time, with increasing the chirp, the pulse length increases such that the overlap of the two pulses in the focal area is increased. This may lead to conversion over a narrow bandwidth (as not all wavelengths in the two chirped pulses overlap), as was observed in experiment, and it also explains the symmetric dependence of efficiency for negative and positive chirps (Fig. 1.15a).

Now we analyse the influence of the delay between the two chirp-free pump pulses (ω and 2ω) inside the filaments on the nonlinear processes studied. Due to the group velocity mismatch between the ω and 2ω waves in the type I BBO crystal, the 790 nm beam is delayed $\{\Delta_{cryst} = l_{cryst}[(n^o_\omega)_{group}/c - (n^e_{2\omega})_{group}/c] \approx 188$ fs for 1 mm thick beta-barium borate$\}$ with respect to the 395 nm beam due to $n^o_\omega > n^e_{2\omega}$ in this negative uniaxial crystal $\{n^o_\omega \approx 1.6605$, $n^e_{2\omega} \approx 1.5681$ [86], $(c/n^o_\omega)_{group}$ and $(c/n^e_{2\omega})_{group}$ are the group velocities for the o and e waves in BBO at 790 nm and 395 nm respectively, and $n_{group} = n - \lambda \times (dn/d\lambda)$ is the group refractive index calculated from the Sellmeier relations$\}$. The duration of the 395 nm pulse is given by $\Delta t_{2\omega} \approx \{(\Delta_{cryst})^2 + 0.5(\Delta t_\omega)^2\}^{1/2}$ [87]. Hence, the 395 nm beam has a longer pulse duration, corresponding to the induced delay and a certain percentage (~50 %) of the 790 nm pulse duration. The latter is because the energy of the fundamental radiation is in general not high enough in the pulse wings to effectively generate the second harmonic. Thus, one can estimate $\Delta t_{2\omega}$ at the output of the 1 mm thick BBO crystal to be about 191 fs. The SH crystal was inserted in the experimental scheme at appropriate position after the focusing lens, so that no impeding processes were observed after propagation of the laser radiation through the crystal. After leaving the crystal, the delay between the two pulses gets reduced due to propagation in air [$n_{\omega}^{air} \approx 1.000274, n_{2\omega}^{air} \approx 1.000283, (n_{\omega}^{air})_{group} \approx 1.000280$ $(n_{2\omega}^{air})_{group} \approx 1.000304$]. In the area of filaments formation, which was approximately $L = 700$ mm away from the BBO crystal, the delay between two pulses decreases by $\Delta_{air} = L[(n_{2\omega}^{air})_{group}/c - (n_{\omega}^{air})_{group}/c] \approx 55$ fs. Thus, the final delay between two pumps inside the filaments became ≈133 fs.

At these conditions (i.e. for chirp-free pulses), no 4H was observed. Although the 48 fs, 790 nm pulse partially overlaps with the trailing part of the 191 fs, 395 nm pulse, the conditions for parametric four wave mixing remained far from optimum. The increase of fundamental pulse duration by introducing the positive or negative chirp led to considerable improvement in the overlapping of two pumps inside the filaments. From the experimental data (Fig. 1.15a), one can define a range of fundamental pulse duration (300–400 fs) at which the most efficient FWM occurs. This range corresponds to the most effective overlap of the chirped fundamental pulse and broadened second harmonic pulse, which led to strongest 4HG. Longer fundamental pulses (>400 fs) caused a decrease of 4H efficiency due to considerable decrease of the laser intensity. Thus, the optimisation of the pulse duration is due to interplay between the overlapping of the pump pulses in the filaments and their relative intensities. These studies showed that the 4HG takes place despite the significant difference between the two pump intensities ($I(\omega)$: $I(2\omega) \cong 12 : 1$), and without any precise temporal and spatial matching of the two pumps. These observations have demonstrated that, by appropriate chirping of the fundamental radiation, one can achieve a considerable enhancement of the 4H yield.

Figure 1.15b shows the spectrum when all three (second, third, and fourth) harmonics, together with partially filtered fundamental radiation, were recorded. The relative intensities of these harmonics cannot be taken into consideration, since the conditions for recording of these spectral components were optimised only for the 4H radiation. The harmonics, other than 4H, were seen in this spectrum since their scattered radiation was quite strong.

The influence of the fundamental and second harmonic intensities on the 4H output was also studied. The yield of the 4H increased almost quadratically with the fundamental wave intensity (Fig. 1.16, scaling parameter $l \approx 2.3$), while the $I_{4\omega}(I_{2\omega})$ plot showed approximately linear dependence ($l \approx 0.8$). Such behaviour is expected in a perturbative regime of the sum mixing [33]. The FWM intensity in that case should follow the dependence $I_{4\omega} \sim (I_{1\omega})^2 \times I_{2\omega}$, which was close to the experimental observations.

Finally, we discuss the results of FWM studies using the variation of polarizations of the pump pulses. Initially, the effect of circularly polarized fundamental radiation on the generation of the 4H was studied. A quarter-wave plate was inserted in front of the focusing lens to change the polarization of the laser beam. In that case also the 4HG was observed, although its intensity was two to three times less compared to that with linearly polarized fundamental beam. The observation of the fourth harmonic generation in the case of circularly polarized incident laser light can be explained

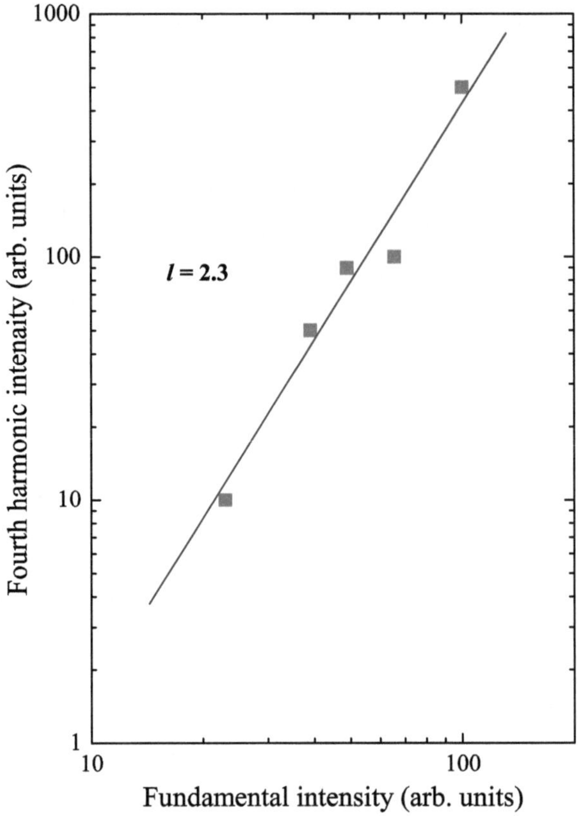

Fig. 1.16 Dependence of the 4H output on the fundamental radiation intensity. Adopted from [83] with permission from American Physical Society

as follows. Circularly polarized light is composed of two orthogonal linear polarizations differing in phase by a quarter wave. Of these, the component, which is an ordinary ray for the BBO crystal (type I phase matched), gets converted to SH, and the other orthogonal component, which becomes extraordinary for the BBO crystal goes through the crystal unconverted (no phase matching). As the refractive indices and group velocities for the ordinary and extraordinary components (of the fundamental) in the BBO crystal are quite different, the two components get separated in time and do not give rise to circularly polarized fundamental, but act as independent radiation pulses. So we have a case of second harmonic (extraordinary) and fundamental (ordinary) orthogonal to each other (two-colour field), and the remaining fundamental field (unconverted extraordinary, single colour field) producing the fourth harmonic radiation independently.

Subsequently, the polarizations of two beams (ω and 2ω) were made parallel by inserting a $\lambda/2$ plate after the BBO crystal. In this case, at the optimally chirped fundamental radiation, the 4HG was decreased, probably due to the group velocity dispersion (GVD) between these two pulses (ω and $2\,\omega$), which get partially separated in time. To confirm the influence of GVD on the 4HG efficiency, optical glasses of different thickness were inserted after the nonlinear crystal. With 10 mm thick glass slides inserted after the BBO crystal, no 4H was observed. 4H started appearing at the thickness of glass slides of 1 mm. With a microslide of thickness of 0.3 mm, the intensity of 4H became stronger (about an order of magnitude higher than in the case of 1 mm thick glass slide), while it further increased (by a factor of about 3) when no optical materials were inserted in the path of fundamental and second harmonic waves. The optimisation of the conditions for strongest 4H output was the same as earlier, i.e. when the negatively chirped orthogonal polarized 300 fs pulses were used.

Apart from the phase modulation, another parameter of a femtosecond laser pulse, which also changes during filamentation of the laser beam, is the polarization of radiation. Initial linear polarization of the laser radiation changes after creation of filaments [88]. Various random directions of polarization appear in this radiation during propagation through the focus of lens, together with the prevailing linear polarization component. The presence of harmonics even after rotation of polarization by quarter- and half-wave plates could be attributed to the simultaneous rotation of randomly rotated polarizations of the parts of laser beam, which can cause the appearance of linearly polarized components in overall pattern of polarization state of the laser beam. These components are still strong enough and can cause the generation of low-order harmonics and/or low-order parametric processes. The harmonic intensity in that case is defined by the intensity of these weak linearly polarized components. The existence of harmonics during introduction of considerable ellipticity and depolarisation is in a stark contrast with conventionally accepted case of a strong influence of the deviation from linear polarization on the harmonic output (both for low- and high-order harmonics).

Measurement of the 4HG conversion efficiency was carried out relative to the THG conversion efficiency. The relative intensities of these two waves were measured to be $I_{3\omega} : I_{4\omega} \approx 10 : 1$, which gives the 4HG conversion efficiency to be $\sim 10^{-4}$. The conversion efficiency of fourth harmonic was also estimated by direct comparison of the intensities of ω and 4ω waves. Their relative intensities were measured using a calibrated photodiode after dispersing the radiation in the fused silica prism. In that case, the 4HG conversion efficiency was estimated to be 2×10^{-4}, which agrees with above-mentioned value (10^{-4}) within a factor of two.

1.5 Low-Order Harmonic Generation During Interaction of Laser Radiation with Surfaces

HG from solid surfaces is caused due to oscillation of the boundary layer ("moving mirror" model [89]) or impulse-like acceleration of electrons [90]. Laser pulse duration is an important parameter, which governs the mechanism and the dynamics of the processes involved in HG. In this section, we present the studies of low-order harmonics from surfaces using picosecond [91] and femtosecond pulses [92].

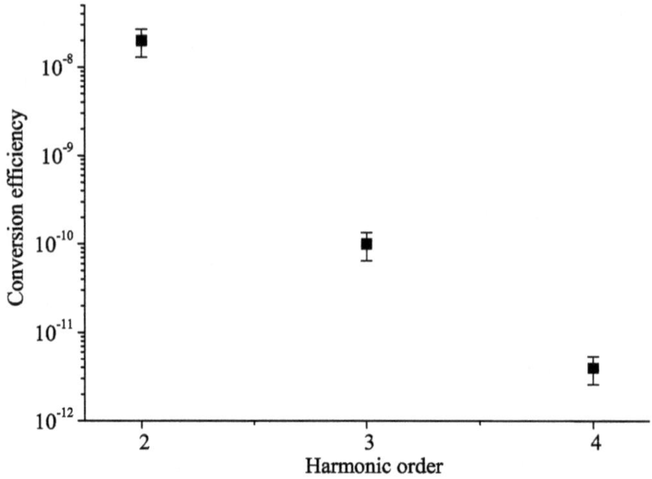

Fig. 1.17 Conversion efficiencies of low-order harmonics for p-polarized pump radiation. Reproduced from [91] with permission from American Physical Society

1.5.1 Picosecond Radiation

The Nd:glass laser provided 27 ps pulses of energy of 2.5 J. The contrast ratio for prepulse and fundamental pulse energies was measured to be smaller than 10^{-5}. The laser beam was focused on an aluminium target placed in a vacuum chamber. The target was prepared by coating of 100 μm thick aluminium film on optically polished planar glass plates. The laser beam was focused on the target using a 750 mm focal length lens at an angle of incidence of 67.5° with a peak intensity of 1.5×10^{15} W cm^{-2}. The target was moved after each shot so that every next pulse of laser radiation interacted with a fresh part of the target. The depolarisation factor of the incident radiation was smaller than 10^{-3}. The Faraday rotator was used to change the polarization of incident radiation from p to s and vice-versa. SH, TH, and 4H radiation were spectrally separated and detected by PMT. Temporal profiles of fundamental radiation and SH were recorded by a streak-camera. Pulse duration of SH radiation (23 ps) was close to that of fundamental radiation (27 ps).

The SHG conversion efficiency was measured to be 2×10^{-8} for p-polarized radiation, and ~10 times smaller for s-polarized radiation. The intensity-dependent measurements can be represented by a power law of $I_{2\omega} \propto I^{1.5}$ for p-polarized pump, and $I_{2\omega} \propto I^{1.7}$ for s-polarized pump. TH and 4H were also observed in those experiments. The intensity scaling exponents for these two harmonics were 1.8 and 3.8, respectively. Conversion efficiencies of various observed harmonics for p-polarized pump radiation are presented in Fig. 1.17.

The increase in the s-polarization induced second harmonic yield with the growth of the fundamental laser intensity was consistent with the increase of Faraday rotation of laser radiation. The spectrum of second harmonic was broadened and red-shifted with respect to the second harmonic wavelength of fundamental radiation. The large difference in the ratio of p- and s-polarization induced second harmonic conversion efficiencies can be understood from the polarization rotation of s-polarized fundamental beam due to Faraday rotation. This occurs due to spontaneously generated mega gauss magnetic fields in laser-produced plasmas, which becomes important for longer duration laser pulses. Faraday rotation measurements were performed for both p- and s-polarized fundamental beams. Laser radiation specularly reflected from the target was analysed with a Glan prism. The calibrated energy meters were used for the measurements of the energies of two orthogonal components. The angle of rotation φ_{rot} was derived from the relative intensities measured by the detectors using the relation $\varphi_{ort} = arctg(E_{ort}/E_{par})$, where E_{par} and E_{ort} are the energies of reflected fundamental radiation of the parallel and orthogonal polarizations. It was shown that no noticeable rotation occurs in the case of p-polarized beam. On the other hand, a rotation of 9° was measured at a laser intensity of $10^{15}\,\mathrm{W\,cm^{-2}}$ for s-polarized laser radiation.

In order to confirm that the above stated result is not affected due to any possible contribution of depolarisation of the reflected fundamental radiation, the measurements were repeated by rotating the Glan prism. The minimum value of the normalized rotation was observed at 10° rotation of the Glan prism, which is in agreement with the aforesaid Faraday rotation angle measurements (9°). This observation confirms that the fundamental radiation does not undergo any significant depolarisation, and there is a definite Faraday rotation for the s-polarized fundamental radiation.

A significant energy of second harmonic radiation generated by s-polarized fundamental radiation can now be explained from the Faraday rotation of the latter. Due to longer pulse duration of 27 ps (in comparison with previous studies where the laser pulses up to 2.5 ps were used), the plasma interaction length and hence the Faraday rotation undergone by the fundamental laser radiation increases. The relative increase in the s-polarization induced SH yield with increase of the fundamental laser intensity is also coincides with the observation of increase in Faraday rotation angle with laser intensity. The latter can be explained by the increase of absorption of the laser energy. It was shown that the reflectivity (~50 %) at small intensities (~$10^{13}\,\mathrm{W\,cm^{-2}}$) decreases up to ~15 % at intensities exceeding $10^{15}\,\mathrm{W\,cm^{-2}}$. This leads to an increase in the spontaneously generated mega gauss magnetic fields in the plasma, which then results in to the increase of polarization rotation.

1.5.2 Femtosecond Radiation

These experiments were carried out using a Ti:sapphire–Nd:glass hybrid laser operating at a wavelength of 1054 nm. The contrast ratio of prepulse and laser pulse was less than 10^{-6}. The laser delivered 200 mJ, 475 fs pulses and produced a maximal peak intensity on the target of $3 \times 10^{17}\,\mathrm{W\,cm^{-2}}$ at a focal spot of 12 μm diameter. The laser pulse was focused onto the surface of a solid target using a 10 cm focal length achromatic lens at an incidence angle of 45° with either s- or p-polarization selected using a $\lambda/2$ wave plate. The target was a 0.1 μm layer of aluminium deposited on a glass substrate. The SH studies were inconsistent with the resonant absorption because the second harmonic light was blueshifted by 1.6 nm for p-polarization. The 5H light was blueshifted by 5.1 nm for p-polarization at the intensity of $I \geq 1 \times 10^{17}$ W cm^{-2}. This shift was much larger than the Doppler shift. Therefore, these results cannot be explained by the hydrodynamic motion of plasma. With increasing laser intensity and electron temperature the role of collisions decrease and the collisionless mechanisms of absorption become dominant.

At a laser intensity of $\sim 1 \times 10^{17}$ W cm^{-2}, the conversion efficiencies of the SHG and 5HG were 1×10^{-6} and 4×10^{-8}, respectively. It was found that for both harmonics the harmonic yield dependence on the laser polarization decreased at laser intensity exceeding 1×10^{17} W cm^{-2}. SH and 5H radiation generated by s-polarized pump were 25 and 6 times smaller compared to that for the p-polarized laser radiation at the intensities below $I = 10^{17}$ W cm^{-2}. The same ratios were decreased to 2.4 and 1.5 at $I > 2 \times 10^{17}$ W cm^{-2} [93]. Different processes can be involved in decreasing of the I_p/I_s ratio with the growth of laser intensity. One of them is the rippling caused by Rayleigh-Taylor-like instability at the critical density surface [94].

References

1. H.-J. Choi, M. Cha, J. Nonlin. Opt. Phys. Mater. **20**, 501 (2011)
2. S.K. Das, C. Schwanke, A. Pfuch, W. Seeber, M. Bock, G. Steinmeyer, T. Elsaesser, R. Ruediger, Opt. Express **19**, 16985 (2011)
3. R. Cisek, V. Barzda, H.E. Ruda, A. Shik, IEEE J. Select. Topics Quantum Electron. **17**, 915 (2011)
4. I.V. Soboleva, S.A. Seregin, A.A. Fedyanin, O.A. Aktsipetrov, J. Opt. Soc. Am. B **28**, 1680 (2011)
5. G.X. Li, T. Li, H. Liu, K.F. Li, S.M. Wang, S.N. Zhu, K.W. Cheah, Appl. Phys. Lett. **98**, 261909 (2011)
6. G.S. Qin, M.S. Liao, C. Chaudhari, X. Yan, C. Kito, T. Suzuki, Y. Ohishi, Opt. Lett. **35**, 58 (2010)
7. M. Oujja, R. de Nalda, M. Lopez-Arias, R. Torres, J.P. Marangos, M. Castillejo, Phys. Rev. A **81**, 043841 (2010)
8. P.A. Franken, A.E. Hill, C.W. Peters, G. Weinreich, Plys. Rev. Lett. **7**, 118 (1961)
9. T. Harimoto, B. Yo, K. Uchida, Opt. Express **19**, 22692 (2011)
10. K. Miyata, V. Petrov, F. Noack, Opt. Lett. **36**, 3627 (2011)
11. J.X. Zhang, L.R. Wang, Y. Wu, G.L. Wang, P.Z. Fu, Y.C. Wu, Opt. Express **19**, 16722 (2011)
12. B. Kulyk, V. Kapustianyk, Y. Burak, V. Adamiv, B. Sahraoui, Mater. Chem. Phys. **120**, 114 (2010)
13. S.E. Mani, J.I. Jang, J.B. Ketterson, Opt. Lett. **34**, 2817 (2009)
14. Y. Zhou, G.L. Wang, Y.C. Yue, C.M. Li, Y.F. Lu, D. Cui, Z.G. Hu, Z.Y. Xu, Opt. Lett. **34**, 746 (2009)
15. A.G. Akmanov, S.A. Akhmanov, B.V. Zhdanov, A.I. Kovrigin, I.K. Podsotskaya, R.V. Khokhlov, JETP Lett. **10**, 154 (1969)
16. I.A. Begishev, R.A. Ganeev, A.A. Gulamov, E.A. Erofeev, S.R. Kamalov, T. Usmanov, A.D. Khodjaev, Quantum Electron. **18**, 224 (1988)
17. R.A. Ganeev, I.A. Kulagin, U.K. Sapaev, T. Usmanov, Opt. Spectrosc. **89**, 308 (2000)
18. Y. Wang, R. Dragila, Phys. Rev. A **41**, 5645 (1990)

19. R.A. Ganeev, T. Usmanov, Appl. Phys. B **65**, 41 (1997)
20. R. Danelius, A. Dubietis, G. Valiulis, A. Piskarskas, Opt. Lett. **20**, 2225 (1995)
21. R.A. Ganeev, T. Usmanov, J. Appl. Spectrosc. **64**, 544 (1997)
22. R.A. Ganeev, J. Opt. A **6**, S3 (2004)
23. H. Hoshi, N. Nakamura, Y. Maruyama, T. Nakagawa, S. Suzuki, H. Shiromaru, J. Achiba, Jpn. J. Appl. Phys. **30**, L1397 (1991)
24. J.S. Meth, H. Vanherzeele, Y. Wang, Chem. Phys. Lett. **197**, 26 (1992)
25. R.A. Ganeev, A.I. Ryasnyansky, N.V. Kamanina, I.A. Kulagin, M.K. Kodirov, T. Usmanov, J. Opt. B **3**, 88 (2001)
26. R.A. Ganeev, A.I. Ryasnyansky, M.K. Kodirov, S.R. Kamalov, V.A. Li, R.I. Tugushev, T. Usmanov, Eur. Phys. J. D **20**, 129 (2002)
27. K.K. Innes, B.P. Stoichev, S.C. Wallace, Appl. Phys. Lett. **29**, 715 (1976)
28. M.N.R. Ashfold, C.D. Heryet, J.D.B. Tutcher, Chem. Phys. Lett. **131**, 291 (1986)
29. A.S. Aleksandrovsky, S.V. Karpov, S.A. Myslivets, A.K. Popov, V.V. Slabko, J. Phys. B **26**, 2965 (1993)
30. R.A. Ganeev, S.R. Kamalov, M.K. Kodirov, M.R. Malikov, A.I. Ryasnyansky, R.I. Tugushev, S.U. Umidullaev, T. Usmanov, Opt. Commun. **184**, 305 (2000)
31. K. Ruegenberg, C.W. Scherr, J. Chem. Phys. **21**, 1565 (1953)
32. T. Kitagava, Molec. Spektrosc. **26**, 1 (1968)
33. J.F. Reintjes, *Nonlinear Optical Parametrical Processes in Liquids and Gases* (Academic Press, London, 1984)
34. A.B. Fedotov, S.M. Gladkov, N.I. Koroteev, A.M. Zheltikov, J. Opt. Soc. Am. B **8**, 363 (1991)
35. R.A. Ganeev, G.S. Boltaev, R.I. Tugushev, T. Usmanov, M. Baba, H. Kuroda, J. Opt. **12**, 055202 (2010)
36. R.A. Ganeev, J. Phys. B: At. Mol. Opt. Phys. **40**, R213 (2007)
37. R.A. Ganeev, H. Kuroda, Opt. Spectrosc. **109**, 921 (2010)
38. M. López-Arias, M. Oujja, M. Sanz, R.A. Ganeev, G.S. Boltaev, NKh Satlikov, R.I. Tugushev, T. Usmanov, M. Castillejo, J. Appl. Phys. **111**, 043111 (2012)
39. R.A. Ganeev, M. Suzuki, M. Baba, H. Kuroda, T. Ozaki, Opt. Lett. **30**, 768 (2005)
40. S.E. Harris, Phys. Rev. Lett. **31**, 341 (1973)
41. R.A. Ganeev, I.A. Kulagin, I.A. Begishev, V.I. Redkorechev, T. Usmanov, Nonlin. Opt. **16**, 109 (1996)
42. J.L. Bobin, M. Decroisette, B. Meyer, Y. Vitel, Phys. Rev. Lett. **30**, 594 (1973)
43. J. Meyer, Y. Zhu, Phys. Rev. Lett. **71**, 2915 (1993)
44. D.A. Russell, D.F. DuBois, Phys. Rev. Lett. **86**, 428 (2001)
45. G.P. Gupta, T.K. Achal, S.V. Gogawale, B.K. Sinha, Plasma. Phys. Control. Fusion **37**, 1277 (1995)
46. L. Veisz, W. Theobald, T. Feurer, H. Schillinger, P. Gibbon, R. Sauerbrey, Phys. Plasmas **9**, 3197 (2002)
47. P.E. Young, B.F. Lasinski, W.L. Kruer, E.A. Williams, K.G. Estabrook, E.M. Campbell, R.P. Drake, Phys. Rev. Lett. **61**, 2766 (1988)
48. R.A. Ganeev, M. Suzuki, M. Baba, M. Turu, H. Kuroda, Appl. Phys. B **78**, 79 (2004)
49. H.C. Pant, K. Eidmann, P. Sachsenmaier, R. Sigel, Opt. Commun. **16**, 396 (1976)
50. W. Seka, B.B. Afeyan, R. Boni, L.M. Goldman, R.W. Short, K. Tanaka, Phys. Fluids **28**, 2570 (1985)
51. A. Simon, R.W. Short, E.A. Williams, T. Dewandre, Phys. Fluids **26**, 3107 (1983)
52. I.B. Foldes, J.S. Bakos, G. Veres, Z. Bakonyi, T. Nagy, S. Szatmari, IEEE J. Sel. Top. Quantum Electron. **2**, 776 (1996)
53. R.A. Ganeev, I.A. Kulagin, T. Usmanov, S.T. Khudaiberganov, Quantum Electron. **12**, 1637 (1982)
54. R.A. Ganeev, V.V. Gorbushin, I.A. Kulagin, T. Usmanov, Appl. Phys. B **41**, 69 (1986)
55. R.A. Ganeev, V.V. Gorbushin, I.A. Kulagin, T. Usmanov, Quantum Electron. **16**, 115 (1986)
56. R.A. Ganeev, T. Usmanov, Quantum Electron. **26**, 903 (1996)
57. R.A. Ganeev, T. Usmanov, J. Opt. A **2**, 550 (2000)

58. H. Yang, J. Zhang, J. Zhang, L.Z. Zhao, Y.J. Li, H. Teng, Y.T. Li, Z.H. Wang, Z.L. Chen, Z.Y. Wei, J.X. Ma, W. Yu, Z.M. Sheng, Phys. Rev. E **67**, 015401 (2003)
59. A.B. Fedotov, N.I. Koroteev, M.M.T. Loy, X. Xiao, A.M. Zheltikov, Opt. Commun. **133**, 587 (1997)
60. N. Aközbek, A. Becker, M. Scalora, S.L. Chin, C.M. Bowden, Appl. Phys. B **77**, 177 (2003)
61. N. Aközbek, A. Iwasaki, A. Becker, M. Scalora, S.L. Chin, C.M. Bowden, Phys. Rev. Lett. **89**, 143901 (2002)
62. A.B. Fedotov, A.N. Naumov, V.P. Silin, S.A. Uryupin, A.M. Zheltikov, A.P. Tarasevitch, D. von der Linde, Phys. Lett. A **271**, 407 (2000)
63. S. Baskus, J. Peatross, Z. Zeek, A. Rundquist, G. Taft, M.M. Murnane, H.C. Kapteyn, Opt. Lett. **21**, 665 (1996)
64. E.V. Vanin, A.V. Kim, A.M. Sergeev, M.C. Downer, JETP Lett. **58**, 900 (1993)
65. F. Théberge, N. Aközbek, W. Liu, J.-F. Gravel, S.L. Chin, Opt. Commun. **245**, 399 (2005)
66. J. Peatross, S. Baskus, J. Zhou, M.M. Murnane, H.C. Kapteyn, J. Opt. Soc. Am. B **15**, 186 (1998)
67. T. Brabec, F. Krausz, Rev. Mod. Phys. **72**, 545 (2000)
68. A. L'Huillier, L.A. Lompre, M. Ferray, X.F. Li, G. Mainfray, C. Manus, Europhys. Lett. **5**, 601 (1988)
69. R.A. Ganeev, M. Suzuki, M. Baba, H. Kuroda, I.A. Kulagin, Appl. Opt. **45**, 748 (2006)
70. Y.-D. Qin, H. Yang, C.-J. Zhu, Q. Gong, Appl. Phys. B **71**, 581 (2000)
71. X. Liu, D. Umstadter, E. Esarey, A. Ting, IEEE Trans. Plasma Sci. **21**, 90 (1993)
72. M.S. Malcuit, R.W. Boyd, W.V. Davis, K. Rzaewski, Phys. Rev. A **41**, 3822 (1990)
73. K.N. Drabovich, I.A. Kulagin, T. Usmanov, Quantum Electron. **15**, 402 (1985)
74. J.L. Ni, J.P. Yao, B. Zeng, W. Chu, G.H. Li, H.S. Zhang, C.R. Jing, S.L. Chin, Y. Cheng, Z. Xu, Phys. Rev. A **84**, 063846 (2011)
75. G.O. Ariunbold, P. Polynkin, J.V. Moloney, Opt. Express **20**, 1662 (2012)
76. H. Xu, W. Chu, Y. Liu, W. Liu, H. Xiong, Y. Fu, J. Yao, B. Zeng, J. Ni, S.L. Chin, Y. Cheng, Z. Xu, Appl. Phys. B **104**, 909 (2011)
77. J.P. Yao, B. Zeng, W. Chu, J.L. Ni, Y. Cheng, J. Modern Opt. **59**, 245 (2012)
78. E. Schulz, D.S. Steingrube, T. Vockerodt, T. Binhammer, U. Morgner, M. Kovacev, Opt. Lett. **36**, 4389 (2011)
79. R.T. Hodgson, P.P. Sorokim, J.J. Wynne, Phys. Rev. Lett. **32**, 342 (1974)
80. F. Théberge, N. Aközbek, W.W. Liu, A. Becker, S.L. Chin, Phys. Rev. Lett. **97**, 023904 (2006)
81. T. Fuji, T. Horio, T. Suzuki, Opt. Lett. **32**, 2481 (2007)
82. T. Fuji, T. Suzuki, Opt. Lett. **32**, 3330 (2007)
83. R.A. Ganeev, H. Singhal, P.A. Naik, J.A. Chakera, M. Kumar, P.D. Gupta, Phys. Rev. A **82**, 043812 (2010)
84. Z. Zhang, X. Lu, T.-T. Xi, W.-X. Liang, Z.-Q. Hao, Y. Zhang, M.-L. Zhou, Z.-H. Wang, J. Zhang, Appl. Phys. B **97**, 207 (2009)
85. M.L. Naudeau, R.J. Law, T.S. Luk, T.R. Nelson, S.M. Cameron, J.V. Rudd, Opt. Express **14**, 6194 (2006)
86. V.G. Dmitriev, G.G. Gurzadyan, D.N. Nikogosyan, *Handbook of Nonlinear Optical Crystals* (Springer, Heidelberg, 1999)
87. G. Lambert, J. Gautier, C.P. Hauri, P. Zeitoun, C. Valentin, T. Marchenko, F. Tissandier, J.P. Goddet, M. Ribiere, G. Rey, M. Fajardo, S. Sebban, New J. Phys. **11**, 083033 (2009)
88. R.A. Ganeev, H. Kuroda, Appl. Phys. Lett. **95**, 201117 (2009)
89. S.V. Bulanov, N.M. Naumova, F. Pegoraro, Phys. Plasmas **1**, 745 (1994)
90. P. Gibbon, Phys. Rev. Lett. **76**, 50 (1996)
91. R.A. Ganeev, J.A. Chakera, M. Raghuramaih, A.K. Sharma, P.A. Naik, P.D. Gupta, Phys. Rev. E **63**, 026402 (2001)
92. A. Ishizawa, R.A. Ganeev, T. Kanai, H. Kuroda, T. Ozaki, Phys. Rev. E **66**, 026414 (2002)
93. R.A. Ganeev, A. Ishizawa, T. Kanai, T. Ozaki, H. Kuroda, Opt. Commun. **227**, 175 (2003)
94. C.A. Popovici, R.A. Ganeev, F. Vidal, T. Ozaki, J. Phys. B: At. Mol. Opt. Phys. **45**, 035601 (2012)

Chapter 2
High-Order Harmonic Generation from Laser Ablation of Various Surfaces

The high-order harmonic generation in laser-produced plasma plumes can be useful for producing an efficient source of short-wavelength ultra-short pulses for various applications and studies of the properties of harmonic emitters. The laser ablation induced high-order harmonic generation spectroscopy is a new method for the studies of material science. We discuss the realization of new ideas, which further improved the frequency conversion efficiency through harmonic generation in specially prepared plasmas and allowed the spectral and structural studies of matter through the plasma harmonic spectroscopy. We also present the current status of plasma harmonic studies, and show new trends in the developments of this filed. In particular, we show new approaches in high-order harmonic generation from various plasmas for the studies of the physical properties of materials.

Coherent short wavelength radiation is of increasing importance for a broad variety of basic and applied research in various fields of physical, chemical, and life sciences. Among them, femtosecond time-resolved coherent diffractive imaging and photo-induced processes at surfaces and nanoparticles, as well as lithography, plasma diagnostics, and materials science processing and diagnostics are of foremost interest. The process of high-order harmonic generation from femtosecond visible laser pulses allows producing coherent radiation in the extreme ultraviolet spectral range. Table-top lasers render these processes possible with the prospect of wide-spread scientific applications. Presently predominantly few gases are employed as target media for HHG. Other efforts rely on the creation of harmonics at solid surfaces at relativistic laser intensities above $10^{18-19}\,\mathrm{W\,cm^{-2}}$ and with extremely high pulse contrast ratio, either by a coherent wake field excitation or, for petawatt class lasers, on a relativistically moving electron gas acting as a plasma mirror. So far, however, only low conversion efficiencies have been obtained, despite the enormous efforts. Other methods include XUV free-electron lasers (FELs) and X-ray lasers.

To promote the use of XUV radiation it seems therefore appropriate to advance laboratory scale sources to a higher application level. Many interesting experiments can be performed by HHG based on laboratory scale femtosecond lasers. These sources easily cover the spectral range between 10 and 100 eV photon energy of harmonics, and with few-cycles laser systems even up to several 100 eV. For practical

R. A. Ganeev, *Nonlinear Optical Properties of Materials*,
Springer Series in Optical Sciences 174, DOI: 10.1007/978-94-007-6022-6_2,

applications of high-order harmonic sources a higher conversion efficiency and thus an increase in the photon flux and also of the maximum photon energy of the harmonic radiation would be beneficial. HHG itself can be used as a spectroscopic tool for analysis of optical, nonlinear optical and structural properties of the emitters of harmonic generation presently comprising on a few noble gases. The generation of high harmonics in laser-produced plasmas on various solid-state targets, being for this purpose a relatively new and largely unexplored medium, promises to yield these advances.

Plasma HHG can open new doors in many unexpected areas of light-matter interaction. Besides considering as an alternative method for generation of coherent XUV radiation, it can be used as a powerful tool for various spectroscopic and analytical applications. A few of them have already emerged during recent years of plasma HHG studies. The application of doubly charged ions for high-order harmonic generation showed a promising extension of the cut-off photon energy in plasma harmonics, without having to rely on few-cycle driving pulses. As it has been shown in the case of low-order harmonics, the conversion efficiency can be strongly enhanced by making use of resonances in atomic or ionic systems. This has been demonstrated in pioneering plasma harmonic experiments using the indium and other metals. For laser-generated plasmas a large variety of materials can be employed, thereby increasing the chance to select such resonances with fixed-frequency Ti:sapphire lasers. Furthermore, it has been shown that two-colour pumping profitably enhances the high-harmonic intensity and significantly influences the output and properties of harmonic spectrum in rare gases. For plasma harmonics, where this technique has recently been adopted, this will be a new approach for the nonlinear spectroscopy of ionic transitions possessing high oscillator strengths.

Most interestingly, recent studies have shown that enhanced high-order harmonics can be generated also from the ablated nanoparticles, which opens the prospects for applications of local field enhancement, broad plasmonic resonances and a more efficient recombination processes for plasma HHG. As a highly interesting perspective an increase of the harmonic output by quasi phase matching in specially prepared plasmas may be considered. For the plasma a different and more flexible technique than used in neutral gases can be applied. The plasma may be spatially modified using a long pulse co-propagating with the fundamental driving pulse, and conditions might be found where quasi phase matching is possible over a long distance in the plasma, while the constructive and destructive interference in such plasmas containing different emitters can provide a new knowledge about the phase-related characteristics of this process.

Thus the above approach can be useful for producing an efficient source of short-wavelength ultra-short pulses for various applications and studies of the properties of harmonic emitters. The laser ablation induced high-order harmonic generation spectroscopy is a new method for the studies of material science and one of the most important applications of HHG. In this chapter, we discuss the realization of new ideas, which further improved the HHG efficiency through harmonic generation in specially prepared plasmas and allowed the spectral and structural studies of matter through the plasma harmonic spectroscopy. We also present the current status of plasma HHG studies, and show new trends in the developments of this filed. In particular, we show new approaches in HHG from various plasmas for the studies of physical properties of materials.

2.1 Current Status of Plasma Harmonic Studies

High-order harmonic generation may presently be considered the simplest and most efficient technique of obtaining coherent short-wavelength radiation in a broad spectral range [1–9]. As it was already mentioned, alternative means in this area are the use of X-ray lasers [10, 11] and FELs [12]. However, unlike sources involving harmonic generation, X-ray lasers have been unable to generate radiation in a broad range of the extreme ultraviolet spectral domain. Other disadvantages of X-ray lasers are their poor spatial coherence and radiation divergence. As regards FELs that generate radiation in the XUV spectral range, there are only a few such lasers, which are limited in number. Furthermore, the application of these lasers is largely restricted by their extremely high development and maintenance cost.

HHG research is actively being pursued due to the availability of new high-power compact laser systems offering high output parameters (high energy and intensity of pulses and a high pulse repetition rate). Two mechanisms are used for HHG: harmonic generation in gases [1, 3–7] and from surfaces [2, 8, 9]. The considerable progress achieved in this area has enabled extending the range of generated coherent radiation toward the spectral region where the radiation can pass through water-bearing components (the so-called water window, 2.3–4.6 nm [6, 7]). This circumstance is attractive from the standpoint of the practical use of coherent short-wavelength radiation in studies of biological objects. However, the data on the generation of such radiation obtained to date with the use of the above techniques have exhibited a low conversion efficiency to the XUV range (10^{-6} and below), which considerably limits their practical use. This is supposedly the reason why in the last few years the emphasis has been placed on the optimisation of another effect discovered in these investigations, the generation of attosecond pulses [13].

The search for ways of increasing the HHG efficiency in the XUV spectral range has long been (and still is) among the most topical problems of nonlinear optics. However, in the majority of cases, the efficiency of conversion to high-order harmonics turns out to be insufficient for using them as real coherent short-wavelength radiation sources in biology, plasma diagnostics, medicine, microscopy, photolithography, XUV coherent diffraction imaging, time-resolved measurements, etc. The feasibility of increasing the intensities of high-order harmonics generated in gas jet sources by using atomic and ion resonances has been studied primarily by theoretical methods [14, 15]. The results of a number of calculations suggest that the intensity of harmonic may be substantially increased when this harmonic is at resonance with transitions in the atomic and ion spectra of gases. This approach, which yet been realized in the gas HHG, may be an alternative (or a complement) to the method of wave phase matching for harmonics and laser radiation [6, 7].

The first experiments on HHG in the passage of laser radiation through the plasma produced at the surface of a solid target carried out during first half of nineties turned out to be much less successful. As noted in a review on the bifurcational properties of harmonics in plasmas, "the effect of harmonic generation (in plasmas) calls for a more careful consideration and an in-depth basic research" [16]. Data obtained with the use of highly excited plasmas containing multiply charged ions revealed several limiting factors, which did not permit generating harmonics of a sufficiently high order and strong intensity [17–22]. Moreover, the harmonic intensity distribution did not correspond to the so-called three-step model of HHG [23], according to which a plateau-like high-order harmonic intensity distribution (i.e., of approximately equal intensity) should be observed. Those studies, which were carried out in the mid-1990s, stopped at the demonstration of relatively low-order harmonics (from the 11th through the 25th). This disadvantage, as well as the low conversion efficiency, led to the erosion of interest in this HHG technique, especially in comparison with the achievements involving gas sources of HHG.

Nevertheless, there is a reason to hope that harmonic intensities may be increased and efficient shorter-wavelength coherent radiation may be obtained using laser-produced plasmas. There are no fundamental limitations here; it only remains to find the optimal conditions for producing a plasma plume to serve as the efficient nonlinear medium for HHG. Laser-produced plasma may be validly used for this process if the effect of the limiting factors (self-defocusing, self phase modulation, and wave phase mismatch of the harmonics and the radiation being converted) is minimized [17, 21, 22].

Among the special features of HHG in laser-produced plasmas, we first of all note a wide range of nonlinear medium characteristics available by varying the conditions of laser plume production on the surface of a solid. This applies to plasma parameters such as the plasma dimension, the density of ions, electrons, and neutral particles, and the degree of their excitation. The use of any elements of the periodic table that exist as solids largely extends the range of materials employed, together with thousands of complex solid-state samples, whereas only a few light rare gases are typically used in gas HHG. Thus the exploration of practically any available solid-state material through the nonlinear spectroscopy comprising laser ablation and harmonic generation can be considered as a new tool for materials science.

In several cases, this method furnishes an opportunity to realize quasi-resonance conditions for increase in the efficiency of single harmonic generation due to the effect of ion transitions on the nonlinear response in the spectral range in question, thus allowing the studies of ionic transitions possessing strong oscillator strengths. This effect can hardly be observed in gas HHG because of a low probability for the coincidence of the atomic transition frequencies of few gases and the frequencies of single harmonics. The advantages of plasma HHG could largely be realized with the use of a low-excited and weakly ionised plasma, because the limiting processes governing the dynamics of laser wavelength conversion would play a minor role in this case. This assumption has been confirmed by several studies concerned with high-order harmonic generation in the plasma media [24–28]. A substantial increase in the highest order of the generated harmonics, the observation of long plateau and emergence of a second plateau in the energy distribution of highest-order harmonics, the high efficiencies obtained with several plasma formations, the realization of resonance enhancement of individual harmonics, the efficient harmonic enhancement from plasma plumes containing clusters of different materials, and other properties revealed in those and other works [29–32] have demonstrated the advantages of using specially prepared plasmas for HHG. The orders of harmonics obtained in plasma media to date range into the sixties and seventies [24, 26, 33, 34]. The highest-order harmonics (the 101st order, wavelength 7.9 nm) have been demonstrated in manganese plasmas [35]. The HHG conversion efficiency in the plateau region amounted to 10^{-5} [36]. In addition to that, the efficiency of conversion to an individual (resonantly enhanced) high-order harmonic approached 10^{-4} [25, 26].

The quest for new plasma media that would favour the enhancement of an individual harmonic allows further enhancement of harmonic conversion efficiency. The production of a single high-intensity harmonic (rather than a group of harmonics of equal intensity in the plateau region) would open up the way to the practical application of these coherent short-wavelength radiation sources. Resonantly enhanced harmonics observed in several plasma media allowed expecting that similar conditions will be discovered for other plasma formations. The generated harmonic wavelength may then be tuned to the transitions with high oscillator strength by wavelength tuning of the driving laser [25, 31], as well as by varying the chirp of the laser radiation [26, 28, 30]. Application of ablated nanoparticles and clusters for HHG can also enhance the yield of harmonics in the XUV range. Further improvements in HHG conversion efficiency and harmonic extension require a systematic study of the influence of various plasma and laser parameters on ablation harmonics. Many new peculiarities of plasma harmonics emerged during last few years [37–73] allow expecting further extension of our knowledge of materials properties using this powerful took of nonlinear spectroscopy.

The future developments in the application of this technique may include such areas as the seeding of plasma resonance harmonics in the XUV free electron lasers, plasma-induced harmonic generation using a few-cycle pulses, application of endohedral fullerenes for plasma HHG, comparative studies of gas- and plasma-induced harmonics, analysis of molecular structures through the study of harmonic spectra from oriented molecules in plasmas, search for quasi-phase matching schemes in plasma plumes, use of single harmonic for the surface science, structural analysis of the multi-dimensional formations in laser plasma, generation of strong combs and single attosecond pulses, quest for quasi-solid-state HHG, application of the double-target schemes for plasma formation, use of rotating targets for improvements of harmonic stability, application of IR (1–3 μm) laser sources for extension of plasma harmonic cutoffs, analysis of plasma components through the HHG, etc.

The reviews on plasma harmonics studies were mostly devoted to the discussion of such specific topics as the application of the nanoparticle-containing plasmas for the HHG [41], resonance-induced enhancement of harmonics [74] and applications of fullerenes as the attractive media for harmonic generation [51], contrary to the two first topical reviews [42, 75], where the whole range of plasma harmonics studies was presented. It seems timely to return back to the practice of showing the broad pattern of various developments in this field [67, 69, 76]. It is also obvious that the comprehensive overview of most recent findings can help in defining the next steps of the development in this relatively new and attractive area of nonlinear optical studies.

Whilst the first stage of these successful studies was entirely focused on the improvements of harmonic yield from plasma, at current stage of knowledge of the high-order nonlinear optical processes in ablation plume one can consider this method as a new tool for material science. Thus the search of the dual role of plasma HHG as a method for efficient coherent XUV light generation and of materials probing is a milestone of further developments in this field. Below, we show new trends emerged during recent years, which demonstrate the attractiveness of this method.

2.2 Harmonic Generation of Picosecond Nd:YAG Laser Radiation in Ablation-Produced Plasmas

As it was already pointed out, the advantages of HHG of laser radiation in a plasma plume could largely be realized with the use of low-excited and weakly ionized plasma, because in that case the limiting processes governing the dynamics of the laser frequency conversion would play a minor role [75]. A search for the best experimental conditions, such as pulse duration of driving laser field, for efficient HHG in different spectral ranges is a way for further enhancement of harmonic yield.

Previous plasma HHG studies were carried out using femtosecond pulses. The spectral range of those experiments was restricted at the longer wavelength side to ~80 nm, which was defined by the registration properties of the commonly used detectors of harmonic spectra (microchannel plates). A limited number of studies of plasma HHG had been performed in the spectral region above this wavelength. A search for resonance enhancement of single harmonic in these conditions can be justified by the presence of strong ionic and neutral transitions in the longer wavelength vacuum ultraviolet range (80–200 nm). Next, the application of longer pulses could avoid the impeding processes restricting the efficiency of harmonic generation in this region. There two main limiting processes, which can decrease the conversion efficiency in the ionic medium. The first one is an excess free electron concentration caused by over-excitation of the target surface. The appearance of a considerable number of free electrons is a result of ionization of both neutral atoms and singly charged particles. This follows with self-defocusing of the propagating probe pulse. The second process is directly related with the first one. The phase matching conditions of the driving and harmonic waves, which are maintained for a moderate number of free electrons appearing during ionization of neutrals, break up due to the increase of free electron concentration. This increase occurs abruptly once the fluence of the heating pulse reaches the threshold level, when the over-excitation of targets leads to the appearance of doubly and higher charged particles. Further, the use of longer pulses can also increase the fluence of harmonic emission at comparable

HHG conversion efficiencies from picosecond and femtosecond sources due to the higher pulse energies available in the former case.

Recent studies have shown the attractiveness of application of the long laser pulses for third-order harmonic generation from laser plasmas [68, 77]. Below we present an analysis of the HHG in various plasmas using 1064 nm, 38 ps pulses [78]. The goal of those studies was a search for the conditions for generation of energetic coherent picosecond pulses in the range of 80–220 nm using various metal-containing plasma plumes.

A passive mode-locked Nd:YAG laser (1064 nm, 1.5 Hz pulse repetition rate) generated a 38 ps pulse. Two-stage amplification of the single pulse was followed by splitting of this radiation into two parts, one (heating pulse) with an energy of 5 mJ, which was used for plasma formation on the target surface, and another (probe pulse) with an energy of up to 28 mJ, which was used after some delay for frequency conversion in various plasma plumes (Fig. 2.1). The heating pulse was focused using a 300 mm focal length lens inside the vacuum chamber containing various targets. The plasma sizes were in the range of 0.5 mm. The heating pulse intensity on the target surface was in the range of 10^{11} W cm^{-2}. The diameter of the single mode probe beam was 4 mm. This beam was focused inside the plasma plume using a 150 mm focal length lens. The probe beam propagated at a distance of 100–150 μm from the surface of the target. The intensity of the probe pulse at the focus was 4×10^{13} W cm^{-2}. The focal spot was inside the plasma plume. The delay between the heating and probe pulses during most of the experiments was maintained at 25 ns, which was optimal for efficient harmonic generation in metal-containing plasmas. The harmonic radiation was analyzed using a vacuum monochromator.

Various metal materials (copper, chromium, zinc, niobium, silver, indium, molybdenum, titanium, tin, lead, tantalum, manganese, tungsten, gold, boron and aluminum) were used as the targets for laser-induced plasma formation. A three-coordinate translating stage allowed movement of the target along the z-axis and control of the interaction zone of the probe radiation with the plasma relative to the target plane.

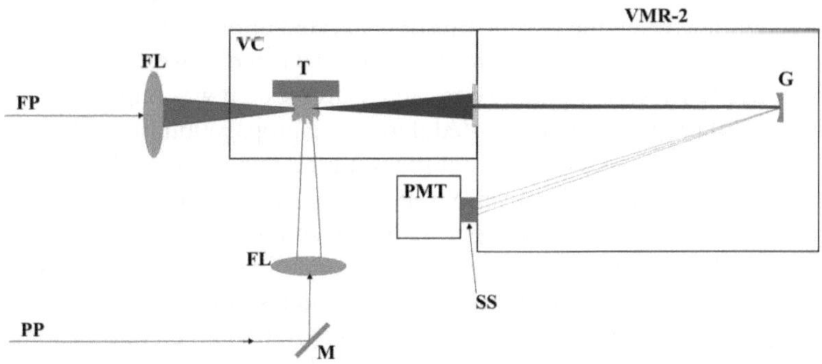

Fig. 2.1 The experimental setup for HHG in a laser plasma using picosecond pulses. *FP*, fundamental probe picosecond pulse; *PP*, heating picosecond pulse; *M*, mirror; *FL*, focusing lenses; *VC*, vacuum chamber; *T*, target; *VMR-2*, vacuum monochromator; *G*, grating; *SS*, sodium salicylate; *PMT*, photomultiplier tube. Reproduced from [78] with permission from IOP Publishing

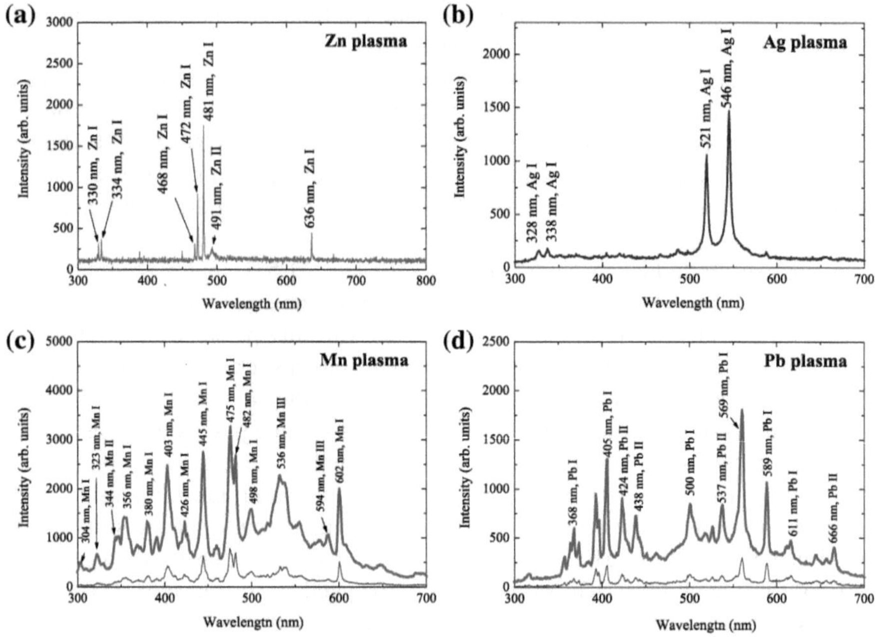

Fig. 2.2 Emission spectra from the (**a**) Zn and (**b**) Ag plasmas used for harmonic generation and plasma emission spectra from the (**c**) Mn and (**d**) Pb targets at weak (*thin lines*, $I = 8 \times 10^{10}$ W cm^{-2}) and strong (*thick lines*, $I = 2 \times 10^{11}$ W cm^{-2}) excitation of target surfaces. Adopted from [78] with permission from IOP Publishing

Analysis of the evolution of the spectra for the diagnosing laser plasma provides important information about the plasma parameters and can be used for multiple applications. In particular, some time-resolved laser-induced plasma spectrometry studies during harmonic studies from gold, silver, manganese and vanadium plasmas [37] allowed for identification of the emissions from native species and optimization of conditions when the plasma mostly consisted of excited neutrals and singly charged ions. It may be noted that most plasma HHG studies were carried out using time-integrated methods of plasma emission analysis, so it was impossible to define exactly what plasma conditions existed during the propagation of the femtosecond pulse through the plume. Below, we discuss a time-integrated analysis of spectral studies of plasma emission from various metal targets used for harmonic generation during propagation of the picosecond radiation through the plasma. These studies were aimed at defining the optimal plasma conditions for efficient HHG in laser plumes and showed that, while, for most plasma plumes, over-excitation during laser ablation leads to drastic decrease of harmonic generation efficiency, in some cases one can achieve the conditions for extension of the harmonic cutoff using picosecond probe pulses.

The spectral studies of atomic and ionic emission from the laser-produced plasmas were carried out in the visible and near ultraviolet ranges (300–700 nm). This spectral region was chosen due to the existence of multiple ionic and atomic transitions of the studied plasma species. To create the ablation, a 1064 nm, 38 ps pulse from a Nd:YAG laser was focused onto a metal target in the vacuum chamber. The spectral characteristics of the laser plasma in the visible and UV ranges were analyzed using a fiber spectrometer (HR4000).

The spectra of plasma emission from the Zn and Ag plasmas are presented in Fig. 2.2a. One can note that these spectra were obtained at the conditions of 'optimal plasma' formation from the point of view of best HHG conversion efficiency. In this way, the maximum yield of harmonics was obtained, especially in the 80–220 nm range, and then the spectral characteristics of the laser plasmas were measured at these conditions. The spectral lines were mostly originated from the excited states of neutral and singly charged ions, when the impeding influence of free electrons on the high-order nonlinear optical processes in the laser plasmas was insignificant.

The increase of heating pulse intensity on the target surfaces above 10^{11} W cm^{-2} led to both the growth of emission intensity for neutral and singly charged ionic species and the appearance of emission lines from the higher charged particles. The appearance of doubly and triply charged ions and a large number of free electrons immediately followed with a considerable decrease of HHG conversion efficiency from almost all the plasma samples. The variations of the plasma spectra in that case are presented in Fig. 2.2b, where one can see the changes of intensity of plasma emission from the Pb and Mn targets. The thin and thick curves correspond to weak and strong excitation, correspondingly. This figure clearly shows the increase of the intensities of the Mn III and Pb II lines, which is correlated with an increase of the concentration of multiply charged ions in the plasma plume and correspondingly the concentration of free electrons.

Previous studies of HHG from plasma plumes have analyzed the UV emission spectra from the plasma to prove that over-excited and over-ionized plasmas could dramatically decrease the harmonic intensity [75]. However, as was mentioned above, these spectral measurements were performed using time-integrated methods that did not allow identification of the plasma state before the interaction with the delayed femtosecond pulse. One can note that the application of time-resolved laser-induced plasma spectrometry allowed previously the definition of the plasma conditions for the extension of harmonic cut-offs [37]. The application of this time-resolved technique, which was not available in the discussed studies [78], could further define better conditions of plasma HHG.

In [78], the restricting features of harmonic generation dynamics during over-excitation of the targets were observed in the case of most of the analyzed plasma plumes. The increase of heating pulse intensity from 1×10^{11} to 3×10^{11} W cm^{-2} during HHG experiments with these targets led to the appearance of strong plasma emission. At these excitation conditions, the harmonic emission from various plasmas was overlapped with the plasma emission. The intensity of the generated harmonics became considerably less than at 1×10^{11} W cm^{-2} excitation. In the case of most of the plasmas, this over-excitation led to a complete disappearance of harmonic emission. The variations of HHG efficiency with the growth of heating pulse intensity are depicted in Fig. 2.3. In particular, the 11th harmonic generated from the Pb plasma started to decrease with growth of the heating pulse energy above 3 mJ (Fig. 2.3a),

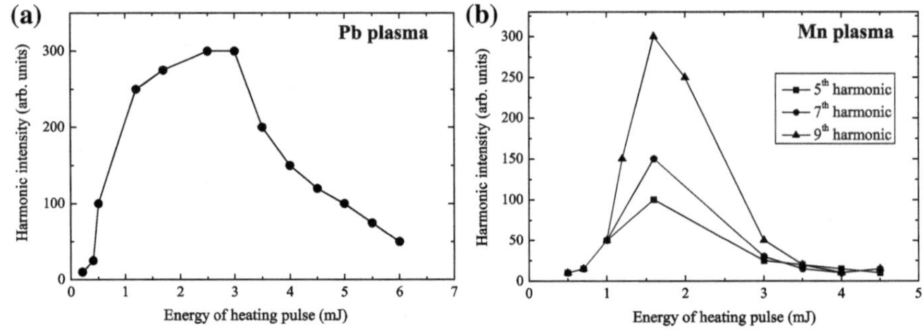

Fig. 2.3 Dependences of harmonic intensity at different heating pulse energies for the (**a**) 11th harmonic generating from the Pb plasma, and (**b**) 5, 7, and 9th harmonics generating from the Mn plasma. Reproduced from [78] with permission from IOP Publishing

which corresponds to an intensity of 1×10^{11} W cm^{-2}. In another case, a decrease of the 5, 7 and 9th harmonics generated in Mn plasma was observed at pulse energies above 1.5 mJ (Fig. 2.3b).

The reason for the decrease of harmonic conversion efficiency is related to over-excitation of the target, which leads to the appearance of an abundance of free electrons in the plasma plume. The latter cause a phase mismatch between the waves of the driving field and the harmonics. This effect is especially important for the lower-order harmonics. One can note that a decrease of harmonic efficiency with over-excitation of plasma has been reported for higher-order harmonics as well [75], although this decrease of converted XUV radiation was less abrupt than that observed in the present studies.

The characteristic pattern of almost all the harmonic spectra from metal plasmas was a featureless sharp decrease of conversion efficiency for lowest orders, which followed with a gradual decrease of higher harmonics (above the 9th order) up to the limit of the registration range of monochromator (50 nm, 21st harmonic). Figure 2.4 shows the characteristic HHG spectra obtained from Cr, Cu and Ti plasmas. Most of these plasmas showed comparable nonlinear optical properties from the point of view of HHG conversion efficiency.

Measurements of the absolute values of the conversion efficiencies of the harmonics generated in the plasmas were carried out using the following procedure. In the first step, the 4th harmonic signal was measured by a 'monochromator + sodium salicylate + PMT' detection system using the known energy of the 4th harmonic of 1064 nm radiation generated in the nonlinear crystals. This allowed calibration of the monochromator at a wavelength of 266 nm. Since the quantum yield of sodium salicylate is equal in a broad spectral range between 40 and 350 nm, calibration of the registration system at 266 nm allowed calculation of the conversion efficiency for the higher harmonics. The monochromator allowed observation of the harmonics down to a spectral range of 50 nm. The conversion efficiency in the plateau range (15th–21st harmonics) was measured to be in the range of $10^{-6} - 10^{-5}$. Table 2.1 summarizes the measured conversion efficiencies for different harmonics in the case of some metal-containing plasma plumes.

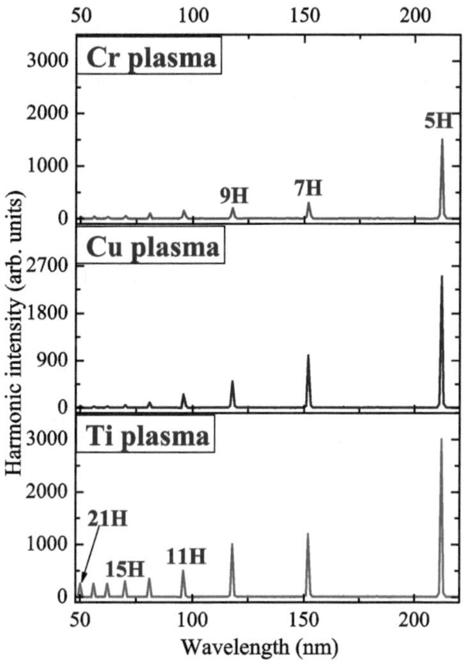

Fig. 2.4 High-order harmonic spectra from the Cr, Cu and Ti plasma plumes. Reproduced from [78] with permission from IOP Publishing

While most of the plasmas demonstrated similar properties, some of them allowed the observation of unusual spectral distributions of harmonics. In particular, in some cases the generation of harmonics was restricted at the 9–15th orders. Among those plasmas were the Al, B and Mo plumes (Fig. 2.5a). A few other plasmas showed even more interesting spectra, where one could distinguish the enhancement of some harmonics with regard to the lower-order ones. These were the In and Nb plasma plumes, which showed enhancement of the 11th harmonics compared with the previous harmonic orders (Fig. 2.5b). Such dependences resemble those observed in previous plasma HHG experiments using 800 nm, few tens of femtosecond pulses in the cases of In, Mn, Cr, Sn and other plasmas [74, 75]. The commonly accepted explanation of these enhancements is related to the closeness of the specific harmonic orders to the resonance ionic transitions of these media. Below we discuss this peculiarity in more detail.

The dependence of the recombination probability on the electron return energy and on the structure of the target is reflected in the HHG spectrum and has been the subject of intensive research in recent years. To enhance the notoriously low efficiency of the HHG process, it appears promising to exploit the effect of resonances, which are known to be of great importance in photoionization. The investigation of resonant peaks in the photoionization cross section has a long history, including studies of autoionising resonances [79], shape resonances [80] and giant resonances [81], but there have been only a few studies on the role of resonances in HHG. The role of

Table 2.1 Harmonic conversion efficiencies ($\times 10^{-6}$) in various plasma plumes

Harmonic order	Wave length, nm	Sn	Zn	Mn	Ti	Cu	W	Au	B	Ta	Ag	Cr	Mo	Al
5	213	100	80	80	70	46	46	40	40	26	20	20	4.2	6
7	152	66	66	20	20	33	40	33	13	10	16	7.1	3.8	2.59
9	118	66	53	10	12	15	38	26	3	3	10	3.8	2.5	1.33
11	97	53	40	6.6	6.6	13	33	20	1.3	1.8	6.6	3.8	2.1	
13	82	26	4	2.1	4.6	2.5	7	1.8	1	0.9	2.6	1.3	0.9	
15	71	16	4	1.75	3.3	0.9	5	1.3		0.9	1.3	0.9		
17	63	6	1.7	1.3	2.1	0.9		1.3		0.9	1	0.9		
19	56	6	1.7	1.3	2.1	0.83		0.9		0.75	1	0.9		
21	51	4	1	1.3	2.1	0.75		0.9		0.75	1	0.9		

Reproduced from [78] with permission from IOP Publishing

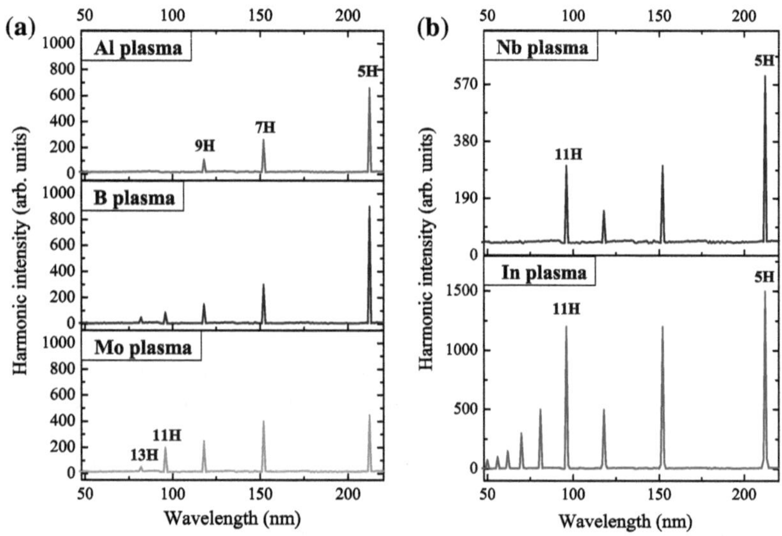

Fig. 2.5 **a** Harmonic spectra from the Al, B, and Mo plasmas producing only low-order harmonics. **b** Harmonic spectra from Nd and In plasmas demonstrating the enhancement of 11th harmonic with regard to the lower-order ones. Reproduced from [78] with permission from IOP Publishing

atomic resonances in increasing the laser radiation conversion efficiency was actively discussed in the framework of perturbation theory in the early stages of the study of low-order harmonic generation (see the monograph [82] and the references therein). In the case of HHG, the increase in the efficiency of the generated harmonics due to resonance processes first came under discussion more than a decade ago, and this approach appears to have considerable promise with the use of ionic and, in some cases, atomic resonances [14, 15, 83–88]. These studies comprised both theoretical treatment of the process and description of the first attempts to form resonance conditions in experiments.

While theoretical estimates testified to the possibility of an efficient enhancement of individual harmonics and groups of harmonics, experimental works revealed the difficulties encountered in HHG in gases. Therefore, the use of plasma media could largely facilitate solution of the problem of resonance harmonic enhancement. Examination of a large group of potential targets allowed identification of some of them as suited for demonstrating this process [75]. The advantages of 'plasma HHG' over 'gas HHG' were amply manifested in this case, because the number of possible media in the former case is far greater than that in the latter case. Some attempts at explanation of the experimental observations of resonant enhancement in plasma harmonics have been reported recently in [89–93].

The observation of resonance enhancement in the case of both femtosecond and picosecond probe pulses is related with the coincidence of harmonic wavelengths and ionic transitions. In the case of femtosecond pulses, this opportunity has more chances to be realized due to the broader bandwidth of the harmonics. However, the conditions of resonance enhancement could be realized in the case of picosecond pulses as well once some of the resonances of Nb and In ions occasionally coincide with the wavelength of the 11th harmonic. These observations were repeated several times during analogous studies of the same plasmas, while none of the other plasma plumes demonstrated these features.

Propagation effects could not explain these observations since in that case we would have to observe the enhancement of at least a few neighboring harmonics as well. One can note that the resonance enhancement of plasma harmonics in the case of narrowband probe pulses is indeed a rarely observed phenomenon. The application of various plumes showed this enhancement only in the two above-mentioned plasmas, while many more plasmas were reported as suitable for resonance enhancement in the case of femtosecond broadband pulses. Overall, Fig. 2.4 shows the tendency of the harmonic distribution, which was common for most of the studied plasma plumes. Figure 2.5 emphasizes the peculiarities of the harmonic distribution observed in some plasma plumes. Those observations showed that the atomic number does not play a significant role in the classification of the harmonic properties of the plasma media. The important parameter here could be a second ionization potential of the species used.

A few properties of the generated harmonics from metal-ablated plasmas were analyzed in depth. The delay between heating and probe pulses is crucial for optimization of the HHG. A typical dependence of the harmonic intensity on the delay between pulses is presented in Fig. 2.6a in the cases of Pb and Mn plasmas. The concentration of particles (neutrals and singly charged ions) is insufficient at the initial stages of plasma formation and spreading out of the target surface, since the species possessing velocities in the range of 5×10^5 cm s^{-1} do not reach the optical axis of propagation of the probe beam (100–150 μm above the target surface). Increase of the delay above 5 ns allowed the appearance of plasma particles along the path of the converting pulse, which led to considerable growth of the HHG conversion efficiency. Further increase of delay led to saturation of the HHG and gradual decrease of the conversion efficiency at longer delays. One can note that there are different optimal delays for the harmonics generated in the Mn and Pb plasmas (20 and 90 ns

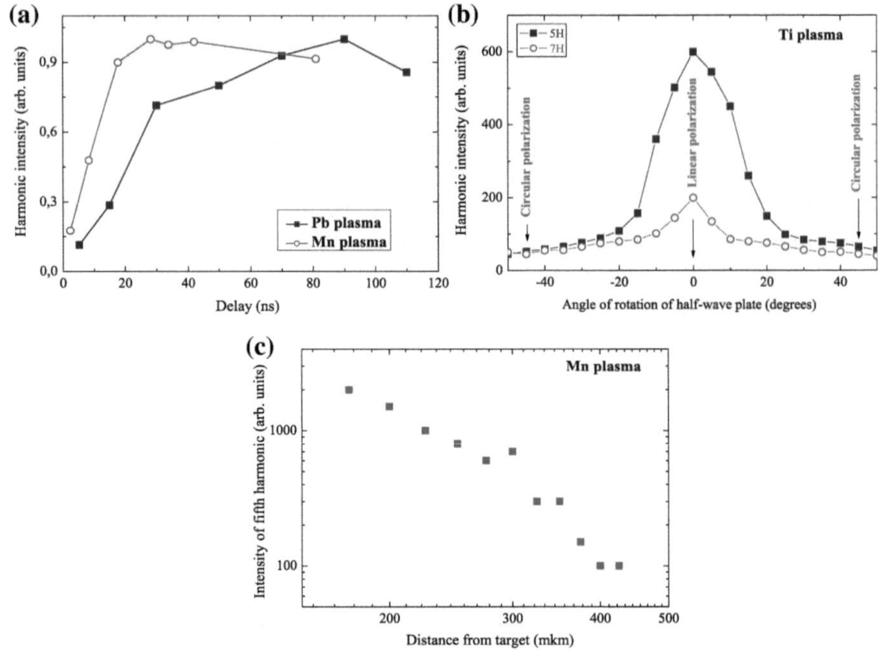

Fig. 2.6 **a** The dependence of the harmonic intensity on the delay between the heating and probe pulses for the Pb and Mn plasmas. **b** The polarization dependences of the 5th and 7th harmonic intensities at different angles of rotation of the half-wave plate in the case of the Ti plasma. **c** The harmonic intensity as a function of the distance between the target surface and the probe beam axis for the 5th harmonic generated in Mn plasma. Reproduced from [78] with permission from IOP Publishing

correspondingly). These observations can be explained by the different velocities of the ablated particles from these targets, which should reach the area of interaction with the probe pulses at different times. In this connection the application of a heavy target (Pb, $Z = 82$) should lead to a longer delay before the bulk amount of ablated material interacts with probe beam compared with a lighter target (Mn, $Z = 25$). The ratio between optimal delays for these plasma samples is approximately coincide with the ratio between the masses of those particles, which should follow from the equal kinetic energies of ablated particles.

Usually, the plasma lifetime is of the order of a few nanoseconds [94]. However, one should consider as another parameter the time when the maximum concentration of particles appears above the target surface in the area of propagation of the driving laser pulse. Since the distance from the target to the axis of the driving pulse propagation is of the order of 100–150 μm, the time when the main cloud of particles reaches this area is of the order of a few tens of nanoseconds and is defined by the velocity of the particles ($\sim 5 \times 10^3$ m s^{-1} for the targets with low Z and $\sim 2 \times 10^3$ m s^{-1} for the targets with high Z). This time is of the order of 20 or 50 ns. So the meaning of the lifetime is incorrect once we consider the process, which depends only on the time required for spreading of the plasma particles towards the area where the driving pulse propagates. In particular, the experiments [68] using nanosecond pulses were optimized at 600 ns delay due to a longer distance from the target to the driving nanosecond pulse (~1 mm). The above estimates allow us to conclude that it is the plasma spreading time, which is more critical than the plasma lifetime. During the plasma HHG the driving pulse interacts with the ion or the atom, because the spreading time is longer than the plasma lifetime.

The important parameter of plasma HHG is the polarization of the probe radiation. Figure 2.6b shows the dependences of the 5 and 7th harmonics generated from the Ti plasma at different angles of rotation of the half-wave plate, which caused variation of the conditions of the driving radiation from linear (at 0°) to circular (at 45°) polarization. A small deviation from linear polarization led to a considerable decrease of the 5 and 7th harmonic intensities, which is typical behavior for high-order harmonics. The application of circularly polarized laser pulses led to complete disappearance of harmonic emission, as should occur assuming the origin of HHG.

The harmonic intensity also considerably depended on the distance between the optical axis of the driving beam and the target surface (Fig. 2.6c). This dependence is induced by the change of plasma concentration above the target. The log–log dependence of the harmonic intensity (I_h) on the distance between the optical axis of the driving beam and the target surface (x) for the 7th harmonic generation in manganese plasma corresponded to $I_h \sim x^{-3.5}$.

We would like to recall that the earliest observations of HHG in gases were also reported using a picosecond Nd:YAG laser [95, 96] (as well as an excimer laser at 248 nm [97]). The harmonics from different gases up to the 21st and 33rd orders of 1064 nm radiation were reported at an intensity of 3×10^{13} W cm^{-2}, which led to an enormous growth of interest in this area of nonlinear optics. The present studies demonstrate that the application of ablated plasma as a nonlinear medium can further amend this HHG technique using picosecond driving pulses, once the metal atoms, ions and clusters become involved as efficient sources of harmonic generation.

2.3 Stable Generation of High-Order Harmonics of Femtosecond Laser Radiation from Laser Produced Plasma Plumes at 1 kHz Pulse Repetition Rate

Almost all HHG studies from weakly ionised plasmas produced during laser ablation of various solid targets were carried out using the 10 Hz pulse repetition rate lasers. Up to now only few studies of HHG from plasmas have been carried out on static targets using 1 kHz class lasers. The ablation process at 1 kHz pulse repetition rate causes a considerable change of the surface properties of the target due to the melting, which deteriorates the plasma plume conditions during laser ablation. Surface degradation of a static target results in an unstable harmonic signal so that movement of the target surface is required to maintain a reasonable stability. The demand in finding

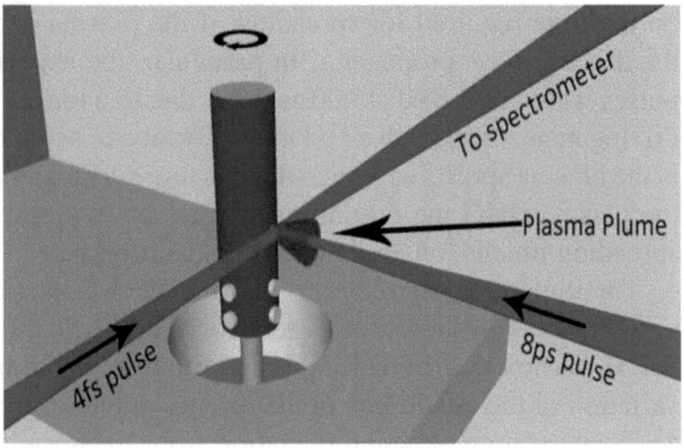

Fig. 2.7 Schematic of the rotating target and HHG configuration. Reproduced from [73] with permission from Optical Society of America

the optimal way for improving the plasma harmonic stability is high due to recent observations of the advanced properties of plasma harmonics over gas harmonics [61, 62, 66]. In particular, in [61], the plasma HHG conversion efficiency was measured one order of magnitude stronger compared with gas HHG efficiency. Analogous features were reported in [66].

The obstacle of all plasma harmonic experiments during earlier studies was an insufficient stability of plasma parameters (density, ion and free electron concentrations, excitation conditions, etc.), which led to the instability of harmonic yield and fast decay of harmonic efficiency during irradiation of the same spot of ablating target. As it was mentioned, most of those early studies were performed using a 10 Hz class lasers. Even at this relatively low pulse repetition rate, the stability of harmonics deteriorated after a few hundred shots on the same spot of the surface and even quicker for powder-like materials (fullerenes, nanotubes, metal nanoparticles, various organic and non-organic powders). One can note that laser ablation of those samples can be considered as an important tool for their structural studies using XUV nonlinear spectroscopy.

The application of soft ablation allows the use of the same target for a much longer period than in the case of earlier studies of over-excited targets during laser ablation. Thus a search of a robust, easy-to-apply method for improving the harmonic stability in the case of plasma HHG could considerably advance a search of the fundamental (structural, orientational, etc.) properties of organic and inorganic atoms and molecules through XUV nonlinear spectroscopy.

The earlier used approaches of a rotating disc geometry [20, 98] are not suitable since the distance between driving femtosecond beam and target surface should be maintained minimal (of order of 100 μm), while the Rayleigh length of the driving beam is maintained at the range of few mm. Below we describe a new method using a motorized rotating rod specifically prepared for the HHG from plasma plumes using high pulse repetition rate lasers, and demonstrate that this target significantly improves the stability of high-order harmonics [73].

Those studies were performed using two laser pulses: one to create the plasma plume and the second to drive the HHG within it. The first (heating) pulse was created by splitting off a portion (200 μJ) of the uncompressed 8 ps laser pulse from a 1 kHz Ti:sapphire chirped pulse amplification laser. The remaining driving pulse was compressed in a prism compressor and then further compressed using a hollow core fibre and chirped mirrors, resulting in 250 μJ, 4 fs pulses. The driving pulse was delayed with respect to the ablation pulse by 35 ns to give the plasma the time to expand away from the target surface to allow the converting pulse to pass through the plasma without being clipped by the target.

Target rotation apparatus consists of three linear stages driven by stepper motors along three axes. The target was attached to an axis of the fourth motor, which provided rotation with a variable speed (from a few rotations per minute (rpm) up to 300 rpm). Rotating the target was sufficient to achieve stable harmonic radiation and an additional vertical movement was not required, though this capability might be useful for future plasma HHG experiments. As the setup requires the target to be positioned very close to the driving beam, it was of paramount importance that the target was carefully aligned to the axis of rotation. Any movement of the target surface due to eccentricity in the radial direction from the driving beam axis would result in an oscillation of the harmonic signal due to variation of the plasma density seen by the driving beam or, in the extreme case, clipping of the laser beam.

The target (cylindrical rode with diameter of 10 mm and length of 30 mm) was positioned as shown in Fig. 2.7, with the probe pulse propagating 100 − 200 μm above the target surface. The picosecond heating pulse was focused onto the surface of the rotating target. In order to efficiently produce high harmonics the plasma must be weakly ionised [75]. To achieve this the target was positioned slightly in front of the focus of the heating pulse using a 50 cm focusing lens, leading to an on target intensity of $\sim 1 \times 10^{10}$ W cm^{-2}. This also had the benefit of increasing the size of the plasma produced from ablating a larger area. The size of the focus at the target surface was measured to be ≈500 μm. The delayed probe pulse was focused through the plasma using a 40 cm spherical mirror. The HHG radiation was analysed by an XUV spectrometer consisting of a flat-field grating and an imaging microchannel plate detector with phosphor screen imaged onto a CCD camera.

Figure 2.8 shows that there is a drastic change in the harmonic signal (integrated over the spectral range of 40–80 nm) when the rotation of the aluminium target is stopped. There is a sharp intensity decrease of more than one order of magnitude over only one thousand shots (or just after one second of ablation using 1 kHz laser). The benefits of the rotating rod are clearly shown in Fig. 2.9a where stable harmonic generation was achieved from the plasma produced on an aluminium target for over 1 million laser shots. Figure 2.9b shows the sharp increase and slow decay in the HHG from silver plasma when the target is moved vertically by one millimetre after about 6×10^5 shots. This caused a quick increase in the harmonic signal from the fresh surface, followed by further decay down to the stable level. The decaying behaviour disappears as the laser creates a rotationally precise micro-channel on the target surface. The reason for the degradation of harmonic signal is clearly due to effects on the surface of the target. Stable harmonics were achieved in a broad range of the speeds of rotation (from 10 rpm and faster). The target rotational speed and the size of the ablation focus imply that the same area of target was undoubtedly used

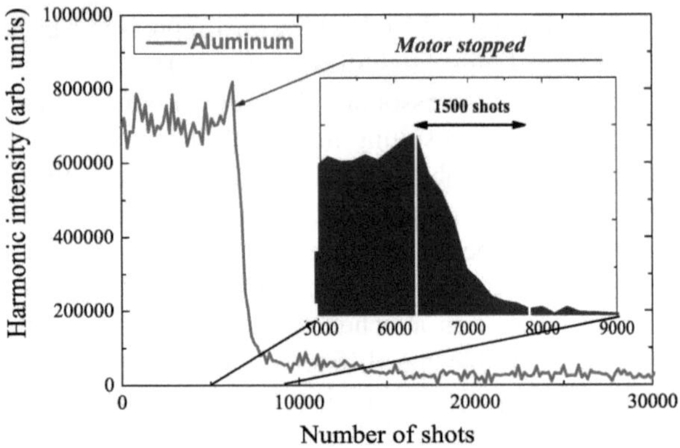

Fig. 2.8 Decay of the harmonics from aluminium plasma after stopping the rotation of the motor. The harmonics were integrated over 40–80 nm spectral range. Reproduced from [73] with permission from Optical Society of America

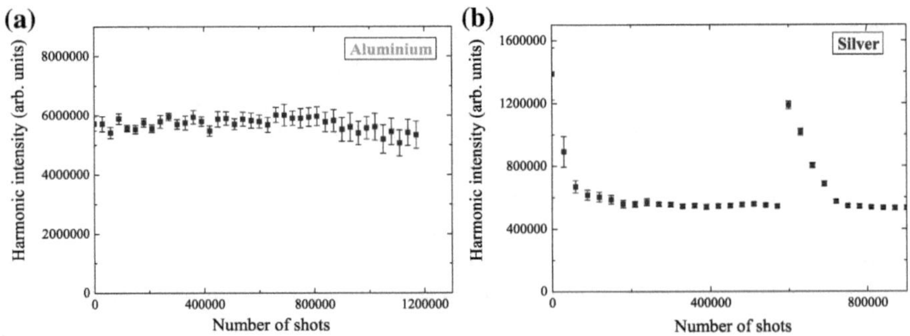

Fig. 2.9 Stability of integrated harmonic signal from (**a**) aluminium and (**b**) silver plasmas. Reproduced from [73] with permission from Optical Society of America

repeatedly for consecutive rotations over the 20 min duration of experiments. This could result in thermal damage issues with this high pulse repetition rate.

It is possible that once the fixed surface is melted the force from a following laser shot and plasma creation could expel some of the liquid target from the ablation area creating a deeper hole out of the focus or with angled side, which would not cause the plasma to be emitted in a direction normal to the surface. These effects are considerably diminished once the target starts to rotate. During rotation, the previously ablated area cools down such that, during the next set of ablation on this spot, the plasma formation occurs at approximately same conditions. To prove that the ablated area cools down with rotation, the target was rotated at different speeds (from 10 to 300 rpm) and no difference in stability of harmonic yield was found. These observations point out the importance of the periodic change of the ablation zone. This also confirms a suggestion that the cooling of the ablation area leads to stable plasma generation.

Characteristics of plasma (density and ionisation state) are the most important parameters to achieve and maintain stable HHG efficiency during an extended period of illumination. The calculations [66] have shown that, in the case of carbon plasma, the concentration of particles in the area of femtosecond laser-plasma interaction at optimal delay between heating and driving pulses ($\sim$40 ns) is $\sim 2 \times 10^{18}\,\mathrm{cm}^{-3}$. The solid surface was considered as the one unheated before the laser ablation. Indeed, after one round of rotation (e.g., after 0.2–2 s), the plasma disappears, the ablated spot cools down, and the next laser shot on the same spot can be considered as a shot on the "fresh" surface. Contrary, in the case of a stationary target, the following shots continue the heating of the same spot.

The novelty of this approach includes the observation of advanced properties of plasma HHG even at extremely small energies of the heating pulses. The efficiency of plasma HHG depends on the possibility to create "optimal" plasma. This can be done using both multi-mJ pulses, as was shown in previous studies [75], and few hundred μJ pulses, as it was demonstrated in the reviewed work [73] and recently published studies [66]. The important point here is the intensity and fluence of the heating pulse on the target surface. The application of a higher energy heating pulse could create the conditions of "optimal" plasma over a longer distance, which could (or could not) increase the harmonic yield depending on the phase relations between the driving and harmonic waves. In addition, it can also lead to the over-heating of the target at 1 kHz ablation. As it was already mentioned above, the rotating speed did not influence the stability of harmonics using 0.2 mJ heating pulses. The use of more energetic pulses at high repetition rate (i.e. of order of few mJ) may require additional optimisation of the rotation target technique (for example, by periodic up and down dragging of the rotating target).

2.4 High-Order Harmonic Generation in Graphite Plasma Plumes Using Ultrashort Laser Pulses: A Systematic Analysis of Harmonic Radiation and Plasma Conditions

The characteristics of laser plasma play a crucial role in determining how efficiently high harmonics can be generated in the plasma plumes. An increase in the free electron density was likely to have been the limiting factor for the harmonic cut-off energy in early experiments with laser plasmas [17, 19, 20]. A search for appropriate target materials, which can provide favourable ablation plasmas for efficient HHG, has motivated the analysis of plasma characteristics at conditions of high harmonic yield. Recent studies have shown that carbon ablation plasmas are promising media to satisfy the above requirements [61–63, 66].

Shot-to-shot stability of the harmonic signal is crucial for any application of the generated radiation and especially for the measurement of the pulse duration of converted XUV radiation. Such temporal measurements were reported in the case of HHG in chromium plasma [61]. Using the "Reconstruction of Attosecond Beating by Interference of Two-photon Transitions" technique [99], the authors have shown that the 11th to the 19th harmonics of a Ti:sapphire laser form, in the time domain, an attosecond pulse train. It was underlined that instability of the harmonic signal in their experiments using a 10 Hz pulse repetition rate laser was the main obstacle for an accurate measurement of the temporal structure of plasma harmonics. Beside its fundamental interest, high-order harmonic generation in plasma plumes could thus provide an intense source of femtosecond and attosecond pulses for various applications.

Optical parametric amplifiers (OPAs) operating in the mid-infrared (MIR) range are promising tools for harmonic cut-off extension and attoscience experiments. The spectral cut-off energy of HHG obeys the scaling law $E_c \sim I\lambda^2$ [23], where I is the peak intensity of the probe field and λ its central wavelength, which allows one to extend the harmonic emission beyond the 100 eV range by using longer wavelength laser sources. Another advantage of mid-infrared optical parametric amplifiers (MIR OPAs) is their wavelength tuneability, which allows one to tune the spectral position of harmonics towards the ionic transitions with strong oscillator strengths. This feature allows the observation of resonance-enhanced harmonics and broadens the range of plasma samples where this phenomenon could be realized compared with the case of ~800-nm lasers of essentially fixed wavelength [69]. Moreover, by using two-colour HHG techniques, the application of MIR OPAs allows the study of complex molecules during their ablation and HHG using the tuneable long-wavelength radiation. These features are interesting for spectroscopic applications of HHG in the MIR range [100, 101].

In the meantime, the use of MIR OPAs for HHG should lead to a reduced harmonic generation efficiency that scales as λ^{-5} [102, 103]. It is of considerable interest to analyse the relative behaviour of plasma harmonics in the cases of 800 nm and MIR lasers and thereby to find the conditions when the reduction of harmonic yield becomes not so dramatic due to some enhancement mechanisms, such as the presence of in-situ produced nanoparticles, which increase the HHG conversion efficiency. It is worth noting that previous studies of plasma HHG in carbon plumes [61, 62] have inferred, through analysis of plasma debris morphology, the formation of nanoparticles during laser ablation of carbon-contained targets.

Atomic carbon is a reactive species, which stabilizes in various multi-atomic structures with different molecular configurations (allotropes). All the allotropic forms of carbon (graphite, diamond, and amorphous carbon) are solids under normal conditions, but graphite has the highest thermodynamic stability. Laser ablation of graphite has been intensively examined during the last ten years to define plasma conditions for the synthesis of carbon structures with unique properties. The physical characteristics of the plasma plume, such as concentration of atoms and clusters, directly affect the properties of the material being formed in the dynamic expansion of the ablated material. The successful synthesis of clusters is strongly dependent on the formation of atomic and molecular species with the required chemistry and aggregation ability. Thus, to select the optimal plasma conditions for HHG, a detailed understanding of the basic physical processes governing the ablation plume composition and reliable methods for controlling of the plume species are needed. The reasons mentioned above and the consideration of recent studies of HHG in carbon plasmas [61, 62], as well as recently reported comparisons of the HHG in graphite-ablated plasmas and argon gas [63, 66], have prompted to systematically analyze the plasma conditions for optimal HHG conversion efficiency in graphite plasmas [104].

2.4.1 HHG in Carbon Plasma at Different Conditions

High-intensity few-cycle pulses (760 nm central wavelength, 0.2 mJ, 3.5 fs, pulse repetition rate 1 kHz) were typically obtained from the Ti:sapphire laser after second stage of compression consisting on hollow fiber filled with neon and bunch of chirped mirrors [105]. The compressed pulses were characterized with a spatially encoded arrangement for direct electric field reconstruction by spectral shearing interferometry. This radiation was used for frequency up-conversion in the specially prepared carbon plasma.

A portion of the uncompressed radiation of this laser (central wavelength 780 nm, pulse energy 120 μJ, pulse duration 8 ps, pulse repetition rate 1 kHz) was split from the beam line prior to the laser compressor stage and was focused into the vacuum chamber to heat the graphite target and create a plasma on its surface (Fig. 2.10). These picosecond heating pulses were focused by a 400 mm focal length lens and created a plasma plume with a diameter of ∼0.5 mm using an intensity on the target surface of $I_{ps} = 2 \times 10^{10}\,\mathrm{W\,cm^{-2}}$. The delay between plasma initiation and femtosecond pulse propagation was fixed at 33 ns. As an alternative ablation, the 10 ns, 1064 nm pulses from a 10 Hz repetition rate Q-switched Nd:YAG laser were used that provided an intensity on the target surface of $1 \times 10^{9}\,\mathrm{W\,cm^{-2}}$. In that case the delay between the 10 ns heating pulses and the 3.5 fs probe pulses was varied in the range of 10–60 ns to maximize the harmonic yield.

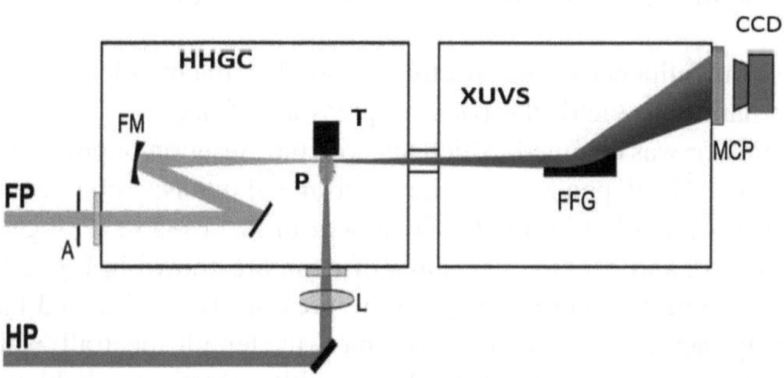

Fig. 2.10 Experimental setup for harmonic generation in plasma plumes. *FP* femtosecond probe pulse, *HP* picosecond heating pulse, *A* aperture, *HHGC* high-order harmonic generation chamber, *FM* focusing mirror, *L* focusing lens, *T* target, *P* plasma, *XUVS* extreme ultraviolet spectrometer, *FFG* flat field grating, *MCP* microchannel plate and phosphor screen detector, *CCD* CCD camera. Reproduced from [104] with permission from IOP Publishing

The 3.5 fs probe pulses, propagating in a direction orthogonal to that of the heating pulse, were focused into the laser plasma using a 400 mm focal length reflective mirror. The position of the focus with respect to the plasma area was chosen to maximize the harmonic signal, and the intensity of femtosecond pulses at the plasma area at these conditions was estimated to be $I_{fs} = 6 \times 10^{14}\,\mathrm{W\,cm^{-2}}$. The 30 fs, 780 nm, 2 mJ probe pulses from another Ti:sapphire laser operating at 1 kHz repetition rate and producing approximately the same intensity inside the laser plasma were also used for HHG. The details of this setup and registration system are presented in [66, 70].

In order to analyse the harmonic yield of the MIR source in the graphite-ablated plasma an OPA pumped by the 30 fs Ti:sapphire laser was used. A beam splitter inserted before the laser compressor of this Ti:sapphire laser allowed to pick off 10 % of the beam (780 nm, 1 mJ, 160 ps, 1 kHz pulses) to generate a plasma plume on the graphite targets, with the remaining 90 % being compressed to 30 fs (7 mJ) to pump a commercial OPA. The OPA was optimised for high conversion efficiency, beam quality and duration of the converted pulses. To achieve high reproducibility of the generated pulses, all the amplification stages were driven to saturation. This device generated 35 fs signal pulses in the 1200–1600 nm range. The idler pulse covered the 1600–2200 nm range. The delay between the heating ablation pulse and MIR pulses from the OPA was set to 35 ns, as this delay was found to be optimal for the efficient generation of extended harmonics.

Since the goal of these studies was to analyse the graphite ablation plasma characteristics at the conditions of efficient HHG of ultrashort laser pulses, this process was firstly optimised by achieving the maximum conversion efficiency and highest harmonic cut-off using the probe radiation from both Ti: sapphire lasers with fixed wavelengths and the tuneable OPA. Then the efforts were concentrated on the analysis of the "optimal" plasma plume using three techniques: optical emission spectroscopy of emitting plasma species in the visible, UV and XUV spectral ranges; scanning electron microscopy for inspection of the deposited plasma debris; and finally time-of-flight mass spectrometry for analysis of the ionic components of the plasma.

To analyse the influence of the spectro-temporal characteristics of the probe radiation on the harmonic yield, the backing pressure of neon in the hollow fiber of second compressor was changed, which allowed the variation of pulse duration from 25 to 3.5 fs [106]. The dependence of the spectral and intensity characteristics of the harmonic images recorded by the CCD camera in the 15–25 eV range at different input pulse spectra and backing pressures of neon are shown in Fig. 2.11. One can clearly see that, with the increase of backing pressure (from 1.2 to 3 bar), the harmonic intensity increases, while the harmonic wavelength spectrally shifts towards the blue. During these experiments the driving pulse energy was held constant.

An interesting feature of the carbon harmonic spectrum from the 10 ns pulse-induced plasma is that the spectral width is about 2–3 times broader than that of harmonics generated in other atom- and ion-rich plasmas at the same fluence and intensity of heating pulse, when using few-cycle pulses. For example, the full width at half maximum for medium-order harmonics was 1.5 nm in the case of graphite

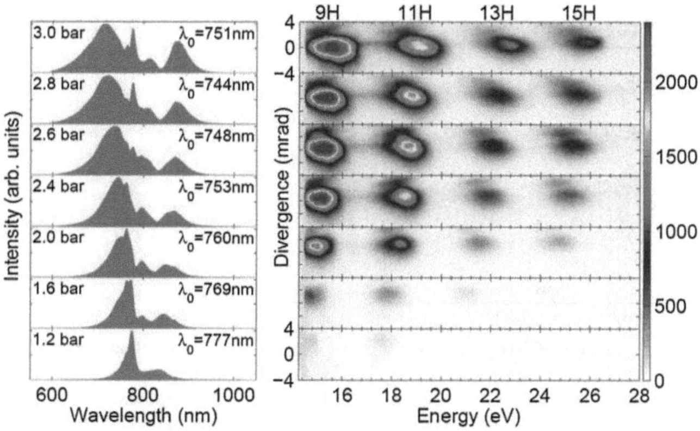

Fig. 2.11 Carbon harmonic spectra as a function of neon pressure in the hollow fiber. The corresponding laser spectra measured in front of the vacuum chamber are presented on the *left* side. The plasma was created using the 10 ns pulses. λ_0 is the central weighted wavelength of the spectral distribution. The colour scale indicates the harmonic intensity. Reproduced from [104] with permission from IOP Publishing

plasma, versus 0.4 nm for different metal (Ag, Al, and Cu) plasmas. The broader width of the harmonics can be explained by self-phase modulation and chirping of the fundamental radiation propagating through the carbon plasma. The presence of nanoparticles in the plasma plume may also contribute to bandwidth broadening of harmonics.

For practical applications of the coherent short-wave radiation generated in graphite plasma using a 1 kHz driving laser, it is necessary to analyse the stability of the plasma characteristics and the generated harmonics. Recently introduced new technique for maintaining a stable ablation plasma for harmonic generation using high pulse repetition rate lasers (1 kHz) based on a cylindrical rotating metal target [73] was described in the previous section. The studies [104] have shown that, in spite of the different properties of metal and graphite targets, the rotating target allowed achieving stable HHG in both metal and graphite plasmas. Figures 2.12a,b show the improved stability over $\sim 10^6$ laser shots of the 11–25th harmonics when using a rotating graphite target and how the harmonic intensity rapidly decays after the target rotation is stopped. The rotating graphite rod allows maintaining a relatively stable harmonic yield well above 1×10^6 laser shots. Harmonics up to the 29th order were routinely observed in these studies using the 3.5 fs pulses.

It is worth noting that harmonic intensity is the same when returning to the same spot after one rotation of the graphite rod. This reveals the unchanged morphological target conditions. Indeed, target analysis by optical inspection has confirmed that there is negligible surface modification due to laser ablation provided the ablation spot continuously moves along the target surface. This means that in graphite, like in metals, under repetitive ablation on the same target position at 1 kHz, the instability of the plasma and generated harmonics is largely related with the unstable conditions

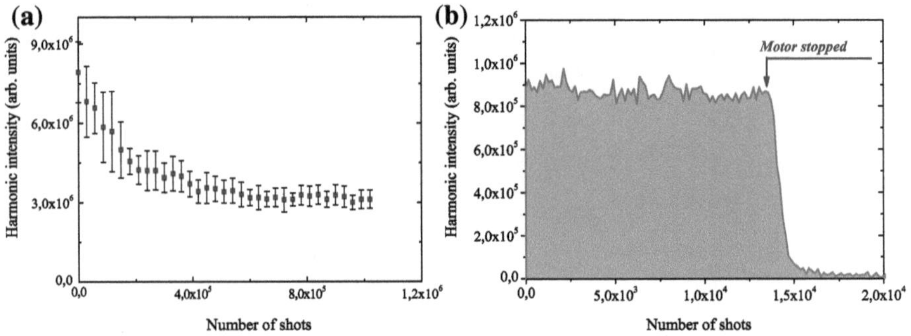

Fig. 2.12 Harmonic generation in graphite plasma using 3.5 fs pulses. **a** Stability of harmonic intensity over 1 million shots on a graphite target integrated over the 11–25th harmonics. **b** Decay of harmonics after stopping the motor rotating the target. Reproduced from [104] with permission from IOP Publishing

of the ablating spot. After moving to a new spot, the previous irradiated target area cools down and becomes again available (one rotation later) for further ablation with the same harmonic output, as in the case of metal rods. Therefore using a rotating graphite rod can significantly improve the stability of the harmonic signal.

Upper panel of Fig. 2.13 shows the harmonic spectrum generated in the case of 1300 nm probe pulses. Harmonics up to the 57th order were observed at the conditions of carbon plasma formation using the heating uncompressed 160 ps pulses from this laser. It is worth noting that application of less intense 1400 nm pulses available by tuning the OPA, while generating weaker harmonics, did not result in a higher harmonic cut-off than in the case of 1300 nm. This observation suggests that the harmonic generation occurred under saturated conditions, with the expectation of even stronger harmonics once the micro- and macro-processes governing frequency conversion are optimised.

Harmonic spectra up to the 29th order in the case of 780 nm, 30 fs probe pulses, are presented in the bottom panel of Fig. 2.13. By comparing with the spectra collected with the 1300 nm driving source (same figure, upper panel), one can clearly see the expected extension of harmonic cut-off in the case of the longer-wavelength driving source. The important peculiarities of these comparative studies are the broadband harmonic spectra in the case of 1300 nm laser and the similar yield of harmonics at the two driving wavelengths. Whilst the former feature depends on the bandwidth of the OPA output, the later observation requires additional consideration. The plasma harmonic yield from the MIR source did not follow the expected $I_h \propto \lambda^{-5}$ rule. In fact, for the intensities of MIR and 780 nm pulses used ($\sim (2\text{–}4) \times 10^{14}\ \mathrm{W\,cm^{-2}}$), the harmonic efficiency of the XUV radiation driven by MIR pulses was higher compared with the case of 780 nm pulses, assuming lower energy of the former pulses (0.2 and 0.54 mJ respectively). One can note that the $I_h \propto \lambda^{-5}$ rule predicts a ~13-fold decrease of conversion efficiency for the MIR (1300 nm) pulses compared with the 780 nm pulses at equal probe pulse intensity.

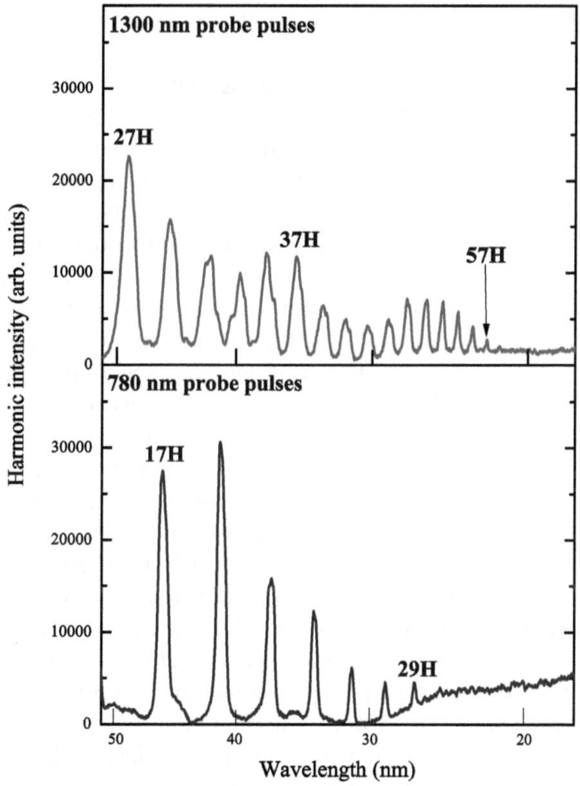

Fig. 2.13 Plasma harmonic spectra using the 1300 nm (*upper panel*) and 780 nm (*bottom panel*) probe pulses. The energies of probe pulses were 0.2 mJ (*upper panel*) and 0.54 mJ (*bottom panel*). Ablation was carried out using 160 ps, 780 nm, 1 kHz laser pulses. Reproduced from [104] with permission from IOP Publishing

2.4.2 Characterization of Optimal Plasma Conditions

This section presents the characterization of the graphite ablation plasma plumes at conditions of maximum HHG conversion efficiency. In graphite, the ablation plasma plume may contain various species of carbon, i.e. neutrals and ions, small molecules, clusters, aggregates, etc., which can contribute to harmonic generation in various extents. It is important to determine their presence in the region where the driving laser pulse interacts with the expanding plasma. In particular, the production of clusters in the laser plasma during laser ablation of various targets has a high probability, while their presence and concentration in the plasma area where the frequency conversion occurs is yet to be confirmed directly. Another issue is how one can define the density of monomers, dimers and clusters and their influence on the HHG yield. Analysis of post-ablation conditions of the deposited debris can provide information on the nature of the nonlinear species, despite the differences between the composition of the plasma in its early stages and the deposited material, due to the influence of conditions of aggregation on the substrate [107]. Another issue of interest is whether the spectral characterisation of the plasma emission in the visible and UV ranges can provide some clues about the plasma conditions, without the simultaneous analysis of the XUV emission.

Plasma characterization through optical spectroscopic measurements in the visible, UV, and XUV ranges at the conditions of different HHG efficiencies were carried out using the above-described XUV spectrometer and a fiber spectrometer. The fluences of heating pulses at which the spectra were recorded corresponded both to optimal and non-optimal conditions of HHG. The acquisition times were set to 1 s, for measurements of XUV spectra, and to 0.5 s, for measurements of visible and UV spectra. Characterisation of the plasma debris collected on silicon wafers placed 4 cm from the ablated target was carried by scanning electron microscopy (SEM).

Cluster composition of the ablation plume produced by nanosecond laser pulses was investigated by time-of-flight mass spectrometry (TOFMS). A brief description of the experimental set up is given here; more details of the TOFMS can be found elsewhere [108]. The laser beam (1064 nm, 5 mJ, 10 ns pulse duration) was focused at normal incidence to a 0.2 mm spot on the surface of the graphite target. The laser intensity was 1.5×10^9 W cm^{-2}, which resulted in the creation of an optimal plasma for efficient HHG using the nanosecond ablation pulses. The target was placed in a vacuum chamber (pumped to $\sim 2 \times 10^{-6}$ bar) between the extracting and accelerating plates of a linear TOFMS. The target surface was parallel to the flight axis of the spectrometer. The target could be rotated and displaced at variable distances from the axis. Positive ions produced in the ablation were deflected along the TOFMS axis by an electric field typically in the range of 300–400 V cm^{-1} and accelerated by a total voltage of 2500 V. A high voltage switch was used to apply the bias voltage at controlled delays with respect to the laser ablation pulse. Ions entered the drift region (flight length $\sim$1 m) and were detected by a microchannel plate. Analysis of neutral species produced in the ablation could also be performed by the use of a second post-ionization laser (F_2 excimer at 157 nm). The post-ionization laser pulse interacted with the ablation plume perpendicularly to the plume propagation axis, at different distances from the target surface and at different delays with respect to the ablation laser.

It has been shown previously that efficient harmonic emission is observed only in the case when the visible and UV plasma emission is dominated by neutral and singly ionized carbon lines [67]. The studies [104] have also confirmed this feature at the laser fluence used to heat the target surface (Fig. 2.14a). The broad features near 470, 515, and 555 nm could be assigned to the bands of excited C_2 molecules. These bands have also been observed early studies of the ablation of graphite (see for instance [109–111]). Other lines in the spectra presented in Fig. 2.12a are attributed to the neutral and singly charged carbon.

The analysis of optical spectra in the visible and UV ranges does not provide information about the presence of highly ionised species, which can be revealed by collecting the plasma emission in the XUV range. The XUV spectrum of carbon plasmas (Fig. 2.14b) collected following excitation by a 8 ps heating pulse at high intensity, without further excitation by the probe pulse, provides some insight into the plasma components prior to interaction with the driving radiation. This spectrum was collected under conditions of considerable decrease in the nonlinear optical

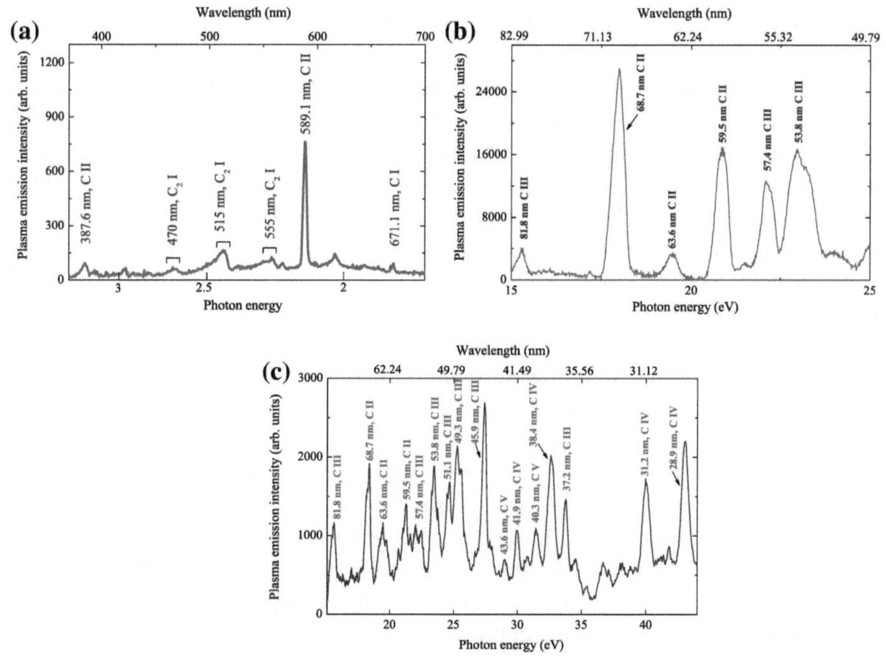

Fig. 2.14 **a** Carbon plasma emission spectrum in the visible and UV ranges at optimal excitation of a graphite target by 8 ps heating pulses (2×10^{10} W cm^{-2}). **b** Spectrum of carbon plasma in the XUV range at over-excitation of the target by 8 ps pulses (5×10^{10} W cm^{-2}). **c** Spectrum of carbon plasma in the XUV range at over-excitation of the target by 10 ns pulses (3×10^{9} W cm^{-2}). Reproduced from [104] with permission from IOP Publishing

response of the medium (i.e. plasma conditions unoptimized for HHG) and revealed the appearance of many emission lines from C II and C III ions. Over-excitation of the target by 10 ns pulses also led to the appearance of emissions from high-charged (C III, C IV) ions (Fig. 2.14c).

It may be noted that these measurements were time-integrated, so one could not say exactly which plasma components existed at the moment of the propagation of the femtosecond beam through the plume. However, the presence of ionic lines from multi-charged species in the last two cases (Figs. 2.14b,c) gives a strong indication of over-excitation of the target and of its negative influence on HHG efficiency. One can note that, at this level of excitation of the graphite plasma, harmonic generation was partially or entirely suppressed. Specifically, a two-fold increase in the intensity of 8 ps pulses (from 2×10^{10} to 4×10^{10} W cm^{-2}) led to a decrease of harmonic intensity by a factor of 2.5. The same can be said about the excitation using longer (10 ns) pulses, though the threshold, at which harmonics started to decay, was considerably lower (2×10^{9} W cm^{-2}). Application of 10 ns pulses with an intensity of 3×10^{9} W cm^{-2} led to a substantial decrease of harmonic efficiency and to the appearance of emission lines from high-charged ionic species (Fig. 2.14c). One can note that, under conditions of efficient HHG, no ion lines appear alongside the harmonic spectra.

To prove the presence of clusters in carbon plasmas under optimal conditions of harmonic generation, the morphology of deposited debris from the graphite-ablated plasma was analysed during ablation using the picosecond and nanosecond pulses. It has already been mentioned that laser ablation of a solid material is a widely accepted technique for the generation of nanoparticles. However, this process has previously been studied without taking into account the role of free electrons and highly excited ions, which destroy the optimal conditions for phase-matched HHG. SEM measurements of the deposited debris were carried out at the laser ablation conditions corresponding to optimal plasma formation for efficient HHG. The substrates (glass plates and silicon wafers) used to collect the deposited material were placed at a distance of 40 mm in front of the ablation area and the debris was further analysed by SEM.

At optimal plasma conditions, when the highest harmonic conversion efficiency from the carbon-containing 8 ps pulses was measured, the SEM images did not reveal the presence of nanoparticles in the deposited debris with sizes above the limit of detection (10 nm) of the microscope (Fig. 2.15a). This was probably due to the small fluence (0.2 J cm^{-2}) of the heating radiation on the target surface ($I_{ps} = 2.5 \times 10^{10}$ W cm^{-2}). It is possible that in the case of the carbon plasmas produced at these conditions, harmonics could also originate from nanoparticles with sizes below the limit of detection. Another pattern was observed upon ablation of the target with 10 ns pulses, where considerably higher heating fluence (10 J cm^{-2}) caused the appearance of nanoparticles on a nearby substrate. At relatively moderate conditions of ablation using 10 ns pulses ($I_{ns} \approx (1 - 2) \times 10^9$ W cm^{-2}), a high density of nanoparticles appeared in the SEM images of the deposits, with sizes mostly distributed in the range between 10 and 200 nm, with a mean size of 50 nm (Fig. 2.15b). One has to reiterate that these debris characteristics were measured at maximum conversion

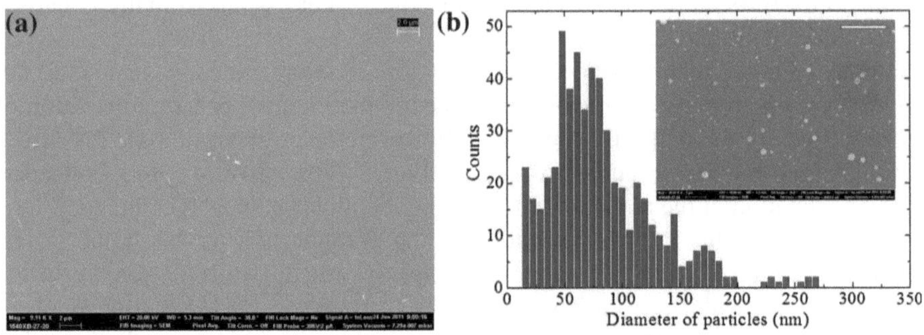

Fig. 2.15 **a** SEM image of the deposited debris after ablation of a graphite target by 8 ps pulses at optimal conditions of HHG. **b** Histogram of size distribution of deposited nanoparticles and corresponding SEM image of deposited debris in the cases of ablation of a graphite target using 10 ns pulses. The size bar on the SEM images is 2 μm. Reproduced from [104] with permission from IOP Publishing

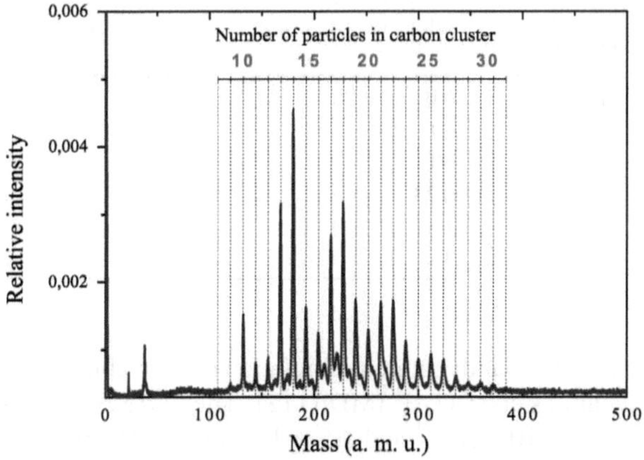

Fig. 2.16 Mass spectrum (in atomic mass units) of carbon plasma obtained during excitation of a graphite target using 10 ns pulses at the conditions close to optimal plasma formation for HHG. Reproduced from [104] with permission from IOP Publishing

efficiency of 15th–23rd harmonics. These morphological studies have confirmed the presence of relatively large nanoparticles deposited on the substrates at conditions of "optimal" ablation using 10 ns pulses. However, some uncertainty still remains about the correlation of these results with the presence of the same nanoparticles in the carbon plasma during harmonic generation, due to the possibility of aggregation of these clusters after deposition. To address this issue, the TOFMS of nanosecond pulse-ablated graphite was carried out.

Figure 2.16 shows the mass-resolved spectrum of carbon plasma after 60 shots of the 10 ns heating pulses. These studies reveal that at plasma conditions close to optimal for HHG, the laser plume contains a group of small, singly ionised carbon clusters (C_{10}–C_{30}). The attempts to find higher mass clusters failed, though the authors of [104] searched for them over a longer range of delays (up to few μs) between the onset of laser ablation and the switching on the triggering pulse in the TOFMS.

To ascertain the presence of neutral species in the ablated plasma, an additional source of ionisation can be used. An F_2 post-ionisation excimer laser ($\lambda = 157$ nm) served that purpose. Photons of this wavelength can induce the ionisation of neutral clusters through single-photon absorption. However, once the 157 nm pulse was focused on the plasma area, no evidence in the mass spectra of neutral clusters was found. This could be explained by the fact that the ionisation potential of carbon ($I_p = 11.2$ eV) is higher than the photon energy of the ionising laser ($E = 7.9$ eV) and by the difficulty of ionising the existing carbon clusters by two-photon absorption of 157 nm radiation, due to low intensity of these pulses in the plasma area.

The concentrations of carbon plasma at the experimental conditions of target ablation (i.e., 2×10^{10} W cm^{-2} in the case of 8 ps pulses and 1×10^{9} W cm^{-2} in the case of 10 ns pulses) allowing efficient harmonic generation were calculated using a

Table 2.2 Calculations of carbon plasma concentration for different intensities of 8 ps and 10 ns pulses heating the graphite target

Intensity, 10^9 W cm^{-2}	Plasma concentration, 10^{17} cm^{-3}	
	8 ps	10 ns
6.6	1.1	
20	2.6	
60	4.0	
0.33		10
1		25
3		37

three-dimensional molecular dynamical simulation of laser ablation of graphite using the molecular dynamics code ITAP IMD [112]. The corresponding concentrations were found to be 2.6×10^{17} cm^{-3} and 2.5×10^{18} cm^{-3}. The results of calculations of carbon plasma concentration for different intensities of heating 8 and 10 ns pulses are summarized in Table 2.2.

2.4.3 Analysis of HHG and Plasma Characterization

A few earlier studies have suggested that the presence of nanoparticles in carbon laser ablation plasmas can explain the observed strong harmonic yield from these media [61, 62]. It was reported that the debris from ablated graphite and carbon lead targets contained nanoparticles with sizes between 100 and 300 nm. The authors of these studies therefore suspected that nanoparticles formed in the plasma by ablation were the source of intense harmonics. Heterogeneous decomposition, liquid phase ejection and fragmentation, homogeneous nucleation and decomposition, and photomechanical ejection are among the processes that can lead to the production and disintegration of nanoparticles [113–115]. A number of different techniques were used in these studies to determine the aggregation state of the evaporated material, including time-resolved emission spectroscopy, CCD camera imaging of the plasma plume, Rayleigh scattering, and laser-induced fluorescence.

In discussed studies [104], SEM for debris analysis and TOFMS were applied for plasma characterization. These two methods have provided useful clues about the conditions and dynamics of plasma plume formed above the target surface. Whilst the former method can provide information about the presence of nanoparticles in the plasma, one has to cautiously consider those results from the following point of view. The deposition process on the substrate happens much later than the time of HHG emission, and the physical process of deposition may lead to further aggregation. Since SEM is an ex-situ method, one cannot exclude the difference between the real composition of clusters in the plasma and the results of SEM measurements, although it clearly proves the presence of clusters in the plasma. TOFMS yields information on the in-situ presence of ionised clusters, although it requires ablation of the target at the same conditions as in the case of HHG experiments and is not well suited for the detection of neutral nanoparticles in the ablated plasma.

TOFMS measurements did not reveal the presence of neutral clusters in the 10 ns pulse produced plasma for the reasons described in the previous section. However, other studies (see for example [110]) have indicated the presence of neutral carbon clusters using two-photon ionisation with an ArF laser (photon energy 6.4 eV). Early TOFMS studies of laser ablation of graphite have revealed the typical characteristics of the expanding plasma species (average velocity 1.5×10^5 cm s^{-1}) and their concentration ($4 \times 10^{18} - 6 \times 10^{19}$ cm^{-3} [116]) for ablation with 532 nm, 10 ns pulses at fluences of the order of 3 J cm^{-2}. The measured mass distribution shown in the spectrum of Fig. 2.16, revealing the presence of C_{10} to C_{30} species, is in good agreement with those observed in previous studies of graphite-ablated plasma at similar excitation conditions [116]. The restriction of cluster sizes to small-sized carbon nanoparticles has also been reported in [117], where it was argued that stronger excitation conditions are necessary to observe clusters larger than C_{32}. In that case, one should expect the appearance of closed cages made of joined five and six member rings. It was confirmed that C_{60} and C_{70} fullerenes, are the most abundant species among the high-mass ions of the carbon plasma plumes at higher ablation fluences. It was also suggested [116] that it is very likely that the plasma is sufficiently dense for cluster growth to occur via ion-molecular reactions. The kinetic mechanism can be responsible for the formation of carbon cluster ions since the supersonic entrainment method is expected to considerably cool down the cluster ions. The growth of clusters is based on the addition of many small carbon neutral species to the ions in a stepwise fashion.

An explanation for strong harmonic generation from nanoparticles compared with single atoms or ions could be the higher concentration of neutral atoms inevitably accompanying the presence of nanoparticles. Unlike single atoms and ions, whose density quickly decreases due to plasma expansion, nanoparticles retain local densities that are close to solid state. The increase of electron recombination cross-section for clusters with respect to atoms can also potentially enhance the HHG efficiency in nanoparticle-contained plasmas. Earlier studies of HHG from gases [101, 118, 119], as well as from plasmas containing various nanoparticles (Ag, Au, $BaTiO_3$, etc) [67, 69], have proven these assumptions by demonstrating the enhanced HHG from clusters as compared with single atoms and ions. Further evidence of the cluster contribution to the enhancement of the harmonic generation process comes from investigations of very intense laser ablation of a silver target [48], which gave clues regarding the participation of in-situ generated nanoparticles.

The observation of a strong extended harmonic plateau in the case of the 1300 nm probe radiation also suggests the involvement of clusters in the HHG process with MIR pulses. Assuming the expected decrease of harmonic intensity from single particle emitters with the growth of driving radiation wavelength ($I_h \propto \lambda^{-5}$, [5, 102, 103, 120]), one can anticipate at least one order of magnitude decrease of harmonic yield from MIR pulses as compared with the yield obtained with 780 nm radiation at other equal conditions, in particular, pulse energy and duration. However, the experiment did not show a considerable difference between the intensities of harmonics originated from these two driving sources (Fig. 2.13). The energy of the 1300 nm pulses in the plasma area (0.2 mJ) was lower than the Ti:sapphire pulse

(0.54 mJ). This suggests the involvement of a mechanism, which compensates for the expected considerable decrease of harmonic efficiency for the longer-wavelength laser. The involvement of a clustered component of the laser plasma in the process of frequency up-conversion may arguably explain the observed inconsistence with the theoretical predictions of the $I_h \propto \lambda^{-5}$ rule defined for atomic species [102, 103].

In principle, the intensity enhancement of the harmonic spectrum from the carbon plume in the 15–26 eV range invokes the involvement of surface plasmon resonances of nanoparticles, analogously to the case of fullerenes [51, 69] in the range of their giant resonance in the vicinity of 20 eV. To prove this in the case of carbon plasma, one should provide evidence of giant absorption in the above range, but this has not been reported yet in the literature. The plasmonic properties of carbon nanoparticles can be responsible for the observed enhancement of carbon harmonics, however their role requires additional study [121]. Another option for explaining the high harmonic generation yield in the carbon plume is the indirect involvement of the clusters in HHG that, while not participating as harmonic emitters, could rather enhance the local field, analogously to recently reported studies using gold nanostructures enhancing gas HHG [122, 123].

As it was mentioned, recent comparative studies of lower-order harmonic efficiency in argon gas and carbon plasmas have revealed stronger conversion efficiency in the carbon plasmas [66]. In this section, we have discussed evidence of the superior properties of graphite ablation for HHG. Some arguments which could explain the enhanced high harmonic yield from this medium are as follows: (a) the graphite target allows easier generation of a relatively dense carbon plasma and the production of adequate phase-matching conditions for lower-order harmonic generation, (b) the first ionization potential of carbon is high enough to prevent the appearance of high concentration of free electrons, a condition that is not necessarily met in metal plasma plumes, (c) neutral carbon atoms dominate in the carbon plume at optimal conditions of HHG before the interaction with the femtosecond laser pulse, and (d) carbon species allow the formation of multi-particle clusters during laser ablation, which can enhance the HHG yield.

2.5 Harmonic Generation in Fullerenes Using Few- and Multi-Cycle Pulses of Different Wavelengths

Fullerenes can be considered as an attractive nonlinear medium for the HHG. Their relatively large sizes and broadband surface plasmon resonance (SPR) in the XUV range allowed the first demonstration of enhanced HHG near the SPR of C_{60} ($\lambda_{SPR} \approx$ 60 nm, with 10 nm full width at half maximum) [32]. The application of laser ablation technique allowed the creation of C_{60}-rich plasma ($\sim 5 \times 10^{16}$ cm^{-3}), in a stark contrast with the density $<10^{14}$ cm^{-3} which can be achieved using the heat oven based methods of production of the fullerene beams.

The theoretical studies of HHG from C_{60} using the multi-cycle pulses include the extension of three-step model [124], analysis of the electron constrained over the surface of a rigid sphere, with geometrical parameters similar to those of the C_{60} [125], and application of dynamical simulations [126]. In the latter work, high-order harmonics were shown to be due to multiple excitations and could be easily generated even with a weak laser field. Those studies reveal how HHG can be used to probe the electronic and molecular structure of C_{60}. At the same time, theoretical investigation of such systems is hampered by the fact that the Hamiltonian of HHG is time dependent and the system consists of many electrons. The investigation of the influence of the electrons on the resonant HHG can be performed by means of a multiconfigurational time-dependent Hartree-Fock (MCTDHF) approach, which has the accuracy of direct numerical solution of Schrödinger equation and is almost as simple as the ordinary time-dependent Hartree-Fock approach. In particular, the computations could be based on the multiconfigurational time-dependent Hartree software packages. In [46], simulations of resonant HHG were performed by means of a MCTDHF approach for three-dimensional fullerene-like systems. The influence of the SPR of C_{60} on the harmonic efficiency in the range of 60 nm ($E = 20$ eV) was analysed and showed the ways of resonant HHG optimisation [127].

The saturation intensities of different charge states of C_{60} are higher compared to isolated atoms of similar ionisation potential. In this connection, it would be interesting to analyse the behaviour of fullerene molecules in the field of few-cycle pulses from the point of view of harmonic generation and compare these studies with those carried out using the multi-cycle pulses. The motivation of studies [128] was to analyse the conditions of efficient HHG from the plasma containing C_{60}, when picosecond radiation ablates the fullerene-containing target at high pulse repetition rate (1 kHz) and then few-cycle pulse ($t = 3.5$ fs) propagates through the fullerene plasma. It would be also interesting to analyse HHG in fullerenes using the longer- and shorter-wavelengths probe sources, in particular using the 1300 and 780 nm multi-cycle (35 and 40 fs) pulses.

The detailed description of experimental setup for plasma HHG is presented in previous sections. Briefly, a small part ($E = 120\,\mu$J) of the uncompressed radiation of a 1 kHz Ti:sapphire laser with central wavelength 780 nm and pulse duration 8 ps was split from the beam line prior to the laser compressor stage and was focused into the vacuum chamber to create a plasma on the C_{60}-containing target using an intensity on the target surface of typically $I_{ps} = 2 \times 10^{10}$ W cm^{-2}. Then the few-cycle pulses were focused into the plasma plume, approximately $150 - 200\,\mu$m above the target surface, to generate high-order harmonics. Two types of targets (C_{60} powder glued on the glass substrates or rotating aluminium rod, and bulk graphite) were used for plasma harmonics generation. These targets were studied from the point of view of comparative intensities of harmonics they can provide. The harmonic spectra from the plasmas produced on the bulk graphite and fullerene powder glued onto the glass surface are presented in Fig. 2.17. The harmonics up to the 29th order were obtained from the fullerene plasma. Note that harmonic efficiency in the case of graphite plasma was a few times stronger compared with the case of fullerene plasma, which probably was caused by higher concentration of plasma in the former case, as it has also been reported in [62].

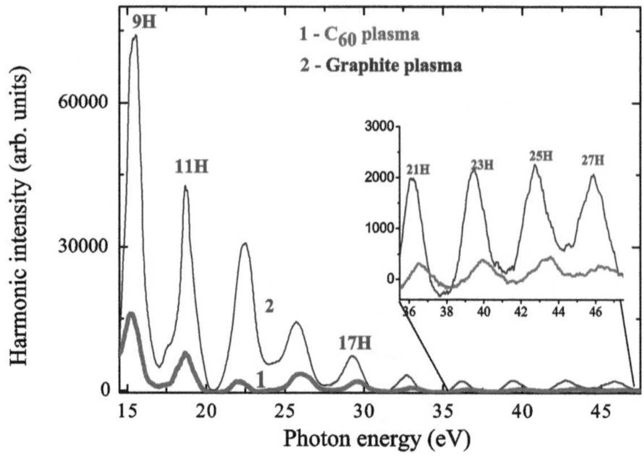

Fig. 2.17 Harmonic spectra from fullerene (*thick curve*) and graphite (*thin curve*) plasmas using the heating 8 ps pulses at identical experimental conditions. Reproduced from [128] with permission from Optical Society of America

The results of comparative studies of the HHG in fullerene plasma using the few-cycle (3.5 fs) and multi-cycle (40 fs) pulses are presented in Fig. 2.18. In the latter case, the Ti:sapphire laser generating the 1 kHz, 4 mJ, 40 fs pulses was used. The conversion efficiency of the 40 fs pulses at the beginning of plateau range in the case of plasma plume containing fullerenes was estimated to be $\sim 5 \times 10^{-6}$ using the comparison with the HHG conversion efficiency in the silver plasma, which has previously been reported at analogous experimental conditions to be 1×10^{-5} [36]. One can see that the cut-off in the case of longer pulses (25th harmonic) was shorter with regard to the few-cycle pulses (29th harmonic).

The problem of using fullerene powder-containing target is a shot-to-shot instability and rapid decrease of harmonic yield, due to abrupt change in the target morphology. This negative feature of harmonics from such targets prevents from using these intense harmonic sources in applications. Nevertheless, currently HHG in laser-produced plasma plumes remains the alone method of studies of the high-order nonlinearities of fullerenes. The alternative could be a creation of long homogeneous

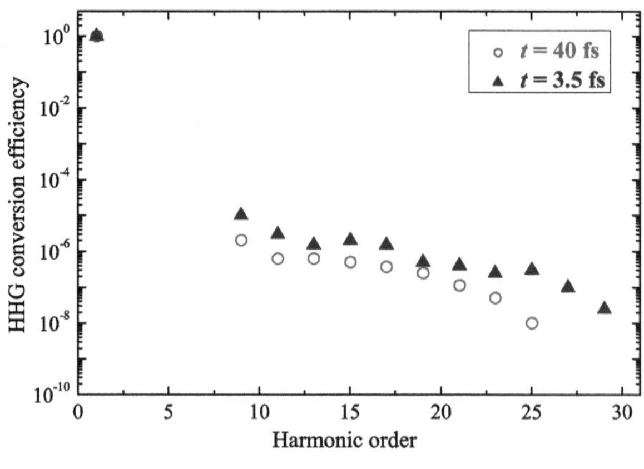

Fig. 2.18 Comparison of HHG conversion efficiencies from fullerene plasma using the 3.5 fs (*triangles*) and 40 fs (*circles*) pulses. Reproduced from [128] with permission from Optical Society of America

tapes containing fullerenes, which continuously move from shot to shot to provide the fresh surface for each next laser pulse. The application of rotating targets containing C_{60} is another method for improvement of the harmonic stability from this medium, which was implemented in these studies. The application of rotating aluminium rods as the substrates on which the fullerene powder was glued considerably improved a stability of harmonics from C_{60}-containing plasmas compared with the case of ablated layer of fullerene powder glued on the fixed glass substrate. One can note that, once the rotating rod stops, the harmonic efficiency from fullerene plasma decreased to almost zero level during 1–2 s. The proposed technique, which was described in the case of bulk targets in previous sections, could be very useful in the case of powder-like targets (fullerenes, metal nanoparticles, organic powder-like samples, etc).

As it was mentioned, the conversion efficiency studies showed advantages of HHG in the case of graphite plasma compared with fullerene plasma [62]. This advantage of graphite plasma over fullerene one can be explained by different concentrations of particles in these plasmas. The concentration of fullerenes can be considered as below 10^{17} cm^{-3} [32], while the estimates of carbon plasma based on a three-dimensional molecular dynamical simulation of laser ablation of graphite using the molecular dynamics code ITAP IMD [112] showed that, in the case of heating by 8 ps pulses, the graphite plasma density can reach 2.6×10^{17} cm^{-3} at the moderate intensity (2×10^{10} W cm^{-2}) of heating radiation [66]. This difference in plasma densities can be instrumental in explanation of the observed higher efficiency of HHG from graphite plasma compared with fullerene one. Another reason for observed advanced features of graphite plasma harmonics could be the cluster formation during laser ablation (see previous section).

In order to analyse the harmonic yield of the mid infrared source in the fullerene-contained plasma an optical parametric oscillator pumped by the 40 fs Ti:sapphire laser was used. Figure 2.19 shows the comparison of fullerene harmonic spectra generated in the case of 1300 and 780 nm multi-cycle probe pulses. Harmonics up to the 41st order (Fig. 2.19, bottom panel) were observed in the case of 1300 nm probe pulses at the conditions of optimal plasma formation using the heating 160 ps pulses. Over-excitation of target by 160 ps pulses (Fig. 2.19, middle panel) led to appearance of plasma emission in the 25–45 eV range of photon energies ($\lambda = 27 - 50$ nm). At these conditions, no harmonics were observed during propagation of femtosecond pulses through such over-excited plasma.

Harmonic spectrum up to the 25th order in the case of 780 nm, 40 fs probe pulses is presented in the Fig. 2.19 (upper panel). One can clearly see the extension of harmonic cut-off in the case of longer-wavelength probe pulses by comparing the harmonic spectra generated using the 780 and 1300 nm probe pulses. The plasma harmonic yields from the 780 and MIR probe pulses followed approximately the expected $I_h \propto \lambda^{-5}$ rule. The harmonic efficiency of the XUV radiation driven by MIR pulses was 7–15 times less compared with the case of 780 nm probe pulses, which is comparable with the expected ratio between harmonic intensities from these sources [$(1300/780)^5 \approx 12.7$] followed from above rule, assuming approximately equal energies of the 780 and 1300 nm pulses (0.53 and 0.5 mJ respectively).

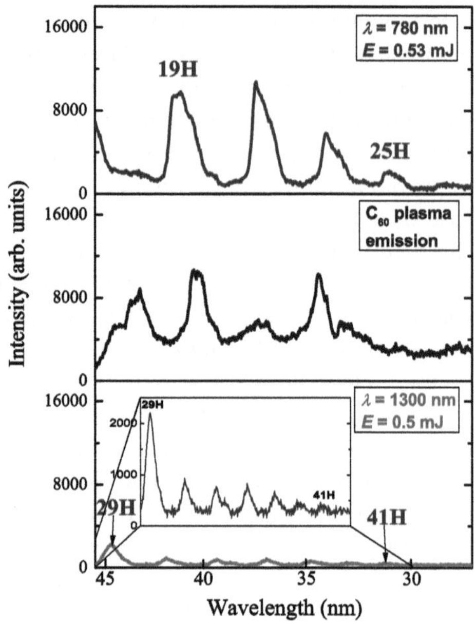

Fig. 2.19 Comparative harmonic spectra from fullerene-contained plume using the 780 nm (*upper panel*) and 1300 nm (*bottom panel*) multi-cycle pulses. Note the equal Y axes for these cases. The middle panel shows C_{60} plasma emission spectrum at over-excitation of target by 160 ps pulses, without further excitation by femtosecond probe pulses. Reproduced from [128] with permission from Optical Society of America

Previous HHG studies in fullerenes were performed using the multi-cycle pulses (30 fs [32], 48 fs [129], and 110 fs [60]). The stability of C_{60} against fragmentation in these multi-cycle laser fields led to fast diffusion of the excitation energy. Even better conditions can occur in the case of few-cycle pulses used for fullerene HHG. In that case fullerenes can withstand the influence of the strong field of few-cycle pulses due to the growth of saturation intensity. This can further increase the diffusion of the excitation energy within the fullerenes due to their very large number of internal degrees of freedom. The increase of energy diffusion should follow with the decrease of disintegration of fullerenes, which enhances the probability of harmonic emission from these clusters.

The carbon, which was compared in these experiments with C_{60}, can easily be aggregated during laser ablation, thus providing the nanoparticles in laser plume (see previous section). In that case the comparison of two clustered species (large 5–20 nm carbon clusters and 0.7 nm C_{60}) can lead to their different nonlinear optical response once the interacting laser pulse becomes compressed from the multi-cycle to few-cycle duration. Above-presented results show that application of shortest pulses changes the dynamics of fullerene behaviour leading to increase of the HHG cut-off, while the intensity of 'fullerene harmonics' became less compared with 'carbon harmonics'. As we mentioned, the additional reason of this could be different concentrations of harmonic emitters from two plasma plumes. It was difficult to maintain equal plasma concentrations in these two cases, since no reliable methods are existed, which could define the exact concentrations of carbon and fullerene plasmas at different 'optimal' conditions of HHG from these media.

The intensity of probe femtosecond pulse is crucial for optimizing the HHG from C_{60}. Increasing the intensity of this pulse did not lead to an extension of the harmonic cutoff from the fullerene plume, which is a signature of HHG saturation in the medium. Moreover, at relatively high femtosecond laser intensities, one can observe a decrease in the harmonic output, which can be ascribed to the phase mismatch caused by propagation effects.

The stability of C_{60} molecules against ionization and fragmentation is of particular importance, especially for their application as a medium for HHG using laser pulses of different duration. Because of this the structural integrity of the fullerenes ablated off the surface should be intact until the probe pulse arrives. Hence, the heating pulse intensity also becomes a sensitive parameter. At lower intensities the concentration of the clusters in the plume would be low, while at higher intensities one can expect fragmentation. This phenomenon is observed when the heating pulse intensity on the surface of fullerene-rich targets is increased above the critical value (Fig. 2.19, middle panel). The abrupt reduction in harmonic intensity in that case can be attributed to phenomena such as fragmentation of fullerenes, increase in free electron density, and self-defocusing, both leading to the phase mismatch of HHG.

2.6 Isolated Sub-Femtosecond XUV Pulse Generation in Mn Plasma Ablation

In this section, we discuss HHG from transition metal plasmas. These are very promising targets in view of the giant resonances found in the photoionization cross sections. For example, the Mn^+ cross section is ~40 Mb at 50 eV photon energy [130], whereas rare gas atoms have cross sections between 1 and 8 Mb at this photon energy [131]. Photorecombination—the third step in the recollision model—is the inverse process of photoionization [132] and therefore HHG and photoionization must exhibit the same resonances. This has been confirmed not only by previous resonance-induced experiments with laser-produced transition metal plasmas but also in a recent study of HHG from xenon gas [133].

Resonance-induced enhancement of a single harmonic of the laser radiation allowed considerable improvement of harmonic efficiency in some specific XUV spectral ranges related with high oscillator strengths of ionic states of metals. This was confirmed in multiple studies following the initial observation of this phenomenon in indium plasma [25]. In particular, the strong enhancement of a single harmonic was reported in Cr [28], and Mn [35] plasmas. The Mn plasma is of special interest since it showed the highest harmonic cut-off energy observed in plasma plumes (101st harmonic [35]). In previous studies multi-cycle (30 [35] and 140 fs [134]) laser pulses were employed and the generation of all harmonics in the plateau was observed together with a strongly enhanced single harmonic.

Recent progress in the generation of few-cycle pulses allowed the observation of various new effects including the realisation of isolated attosecond-pulse generation

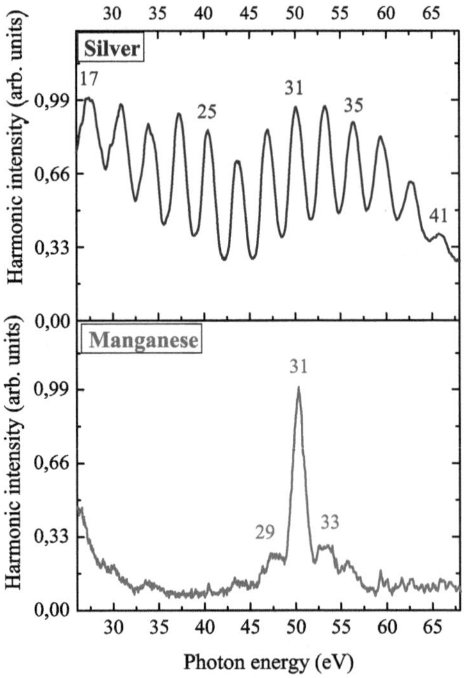

Fig. 2.20 Harmonic spectra from the silver plasma (*upper curve*) and manganese plasma (*bottom curve*). Reproduced from [138] with permission from Optical Society of America

in gas media [135–137]. In this connection it is interesting to analyse resonance-induced processes observed in an ablation plume using the shortest available drive laser pulses. Below, we present the analysis of resonance enhancement in manganese plasmas using 3.5 fs pulses [138]. The most interesting feature observed in those experiments was a suppression of almost all neighbouring harmonics in the vicinity of a resonantly enhanced single harmonic at the photon energy of ~50 eV.

The experimental arrangements were analogous to those presented in Sects. 2.3 and 2.4 of this chapter. The harmonic spectrum in the case of propagation of 3.5 fs pulses through the manganese plasma was strikingly different compared with other plasma samples (for example Ag plasma) analysed in separate experiments. While all other samples studied showed the relatively featureless harmonic spectra with extended cut-offs (Fig. 2.20, upper curve showing the spectrum of harmonics generating in the silver plasma), the Mn plasma allowed the generation of a strong single harmonic substantially enhanced compared with neighbouring ones (Fig. 2.20, bottom curve).

One can note that in earlier work the harmonic spectra from manganese plasmas for 30 fs [35] and 140 fs [134] pulses also showed enhanced harmonics around 50 eV. The assumption of the resonance nature of the enhancement of harmonics of the ~800 nm radiation of Ti:sapphire lasers in this spectral region is supported by the presence of a strong giant resonance in the vicinity of 50 eV confirmed by experimental [130, 139] and theoretical [140] studies. The enhancement of a single harmonic

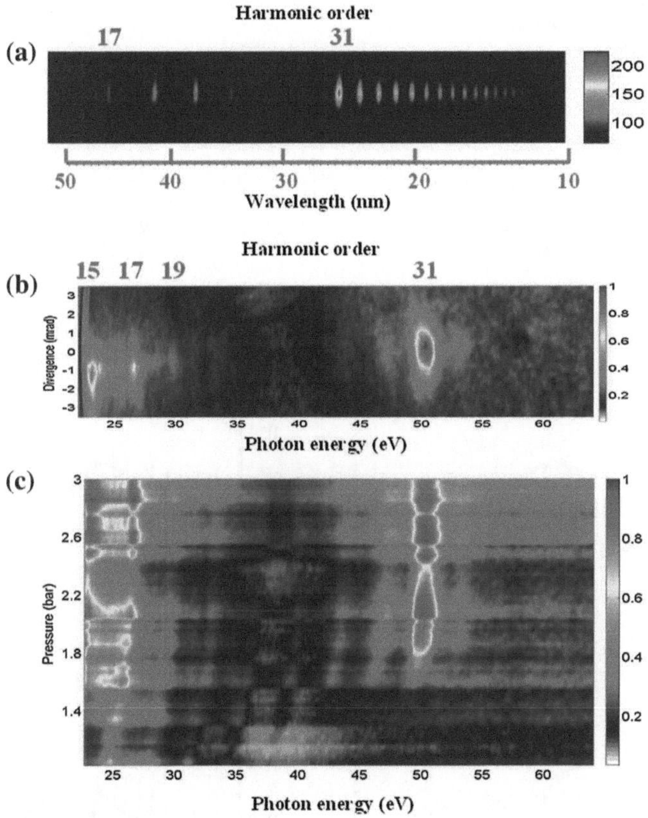

Fig. 2.21 Raw images of harmonic spectra from manganese plasma in the case of (**a**) 40 fs and (**b**) 3.5 fs probe pulses obtained at the same intensity. (**c**) Raw images of harmonic spectra from Mn plasma at different pressures of neon in the hollow fiber obtained at the same energy of probe laser pulses. Reproduced from [138] with permission from Optical Society of America

can be attributed to the broadband resonances of the ions of few metals, such as V, In, Cd, Cr, Cd, and Mn. These "giant" resonances have been experimentally confirmed in the literature [130, 139, 141] and discussed recently in a few theoretical studies [46, 90–93].

However, in previous studies using multi-cycle drive pulses, the intensity of enhanced harmonics was only a few times higher than neighboring harmonic orders. The same features were reproduced in the reviewed studies using 40 fs pulses from another Ti:sapphire laser at similar intensity inside the laser plasma (4×10^{14} W cm^{-2}). The raw image of the harmonic spectrum presented in Fig. 2.21a shows several enhanced harmonics starting from the 31st order followed by an extended second plateau. The extension of the harmonic cutoff exceeding the 71st order is attributed to the involvement of doubly charged Mn ions as the sources of HHG. This feature of Mn plasma harmonics has already been reported earlier [35].

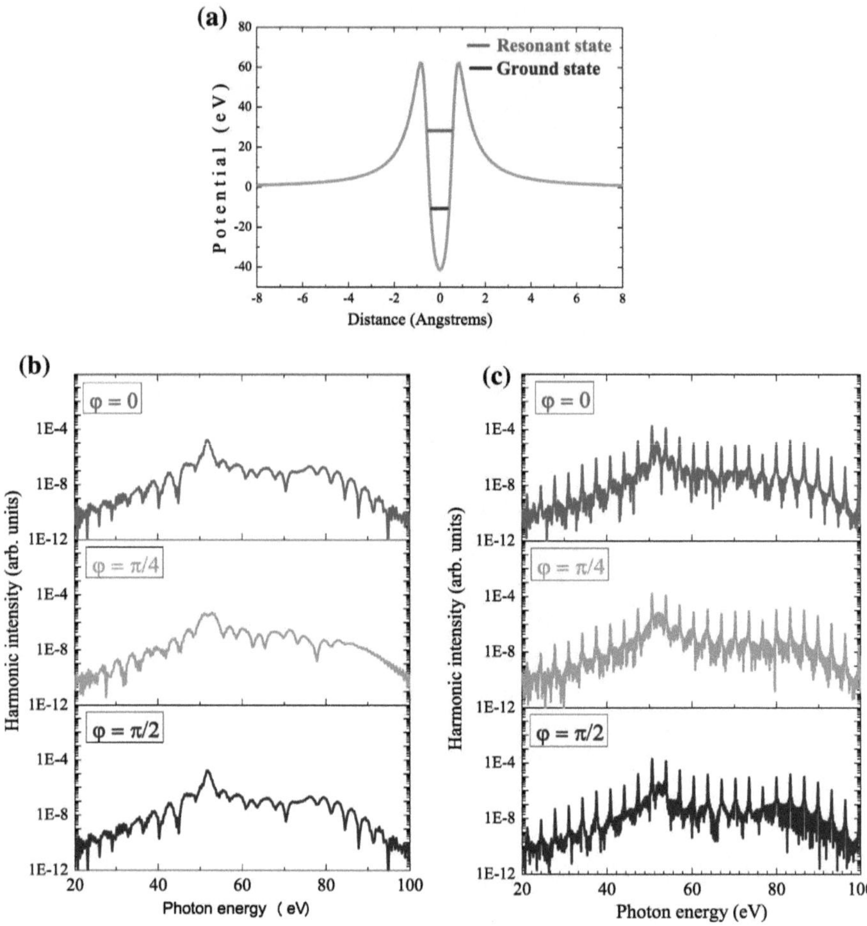

Fig. 2.22 **a** Potential used for the numerical simulations. **b**, **c** Calculated HHG spectra using **b** a long (40 fs) pulse and **c** few-cycle pulses at different values of the CEP ($\phi = 0, \pi/4$, and $\pi/2$). Reproduced from [138] with permission from Optical Society of America

Here also presented a typical image of a Mn harmonic spectrum in the case of 3.5 fs pulses (Fig. 2.21b). No second plateau, which was seen in the case of multi-cycle (40 fs) pulses, is observed for the few-cycle pulse. Most striking was the observation of a single, very strong, broadband (2.5 eV) 31st harmonic. Only two weak neighboring harmonics (around the strong emission) are seen in the 30–65 eV spectral range. The ratio between the intensities of the enhanced harmonic to the weak neighboring harmonics exceeded one order of magnitude. One can note that, at a lower intensity of the femtosecond pulse ($<2 \times 10^{14}$ W cm^{-2}), this strong harmonic disappeared when using both multi- and few-cycle pulses.

The distinctive structure of the harmonic spectra, both for 40 and 3.5 fs pulses, clearly points to the involvement of Mn resonances centred around 50–51 eV. The same can be said about the photoionization or photoabsorption characteristics of Mn^+ plasma, which are due to the 'giant' $3p \rightarrow 3d$ resonance [130]. The laser polarization dependence of this emission was analyzed and it was found that the 50 eV radiation abruptly disappears with the change of the polarization state of the femtosecond probe pulses from linear to circular, which is a clear signature of the emission being due to high harmonic generation.

To analyse the effect of the spectro-temporal characteristics of the femtosecond radiation on the harmonic yield, the pressure of neon in the hollow fiber of second compressor was varied, thus changing the duration of the harmonic drive pulse [106]. The spectral and intensity variations of manganese harmonic spectra in the range of 22–62 eV as the functions of neon pressure in the hollow fiber are shown in Fig. 2.21c. One can clearly see that, with change of pressure (from 1 to 2.3 bar), the single 31st harmonic intensity varies from almost zero to its maximum high value. A blue shift of the harmonics is also evident. Further increase of neon pressure up to 3 bar, at which the experiments with the 3.5 fs pulses were carried out, did not change the harmonic distribution.

In the following, the results of numerical simulations within a one-dimensional model are presented. It was assumed that the main contribution to the resonant peak in the spectrum comes from Mn^+ ions. Note that the ionisation potential of Mn^{2+} ions (33.7 eV) is more than twice higher than the ionisation potential of Mn^+ ions (15.6 eV). The time-dependent Schrödinger equation was solved by means of the split-operator method [90, 93, 142]. The metastable state of this model potential is at 51.8 eV above the ground state (Fig. 2.22a). The laser field is $E(t) = E_0 f(t) \cos(\omega_0 t + \varphi)$, where $f(t)$ is the pulse envelope, φ denotes the carrier envelope phase (CEP) and ω_0 is the laser frequency corresponding to the central wavelength $\lambda = 770$ nm. The laser intensity is $I_0 = 4 \times 10^{14}$ W cm^{-2}. A CEP of $\varphi = 0$ means that the maximum of the envelope corresponds to a maximum of sin $(\omega_0 t)$.

HHG spectra were calculated for pulse shapes with different lengths and for different values of φ (Fig. 2.22b,c). A $\sin^2$ envelope with a total length of 4 full cycles was used to model the 3.5 fs pulse, while an envelope with 13 cycles of constant intensity and 21 cycles of total duration was used to model the 40 fs case. The long pulse led to a HHG spectrum that shows well defined peaks at the odd harmonic orders and that is weakly dependent on the CEP (Fig. 2.22b). Figure 2.22c shows the dependence of the harmonic spectrum on the CEP in the case of the short, few-cycle pulse. In all cases, the resonance dominated the HHG spectrum. The most intense emission was occurred around 51 eV, where the metastable state is located. Although some difference between harmonic spectra for $\varphi = 0$, $\pi/4$, and $\pi/2$ is found, one can note that the CEP dependence is strongest for the spectrum outside the region of the resonance. For random CEP the substructure of the spectrum will average out as confirmed by numerical averaging over 20 values of the CEP in the range from 0 to π. The resonance peak itself depends less on the CEP.

The experiments described above were carried out without CEP stabilisation (that is, for random CEP values). The HHG experiments with Mn plasma using 3.5 fs pulses were also performed with stabilized CEP ($\varphi = 0$ and $\pi/2$) and no specific differences

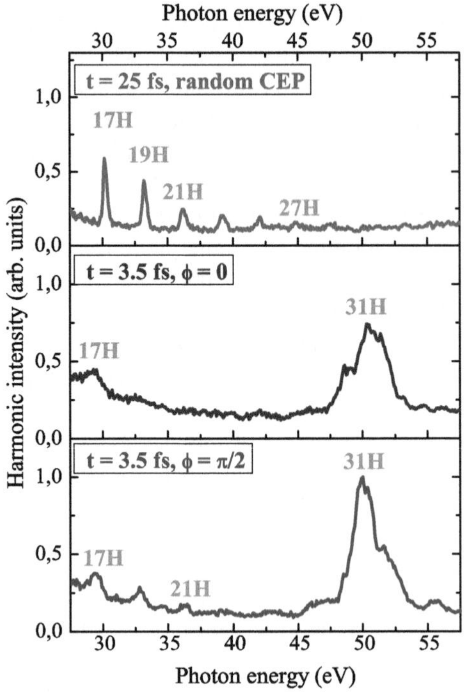

Fig. 2.23 Experimental harmonic spectra generated from manganese plasma in the case of the absence of gas in the hollow fibre compressor ($t = 25$ fs) and random CEP (*upper panel*), and at 3 bar pressure ($t = 3.5$ fs) at fixed CEP ($\varphi = 0$, *middle panel*; $\varphi = \pi/2$, *bottom panel*). Reproduced from [138] with permission from Optical Society of America

were found in that case (Fig. 2.23), though some variation of harmonic distribution was observed for the lower order harmonics (compare the middle and bottom curves of Fig. 2.23). The spectral shapes of the 31st harmonic emission were similar for these two fixed values of CEP, while a considerable difference in harmonic spectra was maintained when comparing to longer pulse duration and lower intensity of the driving pulses. Figure 2.23 shows HHG measurements for 25 fs pulses (upper panel) and 3.5 fs pulses (middle and bottom panels) of the same energy. One can clearly see the absence of harmonic extension and resonance-induced HHG in the case of low-intensity, 25 fs pulses.

The fact that a strong CEP dependence of the harmonic spectra in the case of 3.5 fs pulses was not observed could also be attributed to the presence of a significant density of free electrons in the manganese plasma, which might diminish the difference between the HHG spectra recorded for different values of experimental CEP. The same can be said about other HHG experiments using silver and brass plasmas, which did not show significant differences in harmonic spectra when comparing few-cycle pulses with fixed and random CEP. Comparative studies with gas media under similar experimental conditions were carried out and found a characteristic dependence of the HHG spectra on the CEP. Thus the absence of the strong influence of the CEP on the harmonic pattern generated by few-cycle pulses from the ablation plumes appears to be a common feature of plasma HHG.

References

1. G.D. Tsakiris, K.K. Eidmann, J. Meyer-ter-Vehn, F. Krausz, New J. Phys. **8**, 19 (2006)
2. B. Dromey, M. Zepf, A. Gopal, K. Lancaster, M.S. Wei, K. Krushelnick, M. Tatarakis, N. Vakakis, S. Moustaizis, R. Kodama, M. Tampo, C. Stoeckl, R. Clarke, H. Habara, D. Neely, S. Karsch, P. Norreys, Nature Phys. **2**, 456 (2006)
3. D.H. Reitze, S. Kazamias, F. Weihe, G. Mullot, D. Douillet, F. Aug, O. Albert, V. Ramanathan, J.P. Chambaret, D. Hulin, P. Balcou, Opt. Lett. **29**, 86 (2004)
4. T. Pfeifer, D. Walter, C. Winterfeldt, C. Spielmann, G. Gerber, Appl. Phys. B **80**, 277 (2005)
5. C.A. Froud, E.T.F. Rogers, D.C. Hanna, W.S. Brocklesby, M. Praeger, A.M. de Paula, J.J. Baumberg, Opt. Lett. **31**, 374 (2006)
6. E.A. Gibson, A. Paul, N. Wagner, R. Tobey, D. Gaudiosi, S. Backus, I.P. Christov, A. Aquila, E.M. Gullikson, D.T. Attwood, M.M. Murnane, H.C. Kapteyn, Science **302**, 95 (2003)
7. S. Kazamias, D. Douillet, F. Weihe, C. Valentin, A. Rousse, S. Sebban, G. Grillon, F. Auge, D. Hulin, P. Balcou, Phys. Rev. Lett. **90**, 193901 (2003)
8. U. Teubner, G. Pretzler, T. Schlegel, K. Eidmann, E. Förster, K. Witte, Phys. Rev. A **67**, 013816 (2003)
9. P.A. Norreys, M. Zepf, S. Moustaizis, A.P. Fews, J. Zhang, P. Lee, M. Bakarezos, C.N. Danson, A. Dyson, P. Gibbon, P. Loukakos, D. Neely, F.N. Walsh, J.S. Wark, A.E. Dangor, Phys. Rev. Lett. **76**, 1832 (1996)
10. G.J. Pert, Phys. Rev. A **75**, 023808 (2007)
11. T. Ozaki, R.A. Ganeev, A. Ishizawa, T. Kanai, H. Kuroda, Phys. Rev. Lett. **89**, 253902 (2002)
12. F. Gruner, S. Becker, U. Schramm, T. Eichner, M. Fuchs, R. Weingartner, D. Habs, J. Meyer-ter-Vehn, M. Geissler, M. Ferrario, Appl. Phys. B **86**, 431 (2007)
13. P.B. Corkum, F. Krausz, Nature Phys. **3**, 381 (2007)
14. C. Figueira de Morisson Faria, R. Kopold, W. Becker, J.M. Rost, Phys. Rev. A **65**, 023404 (2002)
15. R. Taieb, V. Véniard, J. Wassaf, A. Maquet, Phys. Rev. A **68**, 033403 (2003)
16. V.P. Silin, P.V. Silin, Phys. Uspekhi **50**, 729 (2007)
17. Y. Akiyama, K. Midorikawa, Y. Matsunawa, Y. Nagata, M. Obara, H. Tashiro, K. Toyoda, Phys. Rev. Lett. **69**, 2176 (1992)
18. S. Kubodera, Y. Nagata, Y. Akiyama, K. Midorikawa, M. Obara, H. Tashiro, K. Toyoda, Phys. Rev. A **48**, 4576 (1993)
19. C.-G. Wahlströom, S. Borgström, J. Larsson, S.-G. Pettersson, Phys. Rev. A **51**, 585 (1995)
20. W. Theobald, C. Wülker, F.R. Schäfer, B.N. Chichkov, Opt. Commun. **120**, 177 (1995)
21. R.A. Ganeev, V.I. Redkorechev, T. Usmanov, Opt. Commun. **135**, 251 (1997)
22. K. Krushelnick, W. Tighe, S. Suckewer, JOSA B **14**, 1687 (1997)
23. P.B. Corkum, Phys. Rev. Lett. **71**, 1994 (1993)
24. R.A. Ganeev, M. Suzuki, M. Baba, H. Kuroda, T. Ozaki, Opt. Lett. **30**, 768 (2005)
25. R.A. Ganeev, M. Suzuki, T. Ozaki, M. Baba, H. Kuroda, Opt. Lett. **31**, 1699 (2006)
26. R.A. Ganeev, H. Singhal, P.A. Naik, V. Arora, U. Chakravarty, J.A. Chakera, R.A. Khan, I.A. Kulagin, P.V. Redkin, M. Raghuramaiah, P.D. Gupta, Phys. Rev. A **74**, 063824 (2006)
27. R.A. Ganeev, M. Baba, M. Suzuki, H. Kuroda, J. Appl. Phys. **99**, 103303 (2006)
28. R.A. Ganeev, P.A. Naik, H. Singhal, J.A. Chakera, P.D. Gupta, Opt. Lett. **32**, 65 (2007)
29. R.A. Ganeev, M. Suzuki, M. Baba, H. Kuroda, Appl. Phys. B **81**, 1081 (2005)
30. R.A. Ganeev, H. Singhal, P.A. Naik, V. Arora, U. Chakravarty, J.A. Chakera, R.A. Khan, P.V. Redkin, M. Raghuramaiah, P.D. Gupta, J. Opt. Soc. Am. B **23**, 2535 (2006)
31. M. Suzuki, M. Baba, R. Ganeev, H. Kuroda, T. Ozaki, Opt. Lett. **31**, 3306 (2006)
32. R.A. Ganeev, L.B. Elouga Bom, J. Abdul-Hadi, M.C.H. Wong, J.P. Brichta, V.R. Bhardwaj, T. Ozaki, Phys. Rev. Lett. **102**, 013903 (2009)
33. R.A. Ganeev, H. Singhal, P.A. Naik, U. Chakravarty, V. Arora, J.A. Chakera, R.A. Khan, M. Raghuramaiah, S.R. Kumbhare, R.P. Kushwaha, P.D. Gupta, Appl. Phys. B **87**, 243 (2007)
34. M. Suzuki, M. Baba, H. Kuroda, R.A. Ganeev, L.B. Elouga Bom, T. Ozaki, Opt. Express **15**, 4112 (2007)

35. R.A. Ganeev, L.B. Elouga Bom, J.-C. Kieffer, T. Ozaki, Phys. Rev. A **76**, 023831 (2007)
36. R.A. Ganeev, M. Baba, M. Suzuki, H. Kuroda, Phys. Lett. A **339**, 103 (2005)
37. L.B. Elouga Bom, J.-C. Kieffer, R.A. Ganeev, M. Suzuki, H. Kuroda, T. Ozaki, Phys. Rev. A **75**, 033804 (2007)
38. R.A. Ganeev, L.B. Elouga Bom, J.-C. Kieffer, T. Ozaki, Phys. Rev. A **75**, 063806 (2007)
39. R.A. Ganeev, M. Suzuki, M. Baba, H. Kuroda, Phys. Rev. A **76**, 023805 (2007)
40. R.A. Ganeev, M. Suzuki, P.V. Redkin, M. Baba, H. Kuroda, Phys. Rev. A **76**, 023832 (2007)
41. R.A. Ganeev, Laser Phys. **18**, 1009 (2008)
42. R.A. Ganeev, Phys. Uspekhi **52**, 55 (2009)
43. R.A. Ganeev, H. Singhal, P.A. Naik, I.A. Kulagin, P.V. Redkin, J.A. Chakera, M. Tayyab, R.A. Khan, P.D. Gupta, Phys. Rev. A **80**, 033845 (2009)
44. R.A. Ganeev, H. Singhal, P.A. Naik, J.A. Chakera, M. Tayyab, M. Baba, H. Kuroda, P.D. Gupta, J. Opt. Soc. Am. B **26**, 2143 (2009)
45. R.A. Ganeev, M. Baba, T. Ozaki, H. Kuroda, J. Opt. Soc. Am. B **27**, 1077 (2010)
46. P.V. Redkin, R.A. Ganeev, Phys. Rev. A **81**, 063825 (2010)
47. R.A. Ganeev, H. Singhal, P.A. Naik, J.A. Chakera, A.K. Srivastava, T.S. Dhami, M.P. Joshi, P.D. Gupta, Appl. Phys. B **100**, 581 (2010)
48. H. Singhal, R.A. Ganeev, P.A. Naik, J.A. Chakera, U. Chakravarty, H.S. Vora, A.K. Srivastava, C. Mukherjee, C.P. Navathe, S.K. Deb, P.D. Gupta, Phys. Rev. A **82**, 043821 (2010)
49. R.A. Ganeev, H. Singhal, P.A. Naik, J.A. Chakera, H.S. Vora, R.A. Khan, P.D. Gupta, Phys. Rev. A **82**, 053831 (2010)
50. P.V. Redkin, M.K. Kodirov, R.A. Ganeev, JOSA B **18**, 165 (2011)
51. R.A. Ganeev, Laser Phys. **21**, 25 (2011)
52. R.A. Ganeev, P.A. Naik, H. Singhal, J.A. Chakera, M. Kumar, M.P. Joshi, A.K. Srivastava, P.D. Gupta, Phys. Rev. A **83**, 013820 (2011)
53. R.A. Ganeev, P.A. Naik, J.A. Chakera, H. Singhal, N.C. Pramanik, P.A. Abraham, N. Rani Panicker, M. Kumar, P.D. Gupta, J. Opt. Soc. Am. B **28**, 360 (2011)
54. R.A. Ganeev, Opt. Spectrosc. **110**, 637 (2011)
55. R.A. Ganeev, H. Kuroda, Appl. Phys. B **103**, 151 (2011)
56. R.A. Ganeev, C. Hutchison, T. Siegel, M.E. López-Arias, A. Zaïr, J.P. Marangos, J. Modern Opt. **58**, 819 (2011)
57. R.A. Ganeev, C. Hutchison, T. Siegel, A. Zaïr, J.P. Marangos, Phys. Rev. A **83**, 063837 (2011)
58. P.V. Redkin, M. Danailov, R.A. Ganeev, Phys. Rev. A **84**, 013407 (2011)
59. R.A. Ganeev, L.B. Elouga Bom, T. Ozaki, Phys. Plasmas **18**, 083101 (2011)
60. R.A. Ganeev, M. Baba, H. Kuroda, G.S. Boltaev, R.I. Tugushev, T. Usmanov, Eur. Phys. J. D **64**, 109 (2011)
61. L.B. Elouga Bom, Y. Pertot, V.R. Bhardwaj, T. Ozaki, Opt. Express **19**, 3077 (2011)
62. Y. Pertot, L.B. Elouga Bom, V.R. Bhardwaj, T. Ozaki, Appl. Phys. Lett. **98**, 101104 (2011)
63. Y. Pertot, S. Chen, S.D. Khan, L.B. Elouga Bom, T. Ozaki, Z. Chang, J. Phys. B: At. Mol. Opt. Phys. **45**, 074017 (2012)
64. S. Haessler, L.B. Elouga Bom, O. Gobert, J.-F. Hergott, F. Lepetit, M. Perdrix, B. Carré, T. Ozaki, P. Salières, J. Phys. B: At. Mol. Opt. Phys. **45**, 074012 (2012)
65. R.A. Ganeev, C. Hutchison, A. Zaïr, T. Witting, F. Frank, W.A. Okell, J.W.G. Tisch, J.P. Marangos, Opt. Express **20**, 90 (2012)
66. R.A. Ganeev, T. Witting, C. Hutchison, F. Frank, P.V. Redkin, W.A. Okell, D.Y. Lei, T. Roschuk, S.A. Maier, J.P. Marangos, J.W.G. Tisch, Phys. Rev. A **85**, 015807 (2012)
67. R.A. Ganeev, J. Modern Opt. **59**, 409 (2012)
68. M. López-Arias, M. Oujja, M. Sanz, R.A. Ganeev, G.S. Boltaev, N.K. Satlikov, R.I. Tugushev, T. Usmanov, M. Castillejo, J. Appl. Phys. **111**, 043111 (2012)
69. R.A. Ganeev, Laser Phys. Lett. **9**, 175 (2012)
70. R.A. Ganeev, V.V. Strelkov, C. Hutchison, A. Zaïr, D. Kilbane, M.A. Khokhlova, J.P. Marangos, Phys. Rev. A **85**, 023832 (2012)
71. M. Suzuki, M. Baba, R.A. Ganeev, L.B. Elouga Bom, H. Kuroda, T. Ozaki, J. Phys. B At. Mol. At. Phys. **45**, 065601 (2012)

72. R.A. Ganeev, P.A. Naik, H. Singhal, J.A. Chakera, M. Kumar, U. Chakravarty, P.D. Gupta, Opt. Commun. **285**, 2934 (2012)
73. C. Hutchison, R.A. Ganeev, T. Witting, F. Frank, W.A. Okell, J.W.G. Tisch, J.P. Marangos, Opt. Lett. **37**, 2064 (2012)
74. R.A. Ganeev, Open Spectrosc. J. **3**, 1 (2009)
75. R.A. Ganeev, J. Phys. B At. Mol. Opt. Phys. **40**, R213 (2007)
76. R.A. Ganeev, Laser Phys. **22**, 1177 (2012)
77. R.A. Ganeev, M. Baba, H. Kuroda, G.S. Boltaev, R.I. Tugushev, T. Usmanov, Eur. Phys. J. D **64**, 109 (2011)
78. R.A. Ganeev, G.S. Boltaev, N.K. Satlikov, T. Usmanov, J. Opt. **14**, 095202 (2012)
79. G. Raşeev, B. Leyh, H. Lefebvre-Brion, Z. Phys. D **2**, 319 (1986)
80. F. Keller, H. Lefebvre-Brion, Z. Phys. D **4**, 15 (1986)
81. M.Y. Amusia, J.-P. Connerade, Rep. Prog. Phys. **63**, 41 (2000)
82. J.F. Reintjes, *Nonlinear Optical Parametric Processes in Liquids and Gases* (Academic, New York, 1984)
83. M.B. Gaarde, K. Schafer, Phys. Rev. A **64**, 013820 (2001)
84. E.S. Toma, P. Antoine, A. de Bohan, H.G. Muller, J. Phys. B At. Mol. Opt. Phys. **32**, 5843 (1999)
85. Z. Zeng, R. Li, Y. Cheng, W. Yu, Z. Xu, Phys. Scr. **66**, 321 (2002)
86. R. Bartels, S. Backus, E. Zeek, L. Misoguti, G. Vdovin, I.P. Christov, M.M. Murnane, H.C. Kapteyn, Nature **406**, 164 (2000)
87. L. Plaja, L. Roso, J. Mod. Opt. **40**, 793 (1993)
88. P.A. Oleinikov, V.T. Platonenko, G. Ferrante, J. Exp. Theor. Phys. Lett. **60**, 246 (1994)
89. I.A. Kulagin, T. Usmanov, Opt. Lett. **34**, 2616 (2009)
90. V. Strelkov, Phys. Rev. Lett. **104**, 123901 (2010)
91. D.B. Milošević, Phys. Rev. A **81**, 023802 (2010)
92. M.Y. Frolov, N.L. Manakov, A.F. Starace, Phys. Rev. A **82**, 023424 (2010)
93. M. Tudorovskaya, M. Lein, Phys. Rev. A **84**, 013430 (2011)
94. J. Yao, B. Zeng, W. Chu, J. Ni, Y. Cheng, J. Mod. Opt. **59**, 245–249 (2012)
95. M. Ferray, A. L'Huillier, X.F. Li, L.A. Lompré, G. Mainfray, G. Manus, J. Phys. B At. Mol. Opt. Phys. **21**, L31 (1988)
96. L.A. Lompré, A. L'Huillier, M. Ferray, P. Monot, G. Mainfray, G. Manus, J. Opt. Soc. Am. B **7**, 754 (1999)
97. A. McPherson, G. Ginson, H. Jara, N. Johann, I.A. McIntyre, K. Boyer, C.K. Rhodes, J. Opt. Soc. Am. B **4**, 595 (1987)
98. A. Borot, A. Malvache, X. Chen, D. Douillet, G. Iaquianiello, T. Lefrou, P. Audebert, J.-P. Geindre, G. Mourou, F. Quéré, R. Lopez-Martens, Opt. Lett. **36**, 1461 (2011)
99. P.M. Paul, E.S. Toma, P. Breger, G. Mullot, F. Augé, P. Balcou, H.G. Muller, P. Agostini, Science **292**, 1689 (2001)
100. R. Torres, T. Siegel, L. Brugnera, I. Procino, J.G. Underwood, C. Altucci, R. Velotta, E. Springate, C. Froud, I.C.E. Turcu, M.Y. Ivanov, O. Smirnova, J.P. Marangos, Opt. Express **18**, 3174 (2010)
101. C. Vozzi, M. Nisoli, J.-P. Caumes, G. Sansone, S. Stagira, S. De Silvestri, M. Vecchiocattivi, D. Bassi, M. Pascolini, L. Poletto, P. Villoresi, G. Tondello, Appl. Phys. Lett. **86**, 111121 (2005)
102. J. Tate, T. Auguste, H.G. Muller, P. Salières, P. Agostini, L.F. DiMauro, Phys. Rev. Lett. **98**, 013901 (2007)
103. K. Schiessl, K.L. Ishikawa, E. Persson, J. Burgdörfer, Phys. Rev. Lett. **99**, 253903 (2007)
104. R.A. Ganeev, C. Hutchison, T. Witting, F. Frank, W.A. Okell, A. Zaïr, S. Weber, P.V. Redkin, D.Y. Lei, T. Roschuk, S.A. Maier, I. López-Quintás, M. Martín, M. Castillejo, J.W.G. Tisch, J.P. Marangos, J. Phys. B At. Mol. At. Phys. **45**, 165402 (2012)
105. T. Witting, F. Frank, C.A. Arrell, W.A. Okell, J.P. Marangos, J.W.G. Tisch, Opt. Lett. **36**, 1680 (2011)

106. J.S. Robinson, C.A. Haworth, H. Teng, R.A. Smith, J.P. Marangos, J.W.G. Tisch, Appl. Phys. B **85**, 525 (2006)
107. M. Sanz, M.E. López-Arias, E. Rebollar, R. de Nalda, M. Castillejo, J. Nanopart. Res. **13**, 6621 (2011)
108. R. Torres, M. Jadraque, M. Martin, Appl. Phys. A **80**, 1671 (2005)
109. M. Anselment, R.S. Smith, E. Daykin, L.F. Dimauro, Chem. Phys. Lett. **134**, 444 (1987)
110. E.A. Rohlfing, J. Chem. Phys. **89**, 6103 (1988)
111. S. Acquaviva, M.L. De Giorgi, Appl. Surf. Sci. **197–198**, 21 (2002)
112. J. Roth, F. Géahler, H.-R. Trebin, Int. J. Mod. Phys. C **11**, 317 (2000)
113. T.E. Glover, J. Opt. Soc. Am. B **20**, 125 (2003)
114. H.O. Jeschke, M.E. Garsia, K.H. Bennemann, Phys. Rev. Lett. **87**, 015003 (2001)
115. A.V. Kabashin, M. Meunier, J. Appl. Phys. **94**, 7941 (2003)
116. W.R. Creasy, J.T. Brenna, Chem. Phys. **126**, 453 (1988)
117. S.C. O'Brien, J.R. Heath, R.F. Curl, R.E. Smalley, J. Chem. Phys. **88**, 220 (1988)
118. T.D. Donnelly, T. Ditmire, K. Neuman, M.D. Perry, R.W. Falcone, Phys. Rev. Lett. **76**, 2472 (1996)
119. J.W.G. Tisch, T. Ditmire, D.J. Frasery, N. Hay, M.B. Mason, E. Springate, J.P. Marangos, M.H.R. Hutchinson, J. Phys. B At. Mol. Opt. Phys. **30**, L709 (1997)
120. A. Altucci, R. Bruzzese, C. de Lisio, M. Nisoli, S. Stagira, S. De Silvestri, O. Svelto, A. Boscolo, P. Ceccherini, L. Poletto, G. Tondello, P. Villoresi, Phys. Rev. A **61**, 021801R (2000)
121. R.A. Ganeev, L.B. Elouga Bom, T. Ozaki, J. Appl. Phys. **106**, 023104 (2009)
122. S. Kim, J. Jin, Y.-J. Kim, I.-Y. Park, Y. Kim, S.-W. Kim, Nature **453**, 757 (2008)
123. A. Husakou, S.-J. Im, J. Herrmann, Phys. Rev. A **83**, 043839 (2011)
124. M.F. Ciappina, A. Becker, A. Jaron-Becker, Phys. Rev. A **76**, 063406 (2007)
125. M. Ruggenthaler, S.V. Popruzhenko, D. Bauer, Phys. Rev. A **78**, 033413 (2008)
126. G.P. Zhang, Phys. Rev. Lett. **95**, 047401 (2005)
127. R.A. Ganeev, L.B. Elouga Bom, M.C.H. Wong, J.-P. Brichta, V.R. Bhardwaj, P.V. Redkin, T. Ozaki, Phys. Rev. A **80**, 043808 (2009)
128. R.A. Ganeev, C. Hutchison, T. Witting, F. Frank, S. Weber, W.A. Okell, E. Fiordilino, D. Cricchio, F. Persico, A. Zaïr, J. W. G. Tisch, J. P. Marangos, J. Opt. Soc. Am. B **30**, 7 (2013)
129. R.A. Ganeev, H. Singhal, P.A. Naik, J.A. Chakera, A.K. Srivastava, T.S. Dhami, M.P. Joshi, P.D. Gupta, J. Appl. Phys. **106**, 103103 (2009)
130. H. Kjeldsen, F. Folkmann, B. Kristensen, J.B. West, J.E. Hansen, J. Phys. B At. Mol. At. Phys. **37**, 1321 (2004)
131. G.V. Marr, J.B. West, At. Data Nucl. Data Tables **18**, 497 (1976)
132. J. Levesque, D. Zeidler, J.P. Marangos, P.B. Corkum, D.M. Villeneuve, Phys. Rev. Lett. **98**, 183903 (2007)
133. A.D. Shiner, B.E. Schmidt, C. Trallero-Herrero, H.J. Wörner, S. Patchkovskii, P.B. Corkum, J.-C. Kieffer, F. Légaré, D.M. Villeneuve, Nat. Phys. **7**, 464 (2011)
134. R.A. Ganeev, M. Suzuki, M. Baba, H. Kuroda, Appl. Phys. Lett. **94**, 051101 (2009)
135. M. Hentschel, R. Kienberger, C. Spielmann, G.A. Reider, N. Milosevic, T. Brabec, P.B. Corkum, U. Heinzmann, M. Drescher, F. Krausz, Nature **414**, 509 (2001)
136. T. Witting, F. Frank, W.A. Okell, C.A. Arrell, J.P. Marangos, J.W.G. Tisch, J. Phys. B At. Mol. Opt. Phys. **45**, 074014 (2012)
137. C. Altucci, J.W.G. Tisch, R. Velotta, J. Mod. Opt. **58**, 1585 (2011)
138. R.A. Ganeev, T. Witting, C. Hutchison, F. Frank, M. Tudorovskaya, M. Lein, W.A. Okell, A. Zaïr, J.P. Marangos, J.W.G. Tisch, Opt. Express **20**, 25239 (2012)
139. E.T. Kilbane, J.-P. Kennedy, P. van Mosnier, J. Kampen, Phys. B At. Mol. At. Phys. **38**, L1 (2005)
140. V.K. Dolmatov, J. Phys. B At. Mol. At. Phys. **29**, L687 (1996)
141. J.B. West, J.E. Hansen, B. Kristensen, F. Folkmann, H. Kjeldsen, J. Phys. B At. Mol. At. Phys. **36**, L327 (2003)
142. M.D. Feit, J.A. Fleck, A. Steiger, J. Comp. Phys. **47**, 412 (1982)

Chapter 3
Nonlinear Optical Refraction and Absorption of Media

Various newly developed materials were recently treated from the point of view of their nonlinear optical properties. Below we briefly show the tendencies in the nonlinear optical studies carried out in the material science. These studies deal with a broad range of media under interest, as well as different nonlinear optical processes.

In [1], 2, 3-butanedione dihydrazone (BDDH) was synthesized via chemical route and the nonlinear optical parameters, such as nonlinear refractive index, nonlinear absorption coefficient, third-order nonlinear optical susceptibility, second hyperpolarizability, and optical limiting behavior were investigated by single beam z-scan technique, which will be described in the following section, for different concentrations of BDDH. Synthesized samples were irradiated by a Q-switched, frequency doubled Nd:YAG laser and found that nonlinear refractive index (n_2 or γ) and nonlinear absorption coefficient (β) increase linearly with increasing concentration. The excited-state absorption cross sections of BDDH were found to be larger than ground-state absorption cross sections, and it leads to reverse saturated absorption. The experimental results are well in agreement with the theory and also establish BDDH as one of the potential candidate materials for optical limiting at 532 nm.

The refractive index of most ion-doped materials increases with the excited state population. This effect was studied in many laser materials, particularly those doped with Cr^{3+} and rare earth ions [2], using several techniques, such as interferometry, wave mixing, and z-scans. It was found that the refractive index variation has an electronic origin and is associated with the difference in the polarizabilites of the Cr_3^+ ion in its excited and ground states.

We will discuss in details the results of experiments on low-order nonlinear optical parameters measurements in the cases of various materials (semiconductors, fullerenes, dyes, liquids, plasma, metals, and crystals). The results of the measurements of the nonlinear refraction coefficients, nonlinear absorption coefficients, saturation intensities, and nonlinear susceptibilities of these media by using laser radiation at different wavelengths and the z-scan technique are presented. Some media are analyzed from the point of view of their application as optical limiters of laser radiation. We also discuss a single-shot method for the measurements of nonlinear optical parameters.

R. A. Ganeev, *Nonlinear Optical Properties of Materials*,
Springer Series in Optical Sciences 174, DOI: 10.1007/978-94-007-6022-6_3,

In [3], $Bi_{1.5}Zn_{1.0}Nb_{1.5}O_7$ (BZN) thin films with good surface morphology were prepared on fused quartz substrates by pulsed laser deposition. The sign and magnitude of both real and imaginary parts of third-order nonlinear susceptibility of the BZN thin films were determined by a single beam z-scan technique performed at 1064 nm with a picoseconds laser. The third-order nonlinear refractive index coefficient the nonlinear absorption coefficient of the BZN thin films measured to be $4.06 \times 10^{-11}\,cm^2\,W^{-1}$ and $-1.45 \times 10^{-7}\,cm\,W^{-1}$, respectively. The real part and the imaginary part of the third-order nonlinear susceptibility of the BZN thin films were calculated to be 6.03×10^{-9} and 1.82×10^{-10} esu, respectively. The mechanisms of nonlinear refraction and nonlinear absorption were discussed. The results suggest that BZN thin films exhibit large third-order nonlinear optical properties and be a promising material for applications in nonlinear optical devices.

Vanadium dioxides (VO_2) thin films, which change from a monoclinic semiconductor phase to a tetragonal metallic structure at the temperature of 29 °C, have been fabricated by reactive ion beam sputtering and studied in [4]. Open aperture and closed aperture measurements of z-scan were used to study the optical absorptive and refractive nonlinearities. Nanostructured VO_2 thin films exhibit two-photon absorption and a negative nonlinear index of refraction when phase transition is induced. It was reported that in that case the optical nonlinearities are due to excitation of electronic subsystem only and without involving of the structural semiconductor-to-metal phase transition.

In [5], Mg-doped PbS/PVA freestanding nanocomposite films were successfully prepared by a simple chemical route. The z-scan results showed that the freestanding nanocomposite films exhibit the large nonlinear refractive index and absorption coefficients with negative sign and they are increased with Mg doping. Third-order nonlinear optical susceptibility and figure of merit were calculated and the maximum value is found to be about 3.4×10^{-5} esu and 2×10^{-9} esu m, respectively, for higher Mg-doped film, which is three orders of magnitude larger than that observed in the substrate-supported films as reported earlier in the literature.

A beam of 33 fs laser pulse with peak power of 15–40 GW was employed in [6] to explore a convenient method to determine the nonlinear refractive index coefficient of an optical glass. It is rare to investigate nonlinearities of optical glass with such an extreme ultrashort and powerful laser pulse. According to their method, only a single beam and a few experimental apparatuses are necessary to measure the nonlinear refractive index coefficient.

Nonlinear absorption of carbon disulfide (CS_2) was investigated in [7] by z-scan technique and time-resolved pump-probe technique with femtosecond pulses at 400 nm wavelength. By the two techniques, they confirmed that the nonlinear absorption of CS_2 arise from a combination of two-photon absorption (TPA) and the excited state absorption induced by TPA under the incident laser pulses with 400 nm wavelength. The coefficient of TPA, the absorption cross-section of low excited state and lifetime of low excited state were obtained by theoretical fitting the experimental results. The results indicated that the CS_2 has good optical limiting capability at 400 nm wavelength.

Nonlinear optical (NLO) materials play a major role in the field of photonics including optical information processing, sensor protector applications, data storage etc. Some organic compounds, such as acids, exhibit large NLO response, in many cases, orders of magnitude larger than widely known inorganic materials. The importance of amino acid in NLO application is due to the fact that all the amino acids, except glycine contain chiral carbon atom and crystallizes in non-centrosymmetric space group. Another added advantage of NLO active material is the presence of namely amino group and carboxyl group, which make it transparent in the UV-visible region. The theoretical aspect of nonlinear optics, solution growth of nonlinear crystals and the analysis of second-order and third-order NLO properties using Kurtz powder and z-scan technique are presented in [8].

A novel functionality was reported in [9] for the ferroelectric-antiferromagnet $YCrO_3$ powder of optical limiting upon illumination by nanosecond laser pulses at 532 nm. The optical limiting properties are investigated using the open aperture z-scan technique. The obtained nonlinearity fits to a three-photon like absorption mechanism. It is proposed that this nonlinearity is caused by two-photon absorption, followed by excited state absorption.

In [10], the optical nonlinearity of hybrid structures composed of CdTe quantum dots and periodical particle array of gold is studied using z-scan method. The optical nonlinearity is dramatically affected by the interaction between exciton in CdTe quantum dots and surface plasmons in Au periodical particle array. When the Au surface plasmon is tuned to be in resonance with the exciton transition in CdTe quantum dots, the largest nonlinear refractive index and the smallest two-photon absorption coefficient ($-0.53\,\mathrm{cm^2GW^{-1}}$ and $25\,\mathrm{cm\,GW^{-1}}$), which are about 8 times larger and 50 times smaller than that of bare CdTe quantum dots, can be achieved.

A closed aperture z-scan technique was used in [11] to study the nonlinear optical refractive index of lanthanum-modified lead zirconate titanate (PLZT) (9/65/35) ceramics. The laser light source was a mode-locked titanium-sapphire laser, the output pulse width was approximately 100 fs at a wavelength of 790 nm, and the repetition rate was 10 kHz. It was found that PLZT had a large, positive nonlinear optical refractive index (3.1×10^{-8} esu), and its third-order susceptibility was calculated to be 9×10^{-9} esu. When an electric field (E) was applied to the PLZT, n_2 increased along in direct proportion to E^2. It was proposed that the increase in n_2 was caused by the more ordered structural arrangement of the PLZT imposed by the applied electric field. The results show that PLZT has potential applications in fast, electric-field-controlled nonlinear optical elements.

Pulsed laser ablations of a silicon rod target in water, hydrogen peroxide water solution and ethanol were performed in [12] by varying the laser parameters. The calculated third-order refractive index and absorption coefficient are found to be weakly affected by the specific ablation process, and hence almost independent from the structural and compositional characteristics of the nanoparticles. The only marked difference is observed in the comparison between silicon/silicon oxide nanoparticle and pure silicon nanoparticles, the latter being produced in ethanol. More specifically, a one order of magnitude higher refractive index is observed with nanoparticles dispersed in ethanol.

The third order NLO properties of thiourea and its metal complexes were measured in [13] using a 532 nm second harmonic of diode pumped Nd:YAG laser (1064 nm, 50 mW) by employing the z-scan technique. The magnitude of nonlinear refractive index, nonlinear absorption coefficient and third-order susceptibility were found to be in the order of $10^{-8}\,cm^2\,W^{-1}$, $10^{-3}\,cm\,W^{-1}$ and 10^{-8} esu respectively. The z-scan reflects the negative nonlinearity of the samples and hence self defocusing nature is responsible for the optical limiting behaviour in the regime of interest. The response time was found to be in the order of milliseconds.

Single crystals of bimetallic MnHg(SCN) are grown by slow cooling method and the second and third order optical nonlinearities are investigated by Kurtz and Perry powder SHG test and single beam z-scan technique respectively in [14]. The influences of SCN ligand in modifying the NLO properties are discussed and the results are compared with other organometallic crystals. The nonlinear refractive index, absorption coefficient and third order susceptibility of this crystal were estimated to be $-1.88 \times 10^{-11}\,cm^2\,W^{-1}$, $8.65 \times 10^{-6}\,cm\,W^{-1}$ and 6.58×10^{-9} esu, respectively.

The nonlinear absorptive characteristics (saturation intensity threshold and effective nonlinear absorption coefficients) and nonlinear refraction in a 50-nm-thick $VO_{(x)}$ thin amorphous film prepared by pulsed DC magnetron reactive sputtering were studied in [15]. The absorptive and refractive nonlinearities were investigated by pump-probe and z-scan techniques. The closed aperture z-scan results reveal self-defocusing characteristics of the amorphous $VO_{(x)}$ thin film for both nanosecond and picosecond pulse durations. Experimental results show that a phase transition does not occur in the range of intensities used for the experiments and the investigated sample can be treated as an amorphous semiconductor structure. The open aperture z-scan curves with nanosecond pulses exhibit saturable absorption for all input intensities. On the other hand, the open aperture z-scan curves with picosecond pulses exhibit nonlinear absorption/saturable absorption for low/high input intensities, respectively. Saturation intensity thresholds were found to be $15.3\,MW\,cm^{-2}$ for 4-ns pulse duration and $586\,MW\,cm^{-2}$ for 65-ps pulse duration.

The optical nonlinear absorption characteristics of the crystalline $Ge_2Sb_2Te_5$ thin films were investigated in [16] by performing z-scan measurements with nanosecond laser pulse. The experimental results showed that the nonlinear saturable absorption coefficient was as large as $-10^{-2}\,m\,W^{-1}$ for the excitation intensity lower than $0.17\,GW\,m^{-2}$. The nonlinear saturable absorption changed to the nonlinear reverse saturable absorption at the excitation intensity of above $0.17\,GW\,m^{-2}$. To explore the internal mechanisms, the first-principle theory was employed to calculate the electronic structure, and a five-level structure suitable for explanation of nonlinear absorption reversal was suggested. The theoretical calculation and analysis indicated that for the excitation intensity smaller than $0.17\,GW\,m^{-2}$, the contribution to the nonlinear saturable absorption is mainly from band-filling effect; for the excitation intensity larger than $0.17\,GW\,m^{-2}$, the nonlinear reverse saturable absorption results from the thermal-induced nonlinearity, which is further confirmed by picosecond laser pulse z-scan measurement and the variable-temperature spectroscopy ellipsometric analysis.

The third-order nonlinear optical susceptibilities of media vary in a broad range from 10^{-17} esu units to a few esu units. Nonlinearities responsible for variations in the refractive and absorbing properties are especially important because they strongly affect the propagation of intense radiation in media. Numerous investigations have been performed in this field due to increasing interest in applications of nonlinear optical effects in optoelectronics, various nonlinear optical devices, optical switching, etc. [17–19].

All above studies, which represent only small part of the nonlinear optical properties of newly developed materials, demonstrate their strong nonlinearities using the z-scan technique. Below we will discuss in details the experiments on low-order nonlinear optical parameters measurements in the cases of different media. In this chapter, we present the studies of the nonlinear refraction and nonlinear absorption in semiconductors, fullerenes, dyes, liquids, plasma, metals, and crystals. The results of measurements of the nonlinear refraction coefficients, nonlinear absorption coefficients, saturation intensities, and nonlinear susceptibilities of these media by using laser radiation at different wavelengths and the z-scan technique are presented. Methods of preparation of nanoparticles based on these media and studies of their nonlinear optical parameters will be discussed in the following chapters. Some media are analyzed from the point of view of their application as optical limiters of laser radiation. Investigations of higher nonlinearities of some media are also considered. We also discuss a single-shot method for the measurements of nonlinear optical parameters.

3.1 Basic Relations and Experimental Methods and Schemes for Analysis of Nonlinear Optical Parameters of Media by the z-Scan Method

The nonlinear response of media can be measured by different methods such as nonlinear interferometry, three- and four-wave mixing, generation of harmonics, z-scan, polarization rotation, probe-beam method, and analysis of distortions of the beam amplitude—phase profile [20–23]. However, most of these methods cannot provide the separation of contributions from different nonlinearities. One of the methods that can do it is the widely used z-scan method, which provides high measurement accuracy [24]. At present, there exist several variants of this method: two-colour, shadow, reflection, autocorrelation, and transmission z-scans [25–28].

The single-beam transmission z-scan method [24] is comparatively simple and offers a number of advantages over interferometric and other methods. The method is based on the analysis of variations in the far-field intensity distribution of a laser beam caused by the nonlinear refraction of a sample displaced in the focal region along the laser beam direction (the z axis in Fig. 3.1). To explain the principle of the z-scan method, we consider the propagation of focused intense laser radiation through a medium under study. Consider nonlinear refraction and assume that the

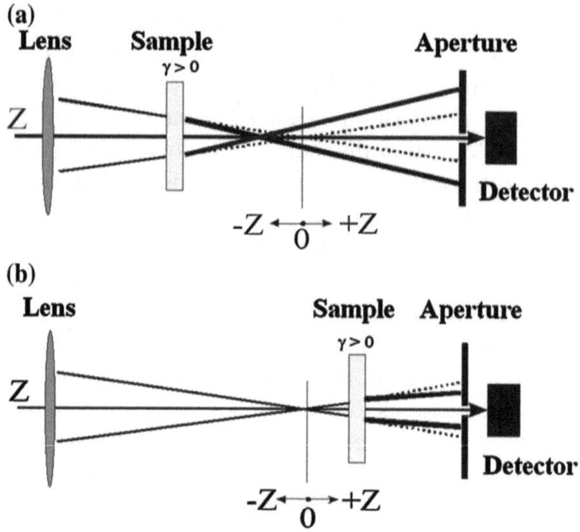

Fig. 3.1 Schemes of propagation of focused intense radiation through a nonlinear self-focusing medium for a sample located **a** in front of the focus and **b** behind it

medium gives a positive nonlinear addition to the refractive index ($\gamma > 0$). If the sample is located at a large distance from the focus (the region of negative values of z), the radiation intensity in the medium is insufficient to excite a noticeable nonlinear refraction, and the transmission of radiation through an aperture mounted in the far-field zone remains invariable and close to unity, as shown in Fig. 3.2. As the sample approaches the focal point, the radiation intensity increases and the self-focusing effect appears in the medium (see solid straight lines behind the sample in Fig. 3.1a). As a result, the far-field radiation has a larger divergence and, hence, transmission through the limiting aperture decreases.

As the sample is scanned along the z axis, transmission will decrease until the point z_1 is reached (see Fig. 3.2), where transmission is minimal. A decrease in the

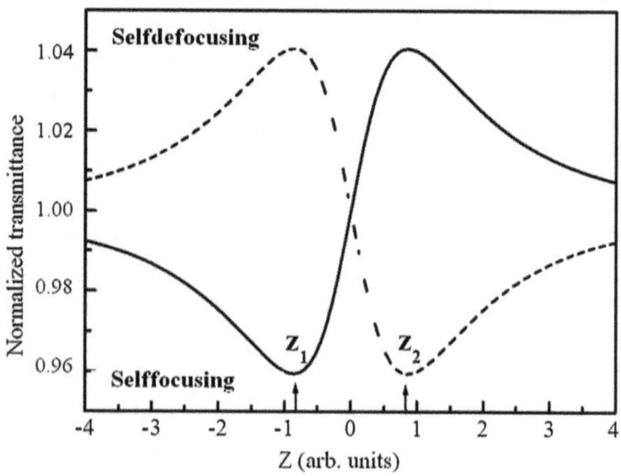

Fig. 3.2 Normalised transmittance in the closed aperture scheme for the media with the positive (*solid curve*) and negative (*dashed curve*) nonlinear refractive indices

curvature of the wave front of a Gaussian beam near the focus causes the general decrease in the far-field radiation divergence, and as the sample further approaches the beam waist, transmission through the aperture increases. After the sample passes through the waist, the following picture is observed. Self-focusing reduces the far-field radiation divergence (see solid straight line behind the sample in Fig. 3.1b). This will continue until the influence of self-focusing on the radiation propagation through the aperture will be maximal (at the point z_2 in Fig. 3.2 corresponding to the maximum of the solid curve). As the sample is further displaced, the influence of this nonlinear optical process on the transmission of radiation through the aperture will decrease because of the decrease in the radiation intensity until the normalized transmission achieves its stationary value close to unity.

Thus, if the maximum of the normalized transmission follows after the minimum during z-scan, the medium has a positive nonlinear refractive index ($\gamma > 0$) and, on the contrary, if the transmission maximum first appears and then minimum, the medium has self-defocusing properties ($\gamma < 0$).

We considered above only one type of nonlinearity (nonlinear refraction), by neglecting the influence of nonlinearities producing variations in the absorption of the medium (such as multiphoton absorption and saturated absorption). The nonlinearity of the first type (multiphoton absorption) will lead to the suppression of the maximum and the increase in the dip depth in the dependence of the transmission $T(z)$ on the sample position. In the case of saturated absorption, the opposite picture will be observed. In the presence of both nonlinear optical processes (nonlinear absorption and nonlinear refraction), we can separate the influence of one effect by subtraction methods and then calculate separately the nonlinear refractive index and nonlinear absorption coefficient. Moreover, taking into account that the z-scan method allows one to determine γ exclusively due to the presence of the limiting aperture (or "closed aperture", as it is often dubbed), the removal of the latter completely eliminates the influence of self-focusing (self-defocusing) on the dependence $T(z)$. In this case, this scheme becomes sensitive only to nonlinear absorption. Therefore, the scheme with an "open aperture" can be used for measuring the nonlinear absorption coefficient.

We did not specify above the order of nonlinearity responsible for the nonlinear optical process. Note that the influence of higher-order nonlinearities (for example, fifth-order nonlinear susceptibilities) becomes substantial for a number of media (in particular, semiconductors in which the appearance of free carriers due to two-photon absorption and their interaction with the intense field leads to the higher-order nonlinear response). In this case, the z-scan method also allows one to determine and separate the contributions of the nonlinearities of different orders by the shape of the dependence $T(z)$ and positions of the transmission minimum and maximum.

It is known that, under real experimental conditions, the precise measurement of the radiation intensity in the focal plane is a nontrivial task due to the difficulties encountered in the determination of its spatial parameters in a medium. Note in this connection an interesting feature of the $T(z)$ dependence. If nonlinear refraction is caused by the third-order nonlinearities, then for the sample length L smaller than the diffraction length z_0 of laser radiation, the distance ΔZ between the transmission maximum and minimum is related to the spatial characteristics of focused radiation in the focal plane by the expression [24]

$$\Delta Z \cong 1.7 z_o. \tag{3.1}$$

Here, $z_0 = 0.5k(w_0)^2$; $k = 2\pi/\lambda$; w_0 is the beam radius at the $1/e^2$ level of the spatial intensity distribution in the focal plane; and λ is the radiation wavelength. Thus, the spatial parameters of focused radiation can be obtained with good accuracy from the dependence of the normalized transmission on z upon scanning the sample. This, in turn, allows one to calculate nonlinear optical parameters with good accuracy.

Variations in the refractive index of a medium in the field of an electromagnetic wave, taking into account only the third-order nonlinearity, are determined by the relation

$$\Delta n(\omega) = \gamma(\omega) I_\omega = n_2(\omega)\frac{|E_\omega|^2}{2}, \tag{3.2}$$

where I_ω and E_ω are the intensity and strength of the electric field of the electromagnetic wave of frequency ω.

The nonlinear refractive indices n_2 and γ describing the same process in different measurement units are related with each other by the expression n_2 [esu units] = $(cn/40\pi)\ \gamma$ [SI units], where n is the refractive index of the medium and c is the speed of light. The value of γ is determined from z-scan experiments in the absence of nonlinear absorption by using the expression [24]

$$\Delta T_{p-v} = 0.404\,(1 - S)^{0.25} \left| \frac{2\pi \gamma I_o \left[1 - \exp(-\alpha L)\right]}{\alpha\lambda} \right| \tag{3.3}$$

where ΔT_{p-v} is the normalized difference of the maximum (peak) and minimum (valley) transmission in the closed aperture scheme, I_0 is the maximum radiation intensity in the waist plane, S is the aperture transmission (fraction of radiation incident on a photocathode); and α is the linear absorption coefficient.

An important advantage of the z-scan method is the possibility to separate processes related to nonlinear absorption and nonlinear refraction when both these processes proceed simultaneously in a sample. In that case, the theoretical dependence of the normalized transmission on z-scan can be written in the form [29]

$$T(z) = 1 + \frac{4x}{(x^2+9)(x^2+1)}\Delta\Phi_o - \frac{2(x^2+3)}{(x^2+9)(x^2+1)}\Delta\Psi_o, \tag{3.4}$$

where $x = z/z_o$, $\Delta\Phi_o = k\gamma I_o L_{eff}$ and $\Delta\Psi_o = \beta I_o L_{eff}/2$ are the parameters characterizing the phase shift in the focus due to nonlinear refraction and nonlinear absorption, respectively, and $L_{eff} = [1 - \exp(-\alpha L)]/\alpha$ is the effective length of the sample. By introducing the notation $\rho = \beta/2k\gamma$, we can obtain the relation between $\Delta\Phi_o$ and $\Delta\Psi_o (\Delta\Psi_o = \rho\Delta\Phi_o)$. In this case, the expression for the normalized transmission can be represented in the form

$$T = 1 + \frac{2(-\rho x^2 + 2x - 3\rho)}{(x^2+9)(x^2+1)} \Delta\Phi_o. \tag{3.5}$$

In the general case, the third-order nonlinear susceptibility is the complex quantity

$$\chi^{(3)} = \mathrm{Re}\,\chi^{(3)} + i\,\mathrm{Im}\,\chi^{(3)}. \tag{3.6}$$

Here, the imaginary part of nonlinearity is related to the nonlinear absorption coefficient by the expression

$$\mathrm{Im}\,\chi^{(3)} = \frac{n^2 \varepsilon_o c \lambda \beta}{2\pi}, \tag{3.7}$$

and the real part is caused by Kerr nonlinearities and is related to γ by the expression

$$\mathrm{Re}\,\chi^{(3)} = 2n^2 \varepsilon_o c \gamma. \tag{3.8}$$

where ε_o is the dielectric constant. The value of $\chi^{(3)}$ in the esu and SI units is represented by the expression $\chi^{(3)}$ [esu units] = $(9 \times 10^8/4\pi)\,\chi^{(3)}$ [SI units].

The real part of the Kerr nonlinear susceptibility in the esu units is related to the nonlinear refractive index by a simpler expression [30]

$$\chi^{(3)}[esu \quad units] = \frac{nn_2}{3\pi}. \tag{3.9}$$

Below, we describe the z-scan scheme used for studying nonlinear optical parameters of various media [31]. A picosecond Nd:YAG laser with a pulse repetition rate of 2 Hz was used in experiments. A single pulse (35 ps) was amplified up to the energy of 2 mJ. Nonlinear optical parameters were studied using the Nd:YAG laser radiation and its harmonics (1064, 532, and 355 nm). Laser radiation was focused by lens (1) (Fig. 3.3). Objects (2) under study were displaced using the translating stage (8) along the optical axis z, by passing through the focal region. The system for sample translation included a micrometer drive and four-phase step motor (9)

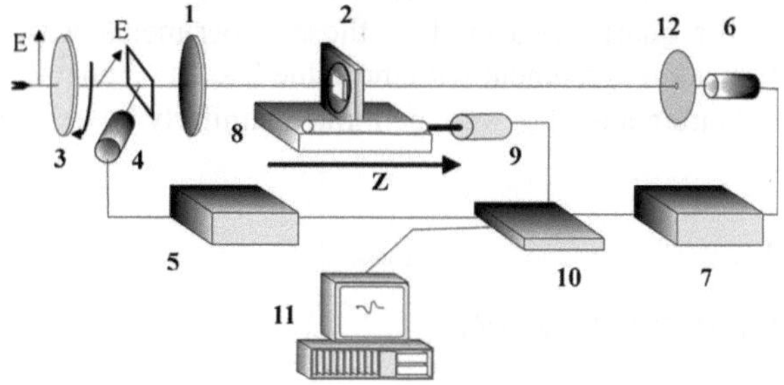

Fig. 3.3 Experimental scheme for z-scans of materials. *1* lens, *2* sample, *3* polarization plate, *4*, *6* photodiodes, *5*, *7* digital voltmeters, *8* moving table, *9* motor, *10* controller, *11* PC, *12* aperture. Reproduced from [31] with permission from Optical Society of America

and provided a translation step of 20 μm per cycle. The focused beam diameter in the waist region was 100 μm. In this case, the maximum radiation intensity was 4×10^{11} W cm^{-2} at 1064 nm, 4×10^{10} W cm^{-2} at 532 nm and 1×10^{10} W cm^{-2} at 354.7 nm, and the used intensity depended on the optical breakdown and multiphoton ionization thresholds of media under study. The laser pulse energy was measured using the calibrated photodiode (4) and recorded with voltage converter (5). The laser radiation energy was varied by using the calibrated neutral filters.

Aperture (12) of diameter 1 mm transmitting 1 % of laser radiation was mounted at a distance of 150 cm from the focal region (the so-called closed aperture scheme). The photodiode (6) was mounted behind the aperture. The output signal of the photodiode was measured by the voltage converter (7). To eliminate the influence of instability of the output energy parameters of the laser on the results of measurements, the signal detected with photodiode (6) was normalized to the signal detected with second photodiode (4). The closed aperture scheme can be used to determine both the sign and values of γ and β as well as the nonlinear susceptibility $\chi^{(3)}$ of materials.

Nonlinear absorption can be also measured by using the so-called open aperture scheme, when the aperture in the z-scan scheme is removed and all radiation transmitted by the sample upon its scanning along the z axis is collimated in photodiode (6). The expression for the normalized transmission in the open aperture scheme has the form

$$T(z) = q(z)^{-1} / \ln[1 + q(z)] \tag{3.10}$$

where $q(z) = \beta I(z) L_{eff} / [1 + (z_o^2/z^2)]$.

The outputs from both voltage converters were analyzed by the controller (10). The results were processed by averaging measurements over the programmable number of points for the given coordinate z and rejecting individual measurements in the case of strong fluctuations on the incident radiation intensity. This reduced considerably the general measurement error of the sample transmission at the given point z. The measurement error were reduced by using a set of 20 individual measurements for a fixed position of the sample.

The measurements of the nonlinear refractive index were calibrated by using media with well-known nonlinear refractive indices (quartz and carbon disulfide). The value of γ for quartz measured in those experiments was $(2.4 \pm 0.8) \times 10^{-16}$ cm^2 W^{-1}, in good agreement with the value 2×10^{-16} cm^2 W^{-1} obtained earlier [32]. The measurements of γ were calibrated similarly by using CS_2 [33].

3.2 Crystals

3.2.1 Photorefractive Crystals

Investigations of higher-order nonlinear optical processes in dielectric crystals involve certain problems due to the closeness of the optical breakdown and multiphoton ionization thresholds to the intensities at which higher-order nonlinearities of these crystals are manifested [34–37]. Due to various applications of photorefractive media [38–45], such crystals as $Bi_{12}SiO_{20}$ (BSO), $Bi_{12}GeO_{20}$(BGO), $Bi_{12}TiO_{20}$ (BTO), $BaTiO_3$, SBN, etc., attract considerable interest. To elucidate the potential possibilities of using photorefractive crystals, it is necessary to study in detail their optical and nonlinear optical parameters. While optical parameters in weak optical fields have been reported in a few work (see [39, 40] and references therein), the nonlinear optical response of these crystals in the high-power IR field has been studied to a lesser extend [46–50]. In particular, picosecond radiation was used to study the efficiency and decay time of a diffraction grating induced in a BSO upon degenerate four-wave mixing. The self-focusing of the 457 nm laser radiation in photorefractive media was first observed in experiments with an SBN crystal [51].

Below we discuss the results of z-scan investigations of self-focusing and other nonlinear optical processes in photorefractive BSO and BGO crystals [52]. These crystals have large linear refraction indices n (2.55 and 2.44 for BSO and BGO, respectively) at a wavelength of 1064 nm. The z-scan experiments were performed by using 55 ps, 0.6 mJ pulses from a 1064 nm Nd:YAG laser and its second-harmonic 532 nm, 0.15 mJ pulses at a pulse repetition rate of 2 Hz. Figure 3.4 presents the dependence of the normalized transmission of the BSO crystal at different positions of z at $\lambda = 1064$ nm in the closed aperture scheme. The nonlinear optical parameters of BSO and BGO crystals measured at wavelengths of 1064 and 532 nm are presented in Table 3.1. According to the commonly accepted practice, the values of γ are presented in $cm^2\,W^{-1}$, β are presented in $cm\,W^{-1}$, and $\chi^{(3)}$, $Re\chi^{(3)}$and $Im\chi^{(3)}$ are presented in the esu units.

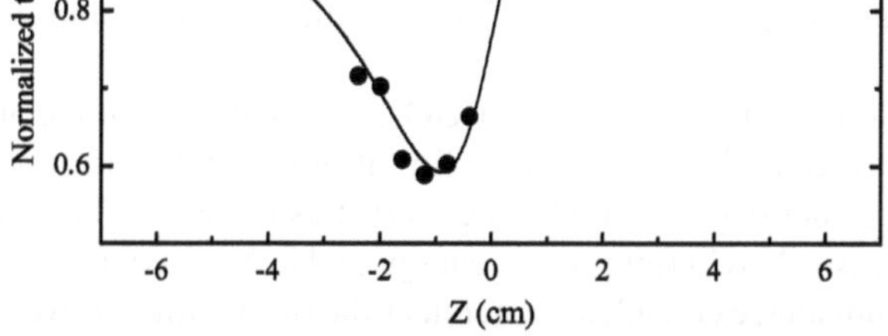

Fig. 3.4 Dependences of the normalised transmission on the BSO crystal position (the ⟨110⟩ orientation) with respect to the focus in case of the closed aperture scheme. *Solid curve* is a theoretical fit, *circles* are the experimental data. Adopted from [52] with permission from Springer Science+Business Media

Table 3.1 Nonlinear optical parameters of BSO and BGO crystals at 1064 and 532 nm Reproduced from [52] with permission from Springer Science+Business Media

Crystal	λ (nm)	L (mm)	γ, 10^{-14} $cm^2\,W^{-1}$	$Re\chi^{(3)}$, 10^{13} esu	$\beta_{2\omega}$, 10^{-9} $cm\,W^{-1}$	$Im\chi^{(3)}$, 10^{13} esu	$\beta_{3\omega}$, $10^{-20}\,cm^3\,W^{-2}$
BSO		10	0.4	1.6			2.5
			0.4	1.7			2.3
BGO	1064	7.3	1	4.3			4.4
		11	1	4.3			4.2
		17	1	4.3			4.3
BSO		10	0.6	2.3	2	20	
			0.7	3.0	1.6	17	
	532	7.3	1.2	5.1	3.7	38	
BGO		11	1.1	4.8	2.9	30	
		17	1.2	5.1	3.6	38	

Earlier [48], the generation of the conjugate wave in BSO and BGO crystals was treated as a result of considerable nonlinear absorption. The open aperture z-scan study performed in [53] has demonstrated the presence of nonlinear absorption, which was the same within the experimental error for different orientations of BSO and BGO crystals. The type of nonlinear absorption was determined by analyzing the width of the dip in the dependence of the normalized transmission on z. In the case of a linear dependence on the intensity I, the absorption coefficient can be written in the form

$$\alpha(I) = \alpha_o + \beta_{2\omega} I, \tag{3.11}$$

where α_0 is the intensity-independent linear absorption coefficient and $\beta_{2\omega}$ is the two-photon absorption coefficient. The theoretical dependence T(z) calculated from (3.11) (dotted curve in Fig. 3.5) differs from the experimental dependence obtained by the open aperture z-scan. One can see that the experimental dip is narrower than the theoretical one. This discrepancy can be explained by assuming that higher-order nonlinear absorption processes (in particular, three-photon absorption) also take place.

The absorption spectra of BSO and BGO crystals show that the absorption edge related to impurities and crystal defects lies at 500 and 450 nm for these crystals, respectively. The energy gap width for these crystals is 3.2 eV. The ratio of the energy gap width to the Nd :YAG laser photon energy (1.17 eV) is equal to 2.7 for BSO and BGO. This circumstance considerably reduces the probability of two-photon absorption in these crystals at 1064 nm. Other possible higher-order mechanisms of nonlinear absorption were considered in [54]. In particular, taking three-photon absorption into account, the dependence of the absorption coefficient a on the intensity can be written in the form [55]

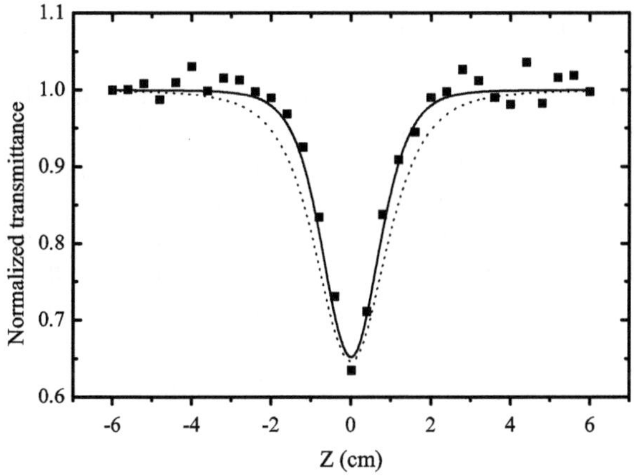

Fig. 3.5 Dependences of the normalised transmission on the BSO crystal position with respect to the focus in the open aperture scheme. *Solid* and *dotted curves* are the theoretical dependences in the case of three- and two-photon absorption, respectively, *squares* are the experiment. Adopted from [52] with permission from Springer Science+Business Media

$$\alpha(I) = \alpha_0 + \beta_{2\omega} I + \beta_{3\omega} I^2, \tag{3.12}$$

where $\beta_{3\omega}$ is the three-photon absorption coefficient.

The theoretical dependence calculated using (3.12) is shown by the solid curve in Fig. 3.5. Good agreement with experimental data suggests that nonlinear absorption is related to three-photon processes. Table 3.1 presents three-photon absorption coefficients calculated using Eq. (3.12). Three-photon absorption coefficients measured in those experiments ($(2.5 \pm 0.8) \times 10^{-20}\,\text{cm}^3\,\text{W}^{-2}$ for BSO and $(4.4 \pm 1.3) \times 10^{-20}\,\text{cm}^3\,\text{W}^{-2}$ for BGO) are close to those coefficients measured for some materials at the same wavelength (in particular, $\beta_{3\omega} = (1.5 \pm 0.75) \times 10^{-20}\,\text{cm}^3\,\text{W}^{-2}$ for CdS). At the same time, these coefficients vary in a rather broad range depending on the medium (from $6.8 \times 10^{-23}\,\text{cm}^3\,\text{W}^{-2}$ for NaCl [34] to $2.0 \times 10^{-19}\,\text{cm}^3\,\text{W}^{-2}$ for InSb [3]).

3.2.2 Nonlinear Crystals for Harmonics Generation

The increase in the intensity of laser pulses due to the development of the techniques for generating and amplifying femtosecond laser pulses requires the consideration of the influence of higher-order optical processes on the second-order optical processes [56]. Despite the fact that nonlinear crystal converters have been long used, their higher-order nonlinearities have been studied inadequately so far. The third-order nonlinear susceptibilities were measured only in some spectral regions and only for certain crystal orientations (see for example [27, 57]). Therefore, analysis of the third-order nonlinear susceptibilities, nonlinear refractive indices, and nonlinear absorption coefficients of KDP and $LiNbO_3$ crystals at 1064 and 532 nm is of current interest [58].

The Kerr nonlinearities are estimated from an empirical model based on a simple assumption about a dominating influence of one efficient optical transition. The transition frequency and the diagonal matrix element of the dipole moment operator (λ_0^{-1} and μ_0) are determined from the values of the refractive index and its dispersion. Unlike approximations [32, 59] used earlier, the values of the refractive index and its dispersion at the same transition frequency are used. In the single-transition approximation, the refractive index is determined by the known expression

$$n_0^2 = 1 + \frac{2}{c\hbar} \frac{f(\mu_0, \lambda_0, \lambda)\mu_0^2}{\lambda_0^{-1} - \lambda^{-1}} N. \tag{3.13}$$

Here, $f(\mu_0, \lambda_0, \lambda)$ is a factor taking into account the local-field correction and N is the density of oscillators. One can obtain from (3.13) the expression for the third-order nonlinear susceptibility responsible for the self-action, which has the form

$$\chi_k^{(3)} = \frac{1}{32\pi^3 c\hbar N} \left(n_0^2 - 1\right) \lambda^2 \frac{\partial n_0^2}{\partial \lambda}. \tag{3.14}$$

Expression (3.14) allows determining the dispersion of the Kerr nonlinearity from the refractive index dispersion. In the general case the Kerr nonlinearity have tensor properties. Although the individual components of the tensor of anisotropic crystals can be estimated from (3.14), the calculation of all components requires an additional analysis. A detailed analysis of the components of the nonlinear susceptibility tensor of a KDP crystal is presented in [60].

In the dispersionless approximation, the third-order Kerr nonlinearities of KDP crystals (the symmetry class 42 m) and $LiNbO_3$ and BBO crystals (the symmetry class 3 m) have four independent components. For the extraordinary wave directed at angles θ and φ to the crystal axes, the Kerr nonlinearity of a crystal of the symmetry 42 m is expressed in terms of these components as

$$\begin{aligned}\chi_k^{(3)} &= \frac{1}{4} [3(\chi_{1111} + \chi_{1122}) + (\chi_{1111} - \chi_{1122}) \cos 4\varphi] \cos^4\theta \\ &\quad + \frac{3}{2}\chi_{2233} \sin^2 2\theta + \chi_{3333} \sin^4\theta,\end{aligned} \tag{3.15}$$

whereas the Kerr nonlinearity for a crystal of the symmetry 3 m has the form

$$\begin{aligned}\chi_k^{(3)} &= 3\chi_{1122} \cos^4\theta + 4\chi_{1123} \left(1 - 4\cos^2\varphi\right) \sin\varphi \sin\theta \cos^3\theta \\ &\quad + \frac{3}{2}\chi_{2233} \sin^2 2\theta + \chi_{3333} \sin^4\theta.\end{aligned} \tag{3.16}$$

It follows from (3.15) and (3.16) that the anisotropy of the Kerr nonlinearity is determined by the relation between its components and can be considerable.

Table 3.2 Nonlinear optical parameters of KDP, BBO, and $LiNbO_3$ at 1064 nm

Crystal	L (cm)	R_d (cm)	Gaussian function-expansion approximation			Paraxial approximation		
			$\chi_k^{(3)}$,10^{-14} esu	n_2,10^{-13} esu	γ,10^{-20} m^2W^{-1}	$\chi_k^{(3)}$,10^{-14} esu	n_2,10^{-13} esu	γ,10^{-20} m^2W^{-1}
KDP ($\Theta - 59°$)	2	0.6	2.0	1.2	3.5	2.6	1.6	4.6
KDP ($\Theta - 59°$)	2	1.6	2.6	1.6	4.6	2.6	1.6	4.6
KDP ($\Theta - 78°$)	1.5	0.6	2.2	1.4	3.8	2.5	1.6	4.4
KDP ($\Theta - 90°$)	1.5	1.6	1.4	0.9	2.5	1.5	0.93	2.6
BBO ($\Theta - 51°$)	0.8	1.3	5.2	3.0	7.4	5.2	3.0	7.4
$LiNbO_3$ ($\Theta - 90°$)	0.8	1.3	6.9	2.9	5.4	7.5	3.2	6.0

Reproduced from [61] with permission from Elsevier

The consideration of the Kerr-nonlinearity dispersion increases the number of independent components up to five (the symmetry class 42 m) and six (the symmetry class 3 m).

In [31, 61], the KDP, $LiNbO_3$ and BBO crystals of length from 0.8 to 2 cm cut at different angles to the optical axis were studied. Measurements were performed for different diffraction lengths R_d. The value of R_d was changed by mounting a telescope in front of a focusing lens. The experimental dependences of the normalized transmission of radiation through a limiting aperture on the sample position were interpreted by the method of the expansion of a Gaussian function, which is widely used to analyze the results of z-scan measurements [24], and by the numerical method in combination with the paraxial approximation. Table 3.2 presents the nonlinear susceptibilities and nonlinear refractive indices of these crystals calculated from the dependence of the normalized transmission of radiation at 1064 nm on the position of samples.

The phase shift in BBO and $LiNbO_3$ crystals is positive, i.e. these crystals have self-focusing properties. Note that the phase shift in a $LiNbO_3$ crystal near $\theta = 90°$ caused by the self-action channel only slightly exceeds the phase shift caused by the cascade process. The nonlinear absorption coefficient was estimated from open aperture z-scan measurements. It was found that nonlinear absorption in this case was determined by the four-photon process. The four-photon absorption coefficient was measured to be 1.7×10^{-32} cm^5 W^{-3}. No nonlinear absorption was observed at 1064 nm in KDP and BBO crystals by the open aperture z-scan [61].

Table 3.3 presents the values of $\chi_k^{(3)}$, n_2 and γ, measured at a wavelength of 532 nm. The diffraction length was the same and equal to ~1 cm in all measurements. The measurements of KDP crystals were performed for the same orientation of the crystals and different intensities of the incident radiation. The nonlinearity of KDP crystals measured at 532 nm was smaller than that measured at 1064 nm. This agrees with calculations based on a simple empirical model.

The nonlinearity of the BBO crystal increased with decreasing wavelength, which also agrees with calculations taking into account the parameters of this crystal. The

Table 3.3 Nonlinear optical parameters of KDP and BBO at 532 nm

Crystal	L (cm)	$\chi_k^{(3)}$, 10^{-14} esu	n_2, 10^{-13} esu	γ, 10^{-16} cm$_2$ W^{-1}
KDP ($\Theta = 41°$)	1.8	1.7	1.1	2.8
KDP ($\Theta = 78°$)	1.9	1.5	0.9	2.5
BBO ($\Theta = 51°$)	0.8	5.7	3.2	8.0

Reproduced from [61] with permission from Elsevier

dependence of the normalized transmission at 532 nm on the crystal position in closed aperture z-scan measurements exhibited asymmetry for all crystals, which is explained by the presence of nonlinear losses. Nonlinear absorption in KDP and BBO crystals is determined by the three-photon processes, whereas nonlinear absorption in $LiNbO_3$ is determined by the two-photon process [61]. The three-photon absorption coefficients were 5.4×10^{-22} cm^3 W^{-2} and 2.1×10^{-21} cm^3 W^{-2} for KDP and BBO crystals, respectively. The two-photon absorption coefficient measured for $LiNbO_3$ crystals (2.1×10^{-10} cm W^{-1}) was smaller than that calculated by expressions from [62] (2.9×10^{-9} cm W^{-1}), but better agrees with the value obtained in [63] (1.5×10^{-10} cm W^{-1}).

3.3 Fullerenes

Most of low-order nonlinear optical studies of fullerenes were performed earlier by the methods of degenerate four-photon mixing z-scan and third harmonic generation [64–70]. The nonlinear susceptibility $\chi^{(3)}$ of fullerenes in films and solutions responsible for phase conjugation at different wavelengths was investigated by the method of degenerate four-photon scattering [47, 71]. The nonlinearities of similar fullerene systems responsible for the second- and third harmonic generation were studied in papers [72–78], respectively. It was found that C_{60} films have high third-order nonlinear susceptibilities at 1064 nm ($\chi^{(3)}(-3\omega; \omega, \omega, \omega) = 2 \times 10^{-10}$ esu and $\chi^{(3)}(-2\omega; \omega, \omega, 0) = 2.1 \times 10^{-9}$ esu).

In fullerenes, the inverse saturated absorption determines an increase in absorption with increasing the laser radiation intensity [66, 79]. It is known that this effect is observed because the absorption cross section in excited states exceeds that from the ground state. Materials with inverse saturated absorption are excellent candidates for using them as optical limiters to protect eyes and other radiation detectors from intense laser pulses [80, 81].

The nonlinear optical parameters of fullerenes C_{60} and C_{70} in polyamide films and toluene excited by 35 ps laser pulses at 1064 and 532 nm were analyzed in [82]. The value of γ for the 0.5 % solution of C_{60} in toluene calculated from experimental data using the 1064 nm radiation was -4×10^{-15} cm^2 W^{-1}. The nonlinear susceptibility of this sample was -2×10^{-13} esu at 1064 nm. Measurements of the nonlinear absorption in samples showed that this process was distinctly observed only in C_{60} solutions. Nonlinear absorption in this case could be caused by the inverse saturated absorption due to excitation of fullerenes to higher-lying states with absorption cross sections exceeding the ground-state absorption cross section. The nonlinear absorption coefficient of the 0.5 % solution of C_{60} in toluene was $1.5 \times 10^{-10}\,\mathrm{cm\,W^{-1}}$ (at $\lambda = 1064\,\mathrm{nm}$).

The inverse saturated absorption was observed in C_{60} and C_{70} films [65, 68] and fullerene solutions [61, 79, 83] at the second harmonic wavelength of Nd:YAG laser (532 nm). The possibility of using fullerene-doped samples for optical limiting at 1064 nm was studied in [69]. The transmission in a cell with the 0.55 % C_{60} solution in toluene did not change with increasing the radiation intensity up to $I = 4 \times 10^{10}\,\mathrm{W\,cm^{-2}}$. Above this intensity, a decrease in transmission was observed. This effect was not observed in C_{70} solutions. Thus, it was shown in [69] that C_{60} solutions can limit the radiation intensity in the IR region as well.

Similar studies of the nonlinearity of fullerene-doped structures were also performed at a wavelength of 532 nm. Table 3.4 presents the values of $\mathrm{Re}\chi^{(3)}$ and γ, measured by different authors under similar spectral conditions [69]. A great difference between the values of $\mathrm{Re}\chi^{(3)}$ and different signs of $\mathrm{Re}\chi^{(3)}$ can be caused by different experimental conditions (pulse duration, laser radiation intensity, etc.). In addition, the influence of the nonlinearities of fullerene-doped matrices (toluene, polymers, and metallorganic compounds) can be also significant. A change in the radiation energy leads to a change in the sign of $\mathrm{Re}\chi^{(3)}$ for some fullerene-containing structures [84]. This can be caused by the influence of additional nonlinear optical processes (thermal effect, Stark effect, etc.) proceeding along with the high-frequency Kerr effect.

The fifth-order nonlinearity of thin fullerene films was studied in [82]. It was shown that, at low laser radiation intensities ($I_0 = 5 \times 10^8\,\mathrm{W\,cm^{-2}}$), the self-defocusing process is determined by third-order nonlinearities. The self-action changed with increasing radiation intensity due to the increasing role of the higher-order nonlinearity. Under these conditions, the self-defocusing properties of a thin (60 nm) fullerene film were demonstrated. The nonlinear refractive indices caused by the third- and fifth-order processes at 532 nm were $2 \times 10^{-9}\,\mathrm{cm^2\,W^{-1}}$ and $-1.4 \times 10^{-19}\,\mathrm{cm^4\,W^{-2}}$, respectively [85].

3.4 Dyes

The practical applications of dyes for various goals of laser physics require the study of their optical parameters [86, 87]. Along with common optical parameters, which have been already investigated to a great extent, an important characteristic of dyes is their nonlinear optical response to the action of short (picosecond and femtosecond) light pulses. It has been shown earlier that the nonlinear susceptibilities of a number of organic dye molecules with conjugated double bonds are comparable with the resonance nonlinear susceptibilities of atoms [87, 88]. Such molecules with double conjugated bonds and delocalized π electrons are very promising as the

Table 3.4 Values and signs of Re$\chi^{(3)}$ of the C_{60} solutions and films

λ (nm)	Sample	Re$\chi^{(3)}$ esu	Experimental method	Pulse duration	Ref.
532	C_{60}/toluene	-1.7×10^{-13}	SFWFM[1]	25 ps	[a][5]
532	C_{60}/toluene	>0	z-scan	19 ps	[b][5]
532	C_{60}/toluene	<0	z-scan	70 ns[2]	[b][5]
532	C_{60}/toluene	<0	z-scan	13 ps	[c][5]
532	C_{60}/toluene	$+5.8 \times 10^{-12}$	ez-scan	15 ns	[d][5]
532	C_{60}/toluene	<0	z-scan	30 ns	[e][5]
<950	C_{60} film (5.2 μm)	<0[3]	z-scan	150 fs	[f][5]
532	C_{60}/toluene	>0[4]	z-scan	23 ps	[g][5]
520	C_{60}/toluene	$-(3.4 \pm 0.7) \times 10^{-11}$	z-scan	15 ns	[h][5]
576	C_{60}/PMMA	<0	Shadow method	12 ns	[i][5]
532	C_{60}/liquid crystal	$+4.86 \times 10^{-8}$ cm^3 erg^{-1}	Holographic recording	20 ns, 400 ps	[k][5]
532	C_{60}/in metal-organic solution	<0	z-scan	8 ns	[l][5]
532	C_{60}/in metal-organic solution	>0	z-scan	8 ns	[l][5]
532	C_{60}/toluene	$+(1.9 \pm 0.38) \times 10^{-13}$	z-scan	55 ps	[m][5]
532	C_{60} film (100 nm)	$+(4.8 \pm 0.96) \times 10^{-8}$	z-scan	55 ps	[m][5]

Reproduced from [69] with permission from Elsevier
[1] SFWFM: stimulated four-wave frequency mixing
[2] Pulse train of duration 19 ps (11 pulses in a train) with a pulse period of 7 ns
[3] $n_2 = 1.9 \times 10^{-12}$ esu
[4] $\gamma = 9.05 \times 10^{-13}$ cm^2 W^{-1}
[5] [a] F. J. Aranda, D. V. G. L. N. Rao, J. F. Roach, et al, J. Appl. Phys. 73, 7949 (1993). [b] T.-H. Wei, T.-H. Huang, T.-T. Wu, et al, Chem. Phys. Lett. 318, 53 (2000). [c] B. Taheri, H. Liu, B. Jassemnejad, et al, Appl. Phys. Lett. 68, 1317 (1996). [d] K. Dou, J. Y. Du, E. T. Knobbe, J. Luminesc. 83-84, 241 (1999). [e] S. R. Mishra, H. S. Rawat, M. P. Joshi, et al, Appl. Phys. B 63, 223 (1996). [f] G. Banfi, D. Fortusini, M. Bellini, et al, Phys. Rev. B 56, R10075 (1997). [g] F. Li, Y. Song, K. Yang, et al, Opt. Commun. 145, 53 (1998). [h] S. Couris, E. Koudoumas, A. A. Ruth, et al, J. Phys. B 28, 4537 (1995). [i] R. V. Goedert, T. A. Whittaker, A. F. Clements, et al, Chem. Phys. Lett. 332, 225 (2000). [k] N. V. Kamanina, Opt. Spectrosc. 93, 639 (2002). [l] G. Fang, Y. Mo, Y. Song, et al, Opt. Commun. 205, 337 (2002). [m] R. A. Ganeev, A. I. Ryasnyansky, V. I. Redkorechev, et al, Opt. Commun. 225, 131 (2003)

nonlinear optical media. Note that the delocalization of π electrons in fullerenes is also responsible for the high nonlinear susceptibility of the latter [65, 77].

The mode locking properties of the polymethine dyes are subjected to irreversible changes with time, which causes, in particular, variations of their nonlinear optical parameters (for example, saturation intensity and nonlinear absorption coefficient). The nonlinear optical parameters of polymethine dyes of different ionic properties (cationic (dyes 1–6), anionic (7), and neutral (8), by classification given in [89]) were analyzed in [90] by the z-scans in the field of picosecond pulses. The z-scan experiments were performed by using a picosecond 1064 nm Nd:YAG laser emitting 55 ps, 2 mJ pulses with a pulse repetition rate of 2 Hz. The dye solutions were z-scanned by using the open aperture scheme. The value of β for the solution of dye 2 (at the concentration $C = 2.5 \times 10^{-3}$ M) was measured to be 3×10^{-13} cm W^{-1}.

Table 3.5 Nonlinear optical parameters of polymethine dyes

Dye	Concentration, 10^{-3} M	γ, $10^{-16}\,\mathrm{cm^2\,W^{-1}}$	β, $10^{-13}\,\mathrm{cm\,W^{-1}}$	I_{sat}, $\mathrm{W\,cm^{-2}}$
1	2.8	2.6	2.3	
2	2.5	5.6	3.0	
3	2	4.5	1.7	
4	2			4×10^8
5	1.3			3.4×10^6
6	1.6	65		
8	4.5	2.7		

Reproduced from [89] with permission from Springer Science+Business Media

Note that calculations performed earlier showed that some of the dyes (tetracene, paraterphenyl, pentacene) have considerable Kerr nonlinearities [91]. Similar measurements were performed for all other dyes. Their nonlinear refractive indices are presented in Table 3.5, where are also given the nonlinear absorption coefficients measured for these dyes at different concentrations. One can see that, having rather low nonlinear refractive indices (at the given concentrations of solutions), some of the dyes exhibit noticeable nonlinear absorption. This can be explained both by inverse saturated absorption and two-photon absorption.

Dye 5 is of certain interest among polymethine dyes. Due to efficient lasing of this dye in polyurethane matrices and alcohol solutions observed in a broad range of 80 nm at 1150 nm upon pumping at 1064 nm [89], this dye is promising not only as a saturable absorber for passive mode locking but also as a laser medium for generating ultrashort IR pulses due to a large width of its lasing spectrum.

The results of the open aperture z-scan measurement of saturated absorption of the dye 5 dissolved in ethanol at different radiation intensities are presented in Fig. 3.6 in the form of the dependences of the normalized transmission of the dye 5 solution on z axis [92]. At low laser radiation intensities, the normalized transmission increases with increasing radiation intensity, which is typical for saturable absorbers. As the radiation intensity was further increased, absorption was saturated, the envelope of the dependence $T(z)$ broadened, and then nonlinear absorption appeared, which was caused by either multiphoton absorption or inverse saturated absorption and manifested in a decrease in the maximum value of the normalized transmission. These dependences can be used to determine the intensity regions where the conditions of the maximum bleaching of the dye are fulfilled and at the same time the influence of other nonlinear optical processes, in particular, nonlinear absorption is absent.

In the trivial case of a linear dependence of the absorption coefficient a on the radiation intensity, this coefficient can be described by Eq. (3.11), where the linear absorption coefficient can be, in particular, negative in the case of saturated absorption. There also exist other models describing saturated absorption. For a two-level system with the inhomogeneous broadening, the equation describing the saturation process could be written in the form [93]

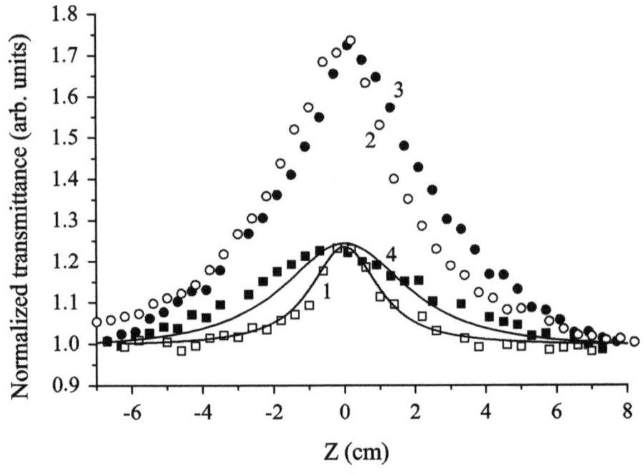

Fig. 3.6 Dependences of the normalised transmission of the dye 5 solution on z in the open aperture scheme for radiation intensities in the focal plane *1* 1.2×10^7, *2* 4.8×10^7, *3* 1.5×10^8, and *4* 1.2×10^9 W cm^{-2}. *Solid curves* are the theoretical calculations for the minimum *(1)* and maximum *(4)* laser radiation intensities; *circles* and *squares* are the experiment. Adopted from [90] with permission from Springer Science+Business Media

$$\alpha = \alpha_0 \frac{1}{(1 + I/I_{sat})^{0.5}}. \tag{3.17}$$

Here, I and I_{sat} are the laser radiation and saturation intensities, respectively. Another ('kinetic') model is used in the case of the depletion of the ground-state concentration and gives the relation [86]

$$\alpha = \alpha_0 \frac{1}{1 + I/I_{sat}}. \tag{3.18}$$

Yet another ('three-level') model giving the expression

$$\alpha = \alpha_0 \frac{1}{1 + (I/I_{sat})^{0.5}} \tag{3.19}$$

was used in [94] to analyze saturated absorption. The experimental dependence of the saturated absorption in the dye 5 solution is well described by the three-level model at low radiation intensities and by the two-level model with the inhomogeneous broadening at high radiation intensities.

3.5 Metals

3.5.1 Organometallic Structures

The doping of the metals into the polymer structures changes both optical and nonlinear optical parameters of these composite materials [95, 96]. In [95], complexes of polymers with metals were studied with the aim of using them for optical switching and optical limiting in the field of nanosecond pulses. In [96], porphyrin polymers doped with zinc were investigated for the purpose of using them for optical limiting. It was shown that these compounds can efficiently limit the intensity of 500 ps pulses at a wavelength of 532 nm. Note that optical limiting in polymer metal-containing complexes was mainly studied in the visible spectral region.

In [97], the nonlinear optical parameters of polyvinyl pyrrolidone (PVP) doped with various metals were investigated. An attractive feature of PVP is that this polymer does not lose its good optical properties (in particular, weak scattering of light) upon doping with molecules and atoms. Note that PVP can be also used as a stabilizing reagent for such nanostructures as metal clusters, which makes it attractive for the development of stable suspensions used in nonlinear optical studies [98]. The aqueous solutions of PVP doped with cobalt at different concentrations (2, 5.3, 6.2 and 13.5 %) and with iron (4.25 %) and zinc (0.85 %) were investigated. One gram of a metallopolymer was dissolved in 100 ml of distilled water at room temperature. The solutions were studied in the cells of thickness of 1 mm. Figure 3.7a presents the normalized transmissions of organometallic polymers at 1064 nm with different cobalt concentrations as functions of z. Figure 3.7b presents similar dependences obtained at 532 nm. Note that the nonlinearity sign is constant in the case of these two wavelengths. One can see from Fig. 3.7 that, for negative values of z, the transmission increases with increasing cobalt concentration. This is explained by the enhancement of defocusing with increasing Co concentration. Experiments with pure water and pure PVP (without cobalt) did not reveal the characteristic nonlinear dependence of $T(z)$.

The open aperture z-scan studies showed that nonlinear absorption at 532 nm was observed for all samples. The nonlinear absorption coefficient of the PVP solution with cobalt at a concentration of 13.5 % was 9.4×10^{-10} cm W^{-1}. Nonlinear absorption at 1064 nm was not observed, although high-power radiation was used. Table 3.6 presents the nonlinear refractive indices, Kerr nonlinear susceptibilities, and nonlinear absorption coefficients of organometallic polymers measured in [97]. A decrease in transmission at 532 nm was caused by reverse saturated absorption, which is manifested in many similar organometallic and polymer structures in the visible spectral range [99–101]. In [100], nonlinear absorption in metalloporphyrins was studied. It was shown that nonlinear absorption at 532 and 600 nm (in the field of picosecond and nanosecond pulses) was caused by reverse saturated absorption. The absence of nonlinear, in particular, two-photon absorption at 1064 nm is probably explained by the fact that the resonance line of metalloporphyrin solutions lies outside the region of the two-photon transition at this wavelength.

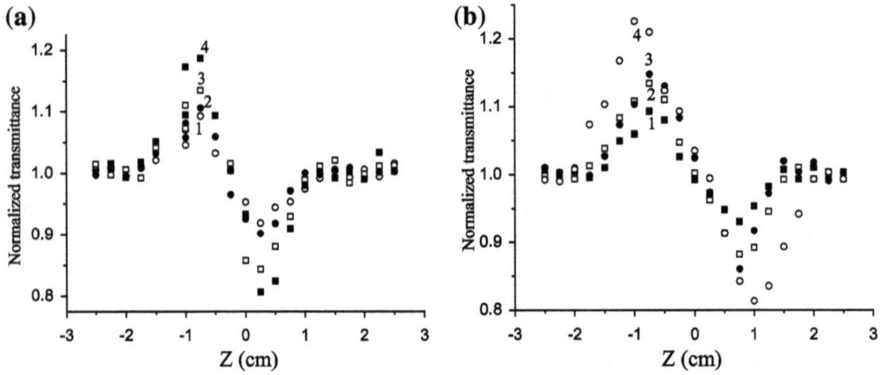

Fig. 3.7 Dependences of the normalised transmission of PVP suspensions with the cobalt concentration *1* 2 %, *2* 5.3 %, *3* 6.2 %, and *4* 13.5 % in the closed aperture scheme at **a** 1064 and **b** 532 nm probe radiation. Reproduced from [97] with permission from Springer Science+Business Media

Table 3.6 Nonlinear refractive indices, nonlinear absorption coefficients, and nonlinear susceptibilities of Co solutions in PVP

Substance	λ (nm)	γ, 10^{-15} cm^2 W^{-1}	$\chi^{(3)}$, 10^{-14} esu	β, 10^{-10} cm W^{-1}
PVP (2 % Co)	1064	−0.9	−6.78	–
PVP (5.3 % Co)	1064	−1.4	−11	–
PVP (6.2 % Co)	1064	−1.8	−13.8	–
PVP (13.5 % Co)	1064	−2.6	−19.7	–
PVP (2 % Co)	532	−1.6	−11.8	2.1
PVP (5.3 % Co)	532	−3.4	−26.5	4.2
PVP (6.2 % Co)	532	−4.6	−35.9	6.1
PVP (13.5 % Co)	532	−6	−47	9.4

Reproduced from [97] with permission from Springer Science+Business Media

One of the mechanisms providing highly efficient nonlinear absorption is, as mentioned above, the reverse saturated absorption. Absorption of this type was successfully described by the so-called five-level model. The reverse saturated absorption can be observed if a medium satisfies the following criteria: (a) the excited-state absorption cross section of the medium should exceed the ground-state absorption cross section, and (b) the excited-state lifetime should be long enough compared to the exciting pulse duration. A number of organometallic materials such as metalloporphyrins [83, 100], metallopolyenes [102], and metallophthalocyanines [99] satisfy these criteria. It was quite reasonable to expect the presence of reverse saturated absorption in structures that were studied (along with other possible optical limiting mechanisms).

3.5.2 Colloidal Metal Solutions

Colloidal metal solutions have high nonlinear optical coefficients and a fast response [103, 104], especially in the frequency range of their surface plasmon resonances. Most of the nonlinear optical studies of colloidal metal solutions have been performed by the method of four-photon mixing. It was shown already in the first experiments on phase conjugation in colloidal gold and silver [105] that the reflection coefficient of the conjugate wave considerably increased near the plasmon resonance.

A number of promising colloidal metal solutions were found for applications in optical limiters. The authors of paper [106] determined the relation between associates of small particles in hydrosols of noble metals and the presence of fractal properties of the latter. It was shown in [103, 107] that colloidal copper solutions have a high nonlinear refractive index. Similar results were obtained for silver [103, 108] and gold [109, 110] solutions. The nonlinear susceptibility of silver and gold hydrosols was investigated in [103, 109]. It was shown that the susceptibility varied from 10^{-11} to 10^{-13} esu for colloidal gold depending on the volume fraction of metal particles in the solvent. At the same time, the nonlinear susceptibilities of metal silver and gold particles themselves were (1–5) $\times 10^{-8}$ esu and (2–4) $\times 10^{-9}$ esu, respectively.

In [110], the nonlinear optical parameters of colloidal solutions of some metals (silver, gold, copper, and platinum) prepared by the chemical method were studied. The absorption spectra of these solutions were changed after irradiation by picosecond laser pulses. The absorption spectrum of gold had the maximum at 525 nm before irradiation, in accordance with data published earlier [111]. However, the frequency of the surface plasmon resonance of gold nanoparticles shifted to the red (525–550 nm) after irradiation at 1064 nm. This shift was caused by the aggregation of gold particles induced by irradiation. A similar picture was also observed in aqueous solutions of some other metals, in particular, silver.

Colloidal silver particles are classical objects for studying optical phenomena in ultradispersion metal systems. The surface plasmon resonance of silver hydrosols (410–420 nm) determines the shape of their optical spectra. Aggregation induced by irradiation, which was earlier observed in colloidal silver solutions [112], resulted both in the surface plasmon resonance frequency shift and corresponding variations in some nonlinear optical parameters. The appearance of the long-wavelength wing in the absorption spectrum of the silver hydrosol was explained in [110] by the aggregation of silver particles to fractal clusters, accompanied by the frequency shift of the intrinsic optical resonances of particles due to dipole-dipole interaction of the photoinduced dipole moments of each particle with the dipole moments of nearest neighbouring particles.

It was found that the nonlinear refractive index at 1064 nm was positive for all colloidal solutions studied except the gold solution. This feature was also preserved for all solutions irradiated by the second harmonic of Nd:YAG laser (532 nm) except the copper solution in which the sign of γ was changed. Analysis of nonlinear absorption in these solutions showed that this process was distinctly observed only in the colloidal gold. Nonlinear absorption in that case could be caused by reverse saturated absorption due to excitation of aggregates to higher-lying levels with larger absorption cross sections compared to the ground-state absorption cross section.

3.5.3 Solid Dielectric Matrices Doped with Metals

Composite materials based on solid dielectric matrices doped with metal nanoparticles attracted interest more than two decades ago [105]. The study of these materials still attracts a great attention [113–116] because they are promising for applications in fast optical switching systems and optical limiters. In [117–120], the nonlinear optical properties of copper and silver nanoparticles in glass matrices were studied in the UV, visible, and IR spectral regions. Silica glasses (SG) containing 100 % of SiO_2 and the SLSG silica glass containing 70 % of SiO_2, 20 % of Na2O and 10 % of CaO were used. These glasses were irradiated by the 60-keV Ag^+ and Cu^+ ion beams with doses 4×10^{16} and 8×10^{16} ions cm^{-2}. The penetration depth of metal nanoparticles was 60 nm. The average size of Cu nanoparticles measured by the method of X-ray reflectometry was 3–5 nm. The size of silver nanoparticles was varied in a broader range (2–18 nm). The formation of metal nanoparticles was also confirmed by the appearance of characteristic absorption lines in the ranges 400–450 nm (for silver-doped glasses) and 550–600 nm (for copper-doped glasses) corresponding to the SPR frequencies of silver and copper nanoparticles.

The nonlinearity of samples in different spectral regions was studied by using the radiation of picosecond Nd:YAG laser and its second and third harmonics. The 1064 nm Nd:YAG laser emitted 55 ps, 1 mJ pulses with a pulse repetition rate of 2 Hz. In Cu:SG and Cu:SLSG samples, self-defocusing and self-focusing was observed, respectively, thereby demonstrating the important role of a matrix. The values of β for SLSG and Cu:SLSG were $2.3 \times 10^{-16}\,\mathrm{cm^2\,W^{-1}}$ and $1 \times 10^{-10}\,\mathrm{cm^2\,W^{-1}}$, respectively. The values of β for Cu:SLSG and Cu:SG ($\lambda = 1064$ nm) were $3.4 \times 10^{-6}\,\mathrm{cm\,W^{-1}}$ and $9 \times 10^{-6}\,\mathrm{cm\,W^{-1}}$, respectively. The nonlinear susceptibility of Cu:SG was 2.4×10^{-8} esu [117]. Self-focusing at 1064 nm was observed for both samples containing silver. As in the case of copper nanoparticles, the SPR frequency of the compounds containing silver nanoparticles depended on the matrix. More details on the low-order nonlinearities of silver nanoparticle-containing matrices will be discussed in Chap. 5.

3.6 Plasma

The characteristics of laser plasma, as a nonlinear medium for high-order harmonics generation, play a crucial role in the restriction of highest achievable harmonic orders and conversion efficiencies. The growth of the concentration of free electrons, probably, led to the restriction of the cutoff energies of harmonics in early experiments with laser plasma [121–124]. It can be assumed that the main contribution to the limitation of harmonic conversion efficiency and cutoff energy was caused by the free electrons-induced self-defocusing of laser beam and phase mismatch between the harmonic and fundamental waves. In the case of gas-jet experiments this process also led to the limitation of cutoff energy (see [125] and references therein).

The rapidly ionizing gas-like medium modifies the temporal structure of the femtosecond laser pulse due to the self-phase modulation [126]. In addition, the spectral structure of high-order harmonics depends critically on the frequency modulation, or chirp, of the driving laser pulses. The propagation of intense chirped laser pulses and harmonic generation from the ionized gas has been analyzed and the spatial, spectral, and temporal characteristics of the laser pulse and of the generated harmonics were defined in [127]. The spatial, temporal, and spectral characteristics of the driving pulse during its propagation in an ionizing medium were estimated, as they proved to influence critically the harmonic generation. The detailed analysis revealed the self-guiding of the laser beam. In addition, the plasma-induced self-phase modulation modifies the initial chirp and affects in a crucial way the spectral response of the single atom.

Since the self-defocusing seems a common limiting factor for both these HHG schemes, it was important to analyze this process and to define the experimental conditions when its influence becomes less significant. Here, we discuss the z-scan studies of the self-defocusing and nonlinear absorption in the laser plasma produced on the surfaces of various targets at the conditions corresponding to the efficient HHG [128]. We also analyze the beam profile in the far field after the propagation of laser radiation through the different plumes.

The pump source used in these studies was a Ti:sapphire laser operated at a 10 Hz pulse repetition rate. A portion of uncompressed radiation (pulse energy $E = 15$ mJ, pulse duration $t = 210$ ps, central wavelength $\lambda = 796$ nm) was split from the main beam by a beam splitter and used as a heating pulse. This radiation was focused by a spherical lens *L2* on a solid target *T* and produced a plume predominantly consisting on neutrals and singly charged ions (Fig. 3.8). The composition of plasma was analyzed using the time-integrated spectral measurements of plume in the XUV,

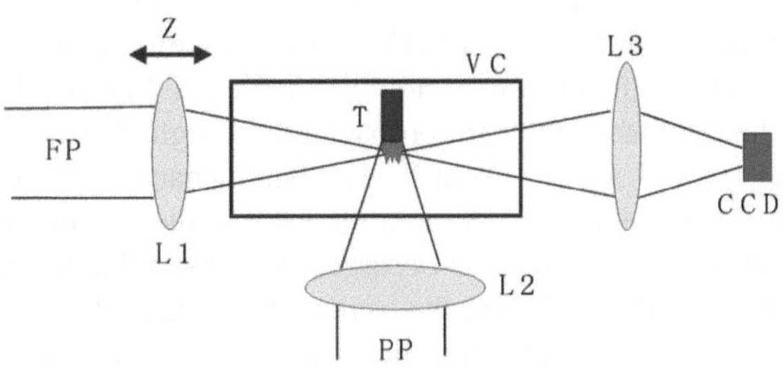

Fig. 3.8 Experimental setup: *T* target, *VC* vacuum chamber, *L1*, *L2*, *L3* lenses, *FP* femtosecond pulse, *PP* picosecond pulse, *CCD* charge-coupled device. Reproduced from [128] with permission from Optical Society of America

UV, and visible ranges. The intensity of picosecond pulse, I_{pp}, at the target surface was varied between 7×10^9 to $8 \times 10^{10}\,W\,cm^{-2}$. After some delay (20–30 ns), a femtosecond pulse ($E = 8\,mJ$, $t = 150\,fs$, $\lambda = 796\,nm$) was focused on the target plasma from the orthogonal direction using 200 or 100 mm focal length lens *L1*. Those experiments were carried out up the maximum intensity of femtosecond probe pulse $I_{fp} = 1.5 \times 10^{15}\,W\,cm^{-2}$.

The shape of the probe beam propagating through the laser plume at different plasma densities (from 5×10^{16} to $2 \times 10^{17}\,cm^{-3}$) and laser intensities was registered in the far field using a charge-coupled device *CCD*. The nonlinear refractive properties of laser plasma were analyzed using the z-scan technique, with a 2 mm aperture placed in the far field and a detector of propagated probe radiation behind it. In the case of conventional z-scan, the medium under investigation moves with regard to the focal plane of focused radiation. In discussed experiments, a focal plane of femtosecond radiation was moved with regard to the plasma area by changing the position of lens *L1*. The conditions of thin medium were fulfilled, since the plasma sizes (0.6 mm) were considerably smaller compared to the confocal parameter of focused radiation (5 mm). The energy of laser pulses prior to the interaction with plasma was measured by a calibrated photodiode and recorded by a digital voltmeter. The aperture was placed at a distance of 900 mm from plasma area to analyze the phase variations of propagated beam using the closed aperture z-scan technique. The nonlinear absorption of plasma was analyzed using both the closed and open aperture schemes.

Initially, the optimal conditions for the HHG produced in various plasmas were defined. The efforts were concentrated on achieving the maximum harmonic order, as well as the maximum conversion efficiency for high harmonics using the low-excited singly-ionized laser plasma. These studies have shown that the highest cutoff energy (101.4 eV) was observed in the case of boron plasma, and the highest conversion efficiency (6×10^{-6}) at the plateau region was achieved in the case of silver plasma. The dependence of the cutoff energy on the probe pulse intensity showed a saturation caused by the free electrons.

During the study of the influence of free electrons on the harmonic generation a shape of probe beam after the propagation through the "optimal" plasma produced on the molybdenum target at different intensities of femtosecond laser pulse was analyzed. The term "optimal" refers to the plasma conditions when the maximum conversion efficiency at the plateau region was achieved. Figure 3.9 shows the spatial profiles of the femtosecond beam propagated through the molybdenum plasma at $I_{fp} = 4 \times 10^{13}$ and $5 \times 10^{14}\,W\,cm^{-2}$. The beam shape remained unchanged at the small intensities of femtosecond radiation even at the conditions of the ionization of neutral atoms (Fig. 3.9a). This observation showed that the concentration of the free electrons generating during the ionization of neutrals was insufficient for the self-defocusing of laser beam. Another beam shape appeared in the case when the pulse intensity exceeded a threshold of ionization of singly charged ions. The barrier suppression intensities for the neutral Mo and Mo^+ were calculated to be $2 \times 10^{13}$and $3 \times 10^{14}\,W\,cm^{-2}$, taking into account the values of first (7.10 eV) and second (16.15 eV) ionization potentials of molybdenum, respectively. The amount of

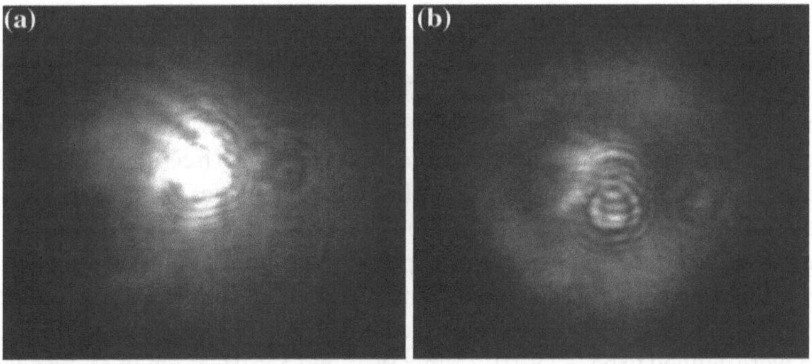

Fig. 3.9 Spatial shapes of femtosecond radiation after the propagation through the Mo plume at different intensities at the focal plane. **a** $4 \times 10^{13}\,\mathrm{W\,cm^{-2}}$, **b** $5 \times 10^{14}\,\mathrm{W\,cm^{-2}}$. Reproduced from [128] with permission from Optical Society of America

the free electrons appearing in the focal area at $I_{fp} > 3 \times 10^{14}\,\mathrm{W\,cm^{-2}}$ became sufficient for the self-defocusing of laser beam. In that case, a ring profile was observed indicating the variation of refractive index in the vicinity of the axis at the focal plane (Fig. 3.9b). Analogous pattern was observed in the case of most plumes. In particular, the dynamics of this process after the propagation of indium plasma showed a gradual variation of spatial distribution from the Gaussian-like shape to the ring-like one.

The closed aperture z-scans were carried out using the femtosecond pulse propagating through the indium plasma. At small laser intensities no variations of the normalized transmittance of femtosecond radiation at the optimal plasma conditions (Fig. 3.10, empty circles, $I_{fp} = 5 \times 10^{13}\,\mathrm{W\,cm^{-2}}$) were observed. With the growth of laser intensity, a characteristic peak-valley shape of normalized transmittance appeared indicating the negative nonlinear refraction inside the plasma (Fig. 3.10,

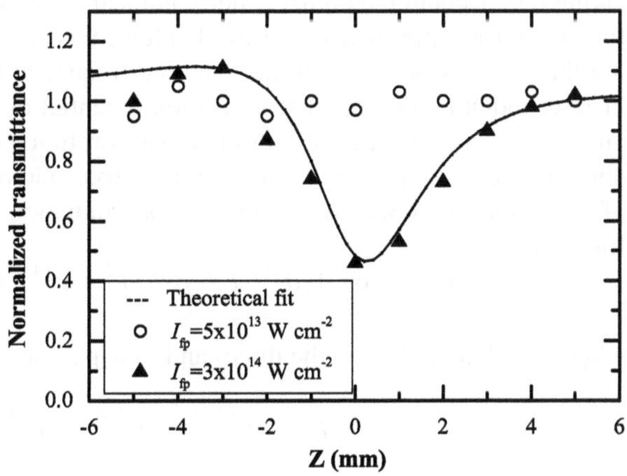

Fig. 3.10 Closed aperture z-scans of indium plasma at low (○) and high (▲) intensities of femtosecond radiation. Fitting *solid line* is a theoretical calculation. Reproduced from [128] with permission from Optical Society of America

filled triangles, $I_{fp} = 3 \times 10^{14}\,\mathrm{W\,cm^{-2}}$). At these intensities a strong nonlinear absorption of femtosecond beam caused by the multiphoton ionization of neutrals and singly charged ions was observed. These studies have shown that the concentration of the free electrons producing by laser ablation was insignificant for the self-defocusing of low-intensity femtosecond beam. The free electrons led to the self-defocusing of laser beam at the probe pulse intensities exceeding the barrier suppression intensity of singly charged ions.

The closed aperture z-scheme was capable determining both the sign and magnitude of the nonlinear refractive index and nonlinear absorption of indium plasma. The solid curve in Fig. 3.10 shows a fitted calculation using Eq. (3.5), from where the γ and β were defined to be $-2 \times 10^{-18}\,\mathrm{cm^2\,W^{-1}}$ and $5 \times 10^{-13}\,\mathrm{cm\,W^{-1}}$, respectively. Note that the last parameter was strongly depended on the intensity of femtosecond pulse.

This analysis is based on the studies of cubic nonlinearities. It mostly refers to the analysis of the nonlinear refraction of plasma. The nonlinear absorption in this case could be used as the parameter depending on the intensity of laser radiation. The nonlinear absorption in plasma was related with the processes of tunnelling and multiphoton ionization using 796 nm radiation. Because of this the value of β was related with the nonlinearities higher than the third-order one. The β measurements have indicated a strong dependence on intensity used, pointing out on the high-order nonlinear processes to be involved. Because of this the analysis of multi-photon absorption at this wavelength was carried out using the open aperture z-scan.

The normalized transmittance for the open aperture z-scheme is given by Eq. (3.10) In the case of the linear dependence of the absorption coefficient on the laser radiation intensity (I) one can use the Eq. (3.11). The theoretical dependencies of the normalized transmittance of laser radiation on the position of plume for the open aperture measurements calculated using this two-photon absorption model have shown the discrepancy with comparing to the experimental data. The experimental dependence (Fig. 3.11) looks considerably narrower on the z scale than the theoretical one (dashed line) calculated by the Eqs. (3.10) and (3.11). This discrepancy can be explained taking into account the higher-order nonlinear absorption process, in particular, the four-photon absorption in the case of In plume that is a reasonable mechanism assuming the comparison between the four-fold frequency ($\lambda = 199\,\mathrm{nm}$, $h\nu = 6.24\,\mathrm{eV}$) of laser radiation and the first ionization potential of In ($U_i = 5.8\,\mathrm{eV}$). Therefore, one can consider the higher multiphoton absorption mechanism, other than the two-photon process. In particular, in the case of four-photon absorption, the dependence of the absorption coefficient α on the laser radiation intensity can be presented as follows:

$$\alpha(I) = \alpha_o + \beta^{(2\omega)} I + \beta^{(3\omega)} I^2 + \beta^{(4\omega)} I^3 = \alpha_o + \beta_{eff} I, \tag{3.20}$$

where $\beta^{(3\omega)}$ and $\beta^{(4\omega)}$ are the three- and four-photon absorption coefficients, and

$$\beta_{eff} = \beta^{(2\omega)} + \beta^{(3\omega)} I + \beta^{(4\omega)} I^2 \tag{3.21}$$

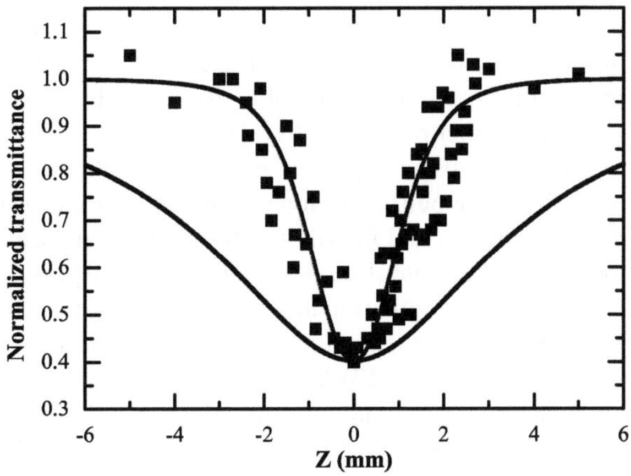

Fig. 3.11 Open aperture z-scan of indium plasma. Fitting *dashed* and *solid lines* are the theoretical calculations for the two- and four-photon absorption mechanisms, respectively. Reproduced from [128] with permission from Optical Society of America

is the "effective" nonlinear absorption coefficient in the case of four-photon absorption. The theoretical dependence (Eq. 3.20) is presented in Fig. 3.11 as a solid line taking into account the Eq. (3.21) and $\beta^{(2\omega)} = \beta^{(3\omega)} = 0$. The comparison of theoretical and experimental results showed a reasonable coincidence that allowed concluding that the four-photon absorption is the dominant one in these experiments. The calculated four-photon absorption coefficient of In plume was found to be $\beta^{(4\omega)} = 5 \times 10^{-42}\,\mathrm{cm}^5\,\mathrm{W}^{-3}$. In the case of Mo ($U_\mathrm{i} = 7.1\,\mathrm{eV}$) the suitable mechanism of nonlinear absorption is a five-photon absorption.

3.7 Liquids

3.7.1 Calibration Measurements of Optical Nonlinearities Using a Standard Medium (Liquid Carbon Disulfide)

The investigations of the nonlinear optical characteristics of various media, which were discussed in previous sections, have shown that the values of these parameters depend strongly on the experimental conditions of their measurement. Due to size effects, materials with chemically identical structures may have different nonlinear refractive indices, nonlinear absorption coefficients, and nonlinear susceptibilities. The difference in the published data for the same materials can also be caused by the methods used for their determination. The most frequently used methods of investigation of the nonlinear optical characteristics are degenerate four-wave mixing (DFWM) [129], nonlinear interferometry [130], and z-scan [24]. All these methods may yield, in some cases, different values for the parameters γ, β, and $\chi^{(3)}$ of the same media. The reason for this discrepancy may be different experimental conditions, in particular different temporal and spectral parameters of the probe pulses used.

The most typical example of such a phenomenon is fullerene-containing structures. The nonlinear optical characteristics of fullerenes previously reported by different authors show considerable discrepancies (see Sect. 3.3). This may be caused by, among other reasons, different pulse durations (from hundreds of femtoseconds to tens of nanoseconds). The measured nonlinear optical parameters may depend strongly on the pulse duration due to the contribution of excited molecules (molecular reorientation and intermolecular interaction), which becomes considerable with increasing duration of pulses. The spectral dispersion of $\chi^{(3)}$ should also be taken into account in comparing the data obtained under different experimental conditions. Measurements of the nonlinear susceptibility of some fullerene-containing media by different methods (DFWM, third harmonic generation, and z-scan) yielded different values for γ, β, and correspondingly $\chi^{(3)}$ of these media. The structure of the fullerenes (C_{60}, C_{70}, and higher), as well as the conditions of their preparation (pure C_{60} films, C_{60}-containing organic films and their derivatives, and C_{60} solutions), also should be taken into account. Concerning the nonlinear refractive index, even the sign of the nonlinear refraction of fullerenes has been the object of numerous studies using various methods. Different signs of γ and considerable discrepancies in the values of $\mathrm{Re}\chi^{(3)}$ can be caused by different experimental conditions (pulse duration, intensity, etc.). The nonlinearities of fullerene-containing matrices (toluene, polymers, and organometallic compounds) can also affect these parameters. A change in the radiation energy alone leads to a change in the sign of $\mathrm{Re}\chi^{(3)}$ in some fullerene-containing structures. This fact can be related to the effect of additional nonlinear optical processes occurring along with the optical Kerr effect (thermal effect, Stark effect, etc.).

In previous sections, we have shown that, over the last two decades, the z-scan technique has proved to be the most frequently used method for determining nonlinear optical parameters owing to its undoubted advantages. One can admit among them the simplicity of use, high accuracy, low cost, and well-developed theoretical basis. However, even this technique requires the attention to the analysis of nonlinear optical processes under different conditions of measurement. An important parameter in such measurements is the duration of laser pulses. Measurements performed in different pulse duration ranges show different values of nonlinear optical parameters (in particular, γ) for some media. Hence, it is necessary to determine the role of the pulse duration in the measurements performed. The most convenient medium for such an analysis may be a liquid carbon disulfide (CS_2). Carbon disulfide has been used as a standard liquid medium for investigation of the nonlinear refractive indices for a long time. Numerous measurements of this parameter at early stages of nonlinear optical investigations showed that γ of CS_2 varies within the range (2–5) $\times 10^{-14}\,\mathrm{cm^2\,W^{-1}}$ (see [131] and references therein). These investigations were mainly performed using pico- and nanosecond pulses. However, the measurements with femtosecond laser radiation showed that this parameter is almost one order of magnitude lower [132–134]. In the meantime, even at the same pulse duration (~400 fs), the data obtained in different studies considerably differ from each other (from 1.3×10^{-16} [135] to 12×10^{-15} [136] and even up to $4 \times 10^{-14}\,\mathrm{cm^2\,W^{-1}}$ [137]).

Below, we analyze the data on the nonlinear optical parameters of CS_2 obtained in the same configuration of the z-scan scheme with the use of laser pulses of various durations (in the femto-, pico-, and nanosecond ranges) [55, 138]. In the experiments, two types of lasers were used. A Ti:sapphire laser (L1) had the following output characteristics: the pulse duration $t = 110$ fs, the wavelength $\lambda = 795$ nm, the single pulse energy $E = 10$ mJ, and the pulse repetition rate $\nu = 10$ Hz. Changing the distance between the diffraction gratings of the compressor of the laser, one can to change the pulse duration from 110 fs to 1.6 ps. This laser also operated in the Q-switched mode with a pulse duration of 8 ns ($E = 10$ mJ) or 75 ns ($E = 0.5$ mJ). The radiation of the oscillator of this laser (Tsunami, $\lambda = 795$ nm, $\nu = 80$ MHz, $t = 100$ fs, the average power $P = 300$ mW) was used to study the thermal self-defocusing in CS_2. As the second laser (L2), a hybrid laser based on neodymium glass and Ti:sapphire regenerative amplifier was used with the following characteristics: $t = 475$ fs, $\lambda = 1054$ nm, $E = 70\,\mu$J, and $\nu = 1$ Hz. The optical nonlinearities in the picosecond range were studied using the uncompressed radiation of this laser ($t = 300$ ps, $E = 100\,\mu$J). The radiation of the oscillator of this laser (GLX-200, $t = 280$ fs, $\nu = 100$ MHz, and $P = 95$ mW) was used for studying the nonlinear absorption and the thermal nonlinear refraction in CS_2 in the near IR region. The 2 mm thick quartz cells filled with carbon disulfide at room temperature were used as the samples under investigation. In the experiments on the influence of the medium length on the measured nonlinear optical parameters of CS_2, cells with lengths of 1, 5, and 10 mm were also used.

3.7.2 Optical Nonlinearities of Carbon Disulfide at 795 nm

The measurements using laser L1 were performed with the shortest pulse duration (110 fs). The dependence of the nonlinear refractive index of CS_2 on the laser radiation intensity was measured using the closed aperture scheme at the intensities ranging from 1×10^9 to 1×10^{12} W cm^{-2}. The nonlinear refraction was observed at the minimum intensity of 2.6×10^9 W cm^{-2} (Fig. 3.12, curve 1). Note that, at low intensities, only nonlinear refraction was observed, which was confirmed by a symmetric form of the dependence T(z) and by a linear dependence of the parameter $\Delta T_{p-v}(\Delta T_{p-v} = |T_p - T_v|$, where T_p and T_v are the maximum (peak) and minimum (valley) transmittances in the dependence T(z)) on the radiation intensity. However, with increasing intensity (at $I \geq 8 \times 10^9$ W cm^{-2}), nonlinear absorption began to manifest itself in the total pattern of the normalized transmittance (Fig. 3.12, curve 3).

These dependences allow concluding that the nonlinear refraction in carbon disulfide is positive, i.e., the self-focusing takes place. In these experiments, the laser radiation intensity was increased up to the saturation of the nonlinear absorption and refraction at the intensities exceeding 1×10^{11} W cm^{-2}. In this case, the nonlinear absorption and the negative addition to the refractive index due to the contribution of free carriers generated at high intensities completely smooth out the peak in the T(z) dependence (Fig. 3.12, curves 4, 5).

In the experiments, the radiation with a nearly Gaussian spatial distribution was used, which was measured using a CCD camera and a spatial analyzer. The beam waist radius ω_0 (at the level $1/e^2$ of the intensity distribution) was $28 \pm 4\,\mu m$. In the case of the influence of third-order optical nonlinearities and the absence of nonlinear absorption (Fig. 3.12, curves 1, 2), the distance between the maximum and the minimum of the dependence $T(z)$ ($\Delta z_{p-v} = 4.5$ mm) relates to the spatial characteristics of the focused Gaussian beam as $\Delta z_{p-v} \approx 1.7 z_0$ [24], where $z_0 = 0.5k(\omega_0)^2$ is the diffraction length of the focused beam and $k = 2\pi/\lambda$ is the wave number. The value of beam radius ω_0 calculated from this relation was $26\,\mu m$, which is in good agreement with direct measurements of this parameter with the use of a Questar QM10 telescope+CCD camera system. Thus, the analysis of the normalized transmittance $T(z)$ allowed determining the spatial sizes of the focused radiation with a high accuracy, which is necessary for the precise calculations of the nonlinear optical characteristics of carbon disulfide.

As it was pointed out in previous sections, one of the advantages of the z-scan technique is the possibility of separating several nonlinear mechanisms if they are simultaneously realized in the medium under study. In the general case, in the presence of both nonlinear refraction and nonlinear absorption, the Eq. (3.5) was used to plot the theoretical dependences $T(z)$ taking into account the experimental conditions. The independent measurements of the parameter β were also performed using the open aperture scheme. In principle, the nonlinear refractive index and the nonlinear absorption coefficient can be measured in the closed aperture scheme. The scheme with an open aperture can be used to compare the obtained values of β.

The solid lines in Fig. 3.12 (curves 2, 3) correspond to theoretical dependences in the absence of saturation. The nonlinear refractive index did not depend on the intensity up to $I = 1 \times 10^{11}$ W cm^{-2}, after which it was observed to decrease with increasing intensity. For intensities not exceeding 1×10^{11} W cm^{-2}, the parameters γ and β for CS_2 were found to be $(3 \pm 0.6) \times 10^{-15}$ cm^2 W^{-1} and $(7 \pm 2.1) \times 10^{-11}$ cm W^{-1}.

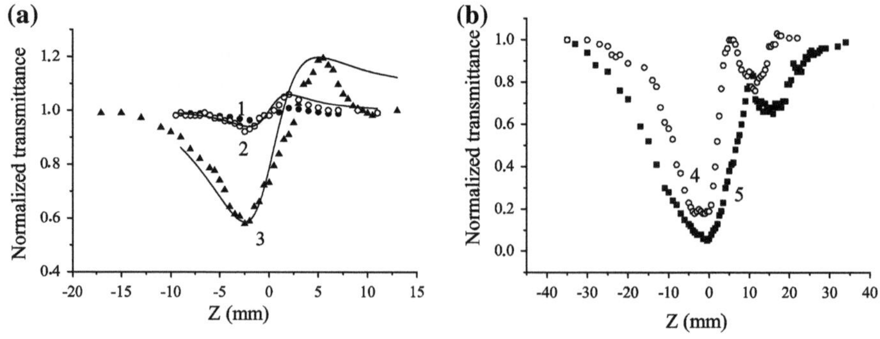

Fig. 3.12 Normalized transmittance in the case of closed aperture scheme at the laser radiation intensities **a** *1* 2.6×10^9, *2* 8×10^9, *3* 3.5×10^{10}, **b** *4* 3×10^{11}, and *5* 9×10^{11} W cm^{-2} (L1, $\lambda = 795$ nm, $t = 110$ fs). *Solid lines* show the theoretical dependences. Reproduced from [138] with permission from Springer Science+Business Media

At higher intensities, the saturation and, probably, the effect of additional nonlinear processes did not allow to correctly calculate these parameters. The measurement error was 20 % for γ and 30 % for β and was mainly related to the error in the calculation of the peak intensity, which, in turn, was determined by the accuracy of the measurement of the radiation energy, the pulse duration, and the beam waist radius. The measurements of the nonlinearities of pure quartz have shown that its γ is $(2 \pm 0.4) \times 10^{-16}\,\mathrm{cm^2\,W^{-1}}$, which agrees well with the data of other authors ($(2.1\text{–}2.3) \times 10^{-16}$ [139] and $(3.5 \pm 1.5) \times 10^{-16}\,\mathrm{cm^2\,W^{-1}}$ [134]) obtained under similar experimental conditions. The threshold intensity of the optical breakdown of the quartz surface was also measured. This threshold intensity in the case of 110 fs pulses was measured to be $(8 \pm 1.6) \times 10^{12}\,\mathrm{W\,cm^{-2}}$, which is in good agreement with the previously reported values of the breakdown threshold at this range of pulse durations ($(1 \pm 0.4) \times 10^{13}\,\mathrm{W\,cm^{-2}}$ [140]).

The uncertainty in the values of the nonlinear refractive index of CS_2 measured at high intensities shows that the nature of the nonlinear refraction and nonlinear absorption is determined not only by third-order nonlinear processes but also by higher order processes induced by intra- and intermolecular interactions. At the same time, one should not exclude the effect of free charge carriers generated in the field of an intense wave. Similar conclusions were previously drawn in studies of various materials in the field of an intense light wave [141–143]. The variations in the nonlinear refractive index with the radiation intensity are an indication of nonlinear processes of a higher (in particular, the fifth) order. In this case, the dependence of the refractive index n on the intensity takes a more complicated form [141, 142]

$$n = n_0 + \gamma_{eff} I = n_0 + (\gamma + \eta I)I, \tag{3.22}$$

where γ_{eff} is the effective value of the nonlinear refractive index and η determines the fifth-order nonlinearity.

From these studies of the nonlinear refractive index at high intensities one can make an assumption that the change (decrease) in γ occurs due to the generation of free carriers. This means that the change in the refractive index caused by these carriers and characterized by the third-order nonlinearity is negative ($\eta < 0$). In the general case, the change in the refractive index can be written as [143]

$$\Delta n = \gamma I + \sigma_r N. \tag{3.23}$$

The second term on the right-hand side of Eq. (3.23) determines the addition to the refractive index due to the effect of free carriers. The parameter σ_r characterizes the change in the refractive index caused by a single charge carrier as a result of photoexcitation and formation of a cloud of free carriers with a density $N(I)$. Note that σ_r for most media is negative and can compensate or even exceed the positive nonlinear addition to the refractive index due to the Kerr nonlinearities of bound electrons. This can lead to variations in the nonlinear optical parameters of the medium and even change the sign of the nonlinear process (from self-focusing to self-defocusing) at high intensities.

The duration of pulses of laser L1 was varied by changing the distance between the diffraction gratings of the compressor. This allowed measuring the nonlinear refractive index of CS_2 at the pulse durations in the range of 110 fs–1.6 ps. The measurements at the pulse durations of 200 and 600 fs showed no noticeable differences from the γ measured at the pulse duration of 110 fs. However, the nonlinear refractive index measured at the pulse duration 1.6 ps was noticeably higher ($\gamma = (1.2 \pm 0.24) \times 10^{-14}\,\mathrm{cm^2\,W^{-1}}$).

In CS_2, three fast processes contributing to the nonlinear optical response of its molecules can occur, namely, intramolecular interaction, orientational Kerr effect, and electronic Kerr effect. All the three processes can contribute to the total change in the refractive index. The slowest of these processes is the orientational Kerr effect. Previously, the response times were reported to be 1.5 ps for the orientational Kerr effect [144] and ~200 fs for the other two processes [136]. The results presented in this section show an increase in γ at $t = 1.6$ ps, which agrees with the assumption about the joint effect of these fast processes as the pulse duration increases from femtosecond to picosecond time scale.

The effect of the length of the cell filled with CS_2 on the nonlinear refractive index was studied using a lens with a long focal distance to meet the conditions of a thin medium. At low intensities, γ did not change with increasing length of the cell. At intensities exceeding $10^{10}\,\mathrm{W\,cm^{-2}}$, the nonlinear refractive index slightly decreased as the cell length changed from 5 to 10 mm. The nature of such variations in γ with increasing length of the nonlinear medium is not quite clear, though one may suppose that it relates to the effect of self-defocusing caused by free charge carriers in an extended medium. In this case, an increase in the medium size can lead to strengthening of the fifth-order rather than the third-order processes.

The following experiments were performed using the laser L1 operating in the Q-switched mode ($t = 8$ ns). Note that all experiments on the study of the nonlinear refractive index and the nonlinear absorption coefficient were performed under the conditions of a thin medium, i.e., under the condition $L < z_0$ [24], when the results of z-scan can be analyzed using Eq. (3.5). The change in the pulse duration of the L1 laser from femtosecond time scale to 8 ns caused a change in the radiation divergence. In the case of nanosecond pulses, the beam divergence was smaller due to the absence of self-modulation, which affected the propagation of femtosecond high-intensity laser radiation. In particular, for the same focusing lens, the diffraction length of the focused beam was changed from $z_0 = 2.6$ mm in the case of femtosecond pulses to $z_0 = 1.8$ mm for nanosecond pulses. In that case, the parameter $r = L/z_0$ was equal to 1.1. The applicability of the z-scan theory for the conditions of an extended medium was analyzed in [145, 146], where the distance between the maximum and the minimum was shown to change from $\Delta z_{p-v} \cong 1.7z_0$ for small r to $\Delta z_{p-v} \cong rz_0 = L$ for large r. The calculations showed that, at $r = 1.1$, this distance changes from $1.7z_0$ to $1.87z_0$ without noticeable changes in ΔT_{p-v}. This means that relation (3.5) of the z-scan theory can be used for $r = 1.1$ as well.

Figure 3.13 shows the results of z-scan of CS_2 using the laser L1 generating the 8 ns pulses in the Q-switched mode. The experiments were performed at various intensities up to $I = 1.5 \times 10^{10}\,\mathrm{W\,cm^{-2}}$. At a 1.5-fold excess of this intensity, we

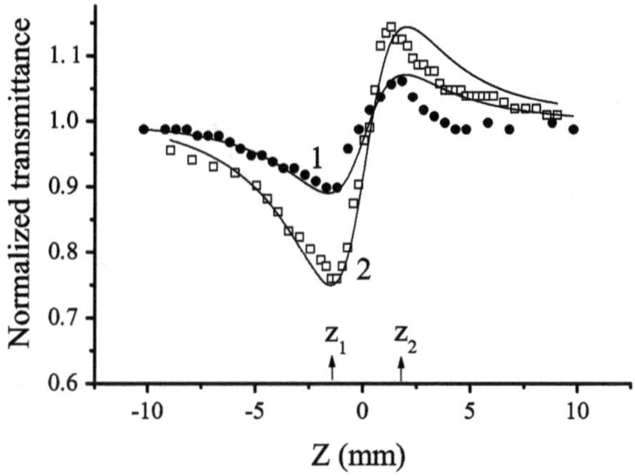

Fig. 3.13 Normalized transmittance in case of closed aperture the scheme at the laser radiation intensities (1) 8×10^8 and (2) $1.7 \times 10^9\,\mathrm{W\,cm^{-2}}$ (L1, $t = 8\,\mathrm{fs}$). *Solid lines* correspond to the theoretical dependences. Reproduced from [138] with permission from Springer Science + Business Media

observed an optical breakdown of the cell surface. This optical breakdown is in good agreement with the data previously reported in [147] ($2.4 \times 10^{10}\,\mathrm{W\,cm^{-2}}$ for 12 ns pulses). The slight asymmetry in the z-scan dependences with respect to the focus (z = 0) was related to the nonlinear absorption.

The theoretical dependences of z-scan (Fig. 3.13) allowed determining the value of γ in the case of nanosecond pulses to be $(3.5 \pm 0.7) \times 10^{-14}\,\mathrm{cm^2\,W^{-1}}$, without noticeable changes in the intensity range from 7×10^8 to $1 \times 10^{10}\,\mathrm{W\,cm^{-2}}$. This value is approximately one order of magnitude higher than the one measured using the femtosecond pulses.

Laser L1 also operated with the pulse duration 75 ns. In that case, the pulse energy was considerably lower (2.5×10^{-4} J) than for 8 ns pulses due to the specific features of the laser. A noticeable nonlinear refraction of CS_2 in this case was observed only under tight focusing conditions. The parameter γ under these conditions was $(4 \pm 2) \times 10^{-14}\,\mathrm{cm^2\,W^{-1}}$.

Previous investigations of the nonlinear refraction caused by the heat accumulation in CS_2 in the field of femtosecond pulses at a high pulse repetition rate (tens of megahertz [132, 134]) showed the importance of this process. Similar studies were performed using the Tsunami oscillator of laser L1 operating at a high pulse repetition rate. Figure 3.14 shows the results of z-scan using the closed aperture scheme with the use of 100 fs pulses with the repetition rate 80 MHz, the power 260 mW, and the wavelength 795 nm, which demonstrate the self-defocusing of laser radiation in carbon disulfide. The γ in this case was equal to $-(3 \pm 0.6) \times 10^{-14}\,\mathrm{cm^2\,W^{-1}}$. No positive nonlinear refraction was observed under these conditions due to the low intensity of pulses. The negative sign of γ observed in the experiments with the pulse repetition rate 80 MHz could be attributed to the thermal lens effect.

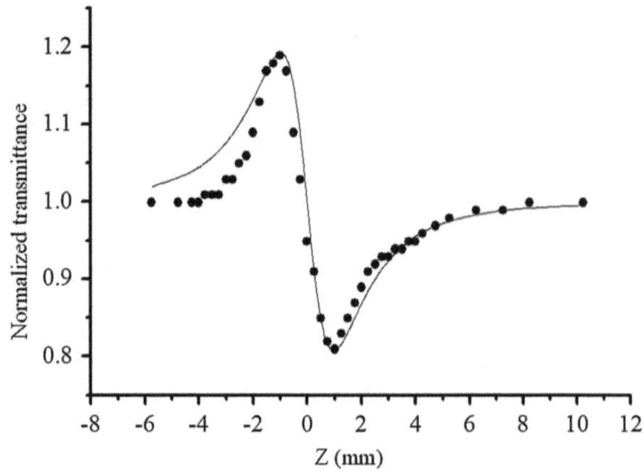

Fig. 3.14 Normalized transmittance at a high pulse repetition rate in the case of closed aperture scheme (L1, t = 100 fs, ν = 80 MHz). Reproduced from [55] with permission from Elsevier

The fact that the thermal effect is observed even at a low intensity (1×10^9 W cm^{-2}) indicates that the dominant mechanism of this effect is the heat accumulation. Under this intensity and a low repetition rate of femtosecond pulses, no nonlinear absorption was detected (10 Hz, Fig. 3.12). It began to manifest itself only at a much higher intensity, 8×10^9 W cm^{-2}. This means that, even at a comparatively weak nonlinear absorption, the heat accumulation during the time of propagation of $\tau/\Delta t \cong 10^5$ pulses results in a pronounced thermal effect ($\Delta T_{\text{p-v}} \sim 0.4$, Fig. 3.14). Here τ is the time required for thermal conduction effects to become significant (τ = 1 ms [55]) and Δt is the distance between pulses in the case of 80 MHz laser (Δt = 12 ns).

The measurements using laser L1 (t = 110 fs, $I \leq 3 \times 10^{11}$ W cm^{-2}, ν = 10 Hz) in the scheme with an open aperture showed that the nonlinear absorption coefficient decreases with increasing intensity at $I > 8 \times 10^{10}$ W cm^{-2}. These independent measurements confirmed the conclusion about the effect of additional nonlinear processes at high intensities (Fig. 3.15). This feature can be related to both the saturation of the nonlinear absorption of CS_2 molecules in the vicinity of 397 nm and the effect of stimulated Raman scattering.

In the case of z-scan in the scheme with an open aperture, the normalized transmittance can be followed with Eq. (3.10). The theoretical dependences calculated by Eq. (3.10) are shown by solid lines in Fig. 3.15. The value of β was $(5 \pm 1.5) \times 10^{-11}$ cm W^{-1}, remaining unchanged with increasing intensity up to $I_0 = 3 \times 10^{10}$ W cm^{-2}. A decrease in the nonlinear absorption coefficient due to the saturation and other nonlinear optical processes at high intensities was determined from the theoretical dependences.

Analyzing the results on the nonlinear absorption and refraction, one can estimate the values of the real and imaginary parts of $\chi^{(3)}$. The main contribution to the absolute value of $\chi^{(3)}$ in the case of 110 fs pulses is made by the real part of $\chi^{(3)}$ ($\text{Re}\chi^{(3)} = 5 \times 10^{-14}$ esu, $\text{Im}\chi^{(3)} = 1 \times 10^{-15}$ esu).

Previously, the values of β for CS_2 were reported to be 10^{-11} cm W^{-1} (800 nm) [134] and $(4.5 \pm 1.0) \times 10^{-13}$ cm W^{-1} (770 nm) [132], while measurements at the

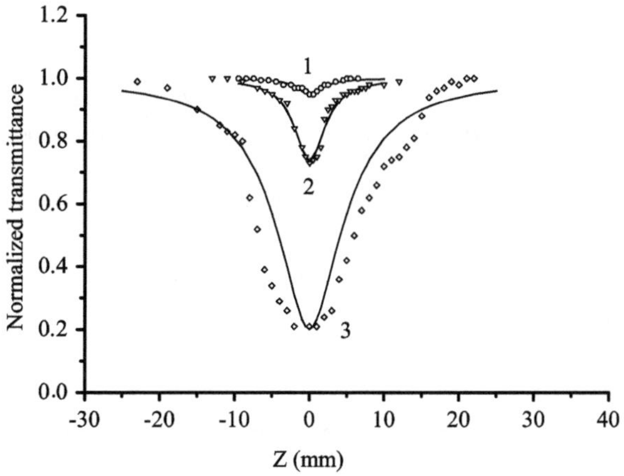

Fig. 3.15 Normalized transmittance using the open aperture scheme at the laser radiation intensities (1) 4×10^9, (2) 2.2×10^{10}, and (3) 3.8×10^{11} W cm^{-2} (L1, $t = 110$ fs, $\nu = 10$ Hz). *Solid lines* are the theoretical dependences. Adopted from [55] with permission from Elsevier

wavelength 532 nm yielded much higher values of this parameter (8.7×10^{-11} cm W^{-1} [148]). The two-photon absorption near 266 nm is comparable with that in the region of 397 nm since these two regions lie near a broad absorption band of CS_2 in the region 290–390 nm.

3.7.3 Optical Nonlinearities of Carbon Disulfide at 1064 nm

Figure 3.16 presents the dependences of the normalized transmittance in the open aperture scheme for the radiation of laser L2 with $\lambda = 1054$ nm, $E = 7 \times 10^{-6}$ J, $t = 475$ fs, and the pulse repetition rate 1 Hz that characterize the self-focusing in carbon disulfide at this wavelength. The effect of nonlinear absorption was clearly seen at high intensities (curves 2, 3). The parameters γ and β measured with laser L2 were equal to $(3.5 \pm 0.7) \times 10^{-15}$ cm^2 W^{-1} and 8×10^{-14} cm W^{-1} (the latter parameter was calculated at the intensity $I_0 = 3 \times 10^{10}$ W cm^{-2}). The self-focusing in CS_2 was also observed when the nonlinear refraction using uncompressed pulses of laser L2 ($t = 300$ ps) was studied. These measurements of γ were performed at the detection threshold of this process because of the low radiation intensity. In that case, the value of γ was measured to be $(3.2 \pm 1.6) \times 10^{-14}$ cm^2 W^{-1}.

The attempts were made, as in the case with laser L1, to detect the self-defocusing of laser radiation due to the heat accumulation using the radiation of the oscillator of laser L2 (GLX-200, $t = 280$ fs, $P = 100$ mW) operating at 100 MHz pulse repetition rate. However, the peak intensity of a single pulse in this case was too low (3×10^8 W cm^{-2}) for the nonlinear absorption to be efficient and did not allow detecting any nonlinearity caused by thermal processes.

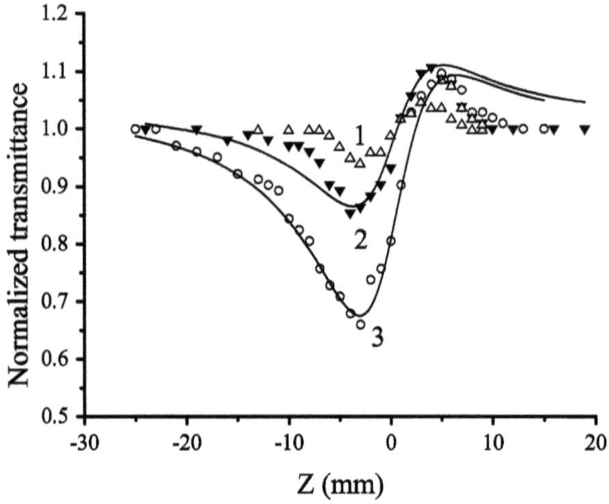

Fig. 3.16 Normalized transmittance using the closed aperture scheme at the laser radiation intensities (1) 3×10^9, (2) 1.4×10^{10}, and (3) 3×10^{10} W cm^{-2} (L2, $\lambda = 1054$ nm, $t = 475$ fs). Reproduced from [138] with permission from Springer Science+Business Media

The experiments using laser L2 with $t = 475$ fs showed no noticeable dependence of γ on the laser radiation intensity but revealed a strong intensity dependence of the nonlinear absorption, which points out higher-order nonlinear absorption processes. The independent measurements of the nonlinear absorption coefficient were also carried out in the scheme with an open aperture at the same wavelength ($\lambda = 1064$ nm) (Fig. 3.17). In these experiments, a nonlinear optical process different from the two-photon absorption was observed.

In the case of a linear dependence of the linear absorption coefficient α on the intensity I, the Eq. (3.11) was used. The theoretical dependence T(z), taking into account the two-photon absorption, noticeably differs from the experimental data (Fig. 3.17, the thin line for $I_0 = 4 \times 10^{11}$ W cm^{-2}). The experimental dependence looks narrower along the z axis than the theoretical dependence, which takes into account only the two-photon absorption (see Eq. (3.11)). This discrepancy may be caused by the effect of a higher-order nonlinear absorption, in particular, three-photon absorption, which seems to be realistic taking into account the correspondence of the energy of three infrared photons ($\lambda/3 = 351$ nm) and the absorption band of CS_2 in the ultraviolet region.

Investigations of the higher-order nonlinear absorption in various media have shown that these processes occur at intensities close to the optical breakdown and multiphoton ionization thresholds. The three-photon absorption coefficients in various crystals were measured in [34–37, 53]. The interest has also arisen in studies of liquid media. Data on the multiphoton absorption of some liquids are reported in [149–152]. Most of these studies were aimed at determining the possibility of using such media as optical limiters. Liquids show a wide spread in the three-photon absorption coefficients, from 10^{-20} to 10^{-24} cm^3 W^{-2}.

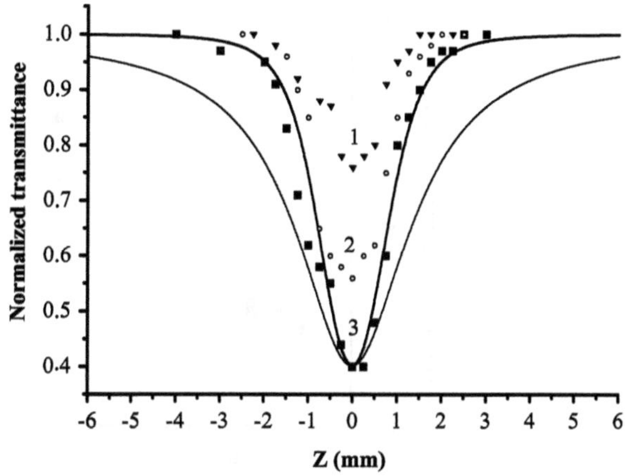

Fig. 3.17 Normalized transmittance using the open aperture scheme at the laser radiation intensities (1) 2.2×10^{11}, (2) 3.2×10^{11}, and (3) $4 \times 10^{11}\,\mathrm{W\,cm^{-2}}$ (L2, t = 475 fs). *Thin line* shows the theoretical dependence accounting for two-photon absorption; the *thick line* corresponds to the theoretical dependence for three-photon absorption at the intensity $4 \times 10^{11}\,\mathrm{W\,cm^{-2}}$. Reproduced from [55] with permission from Elsevier

The analysis of the CS_2 absorption spectrum revealed that, in the case of 1054 nm radiation, the probability of two-photon absorption is very low due to the absence of a pronounced absorption of CS_2 in the region of 527 nm. However, as was shown, under intense infrared radiation, this medium may exhibit three-photon absorption. In that case, the dependence of the coefficient α on the laser radiation intensity can be written in the form of Eq. (3.12). The thick line in Fig. 3.17 corresponds to the theoretical dependence that takes into account three-photon absorption. This theoretical dependence agrees well with the experimental data. The coefficient $\beta_{3\omega}$ for carbon disulfide was found to be $(2.8 \pm 0.8) \times 10^{-23}\,\mathrm{cm^3\,W^{-2}}$.

Figure 3.18 summarizes the measured data of the nonlinear refractive index of CS_2 at various durations of laser pulses. It includes no data on thermal nonlinearities because it was aimed comparing the measured γ related to the fast processes (the orientational and electronic Kerr effects and intermolecular interaction). These fast processes were previously considered in [144] as the main effects determining the optical nonlinearities of such molecular liquids. Figure 3.18 also presents data of the γ of carbon disulfide measured using the second harmonic of a Nd:YAG laser (λ = 532 nm, t = 20 ns). These measurements showed that the contribution of the orientational Kerr effect is insignificant in the range of femtosecond pulses because the response time of this process exceeds few picoseconds but this process makes the main contribution to the total value of the nonlinear refractive index in the case of multi-picosecond and nanosecond pulses. One can also note a weak spectral dependence of this parameter in the near-IR and visible spectral ranges.

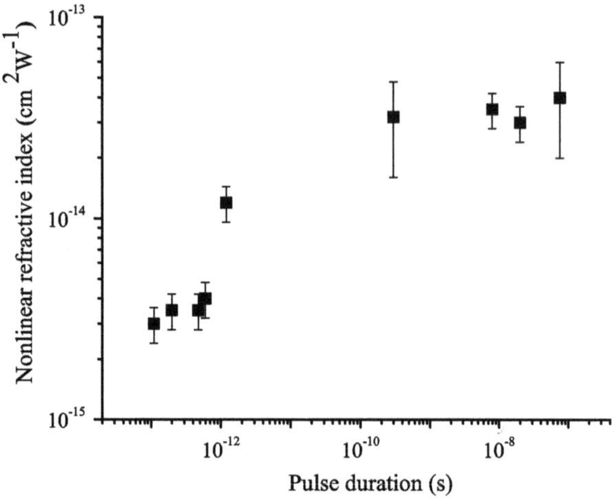

Fig. 3.18 Dependence of the nonlinear refractive index of CS_2 on the pulse duration. Reproduced from [138] with permission from Springer Science+Business Media

3.8 Measurements of Nonlinear Optical Parameters of Transparent and Nontransparent Materials Using Single-Shot Techniques

3.8.1 Single-Shot Reflection z-Scan for Measurements of the Nonlinear Refraction of Non-Transparent Materials

As it was already mentioned in Sect. 3.1, the z-scan technique for the measurements of the nonlinear optical parameters of media, which currently can be considered as most versatile one, was realized as a conventional transmission z-scan, off-axis z-scan, time-resolved z-scan, eclipsing z-scan, and reflection z-scan). Most of these modifications, excluding the latter one, are based on the propagation of the transparent sample through the focal plane of focused laser radiation and the analysis of the phase and amplitude characteristics of transmitted radiation in the far field.

The reflection z-scan (rz-scan) is used in the case of the opaque materials or the samples with limited transparency. The principles of this method are based on the analysis of the phase and amplitude characteristics of the beam reflecting from the non-transparent sample during its propagation through the focal plane of the focused radiation [25, 27, 153–155]. A drawback of this technique is a necessity of multiple measurements during the reflection of strong laser pulse from the surfaces of the materials propagating through the focal plane. The instability of the output intensity of lasers requires the averaged measurements of the reflectance of the samples under investigation, when, at the same position of sample, one has to carry out the multiple measurements. These averaged measurements have to be repeated at least few tens of times to define the normalized reflectance of the sample. This can lead, in some cases, to the damage of the sample, since one has to apply the strong laser fields to achieve the measurable nonlinear optical response from the surface. In that case one has to repeat all these measurements at the fresh position of the sample.

These drawbacks of rz-scan technique could be diminished by using some new approaches. The application of single-shot scans would allow overcoming the above difficulties. In this subsection, a new rz-scan method for defining the nonlinear optical refraction during single shot of laser radiation is discussed [156]. The method is based on the analysis of the spatial shape of laser beam after the reflection from the tilted sample, which is placed at the focal plane of the cylindrical lens. GaAs and Si plates were used to validate this approach. The proposed method is advantageous due to lower stress imposed on the fragile materials, since a single shot of laser pulse is sufficient for the definition of their nonlinear refractive indices. Another advantage is that it is no longer necessary to average a large number of pulses. It also needs no displacement of the medium under investigation, unlike the case of conventional rz-scan method.

The experimental scheme of this method is extremely simple. A broad (18 mm diameter FW1/e^{2}M) laser beam (Ti:sapphire laser, $\lambda = 790\,\text{nm}$, $t = 120\,\text{fs}$) was focused by a cylindrical lens ($f = 150\,\text{mm}$) on a sample tilted at the angle of 45° (Fig. 3.19). The focusing occurs in such a manner that the laser beam size does not change along the plane of drawing of this scheme. A sample was placed at such a position of z-axis when the central part of sample approximately coincided with the focal plane of the cylindrical lens, as shown in the picture. A CCD camera registered the shape of reflected radiation.

Two sets of acquisitions are need for these measurements. Firstly, a dense filter [or a 99.7 % reflecting mirror, which can leak a small (~0.3 %) amount of radiation] is placed in front of the cylindrical lens. The obtained image of the laser beam leaked through this strong filter (or mirror) is used just for confirmation of the approximately homogeneous radiation distribution along the z-axis, especially close to the focal plane. Secondly, a strong filter, or mirror was moved from the previous position

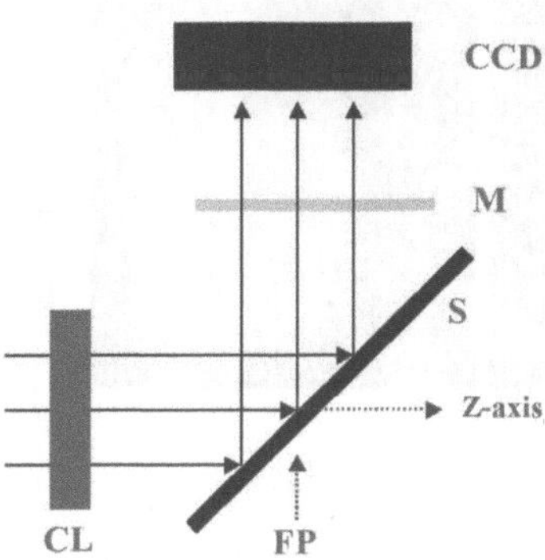

Fig. 3.19 Experimental scheme for the SSSR. *CL* cylindrical lens, *S* sample, *M* 99.7 % reflection mirror, *CCD* charge-coupled device, *FP* focal plane of the cylindrical lens. Reproduced from [156] with permission from Springer Science+Business Media

toward the position in front of the CCD camera (as shown in Fig. 3.19). In that case, the intensity of focused radiation becomes strong enough to induce the nonlinear optical processes influencing the shape of reflected beam along the z-axis.

This method can be dubbed as "Single-Shot Scan by Reflection from the tilted samples" (SSSR). The proposed scheme requires, after confirming the homogeneity of the beam shape reflecting from the sample at low-intensity conditions, only one shot to define the variations of the spatial distribution along the z axis, which further could be processed to define the nonlinear refractive index of the sample.

GaAs and Si plates were used for the definition of their γ at the wavelength of $\lambda = 790$ nm using the SSSR. This wavelength is below the band gaps of the GaAs (890 nm) and Si (1170 nm), so they were fully non-transparent at this spectral range and could not be analyzed using the conventional transmission z-scan technique. The 2 mm thick plate of intrinsic, polished crystalline Si cut with normal axis along the $\langle 001 \rangle$ crystal direction and the 0.5 mm thick plate of polished GaAs were used in these studies.

Figure 3.20 shows the beam shapes reflected from the GaAs plate in the cases of weak and strong irradiation of the focused pulse. At weak irradiation (i.e., when the 99.7 % reflecting mirror was placed in front of the cylindrical lens), the spatial distribution along the z-axis remains unchanged with regard to the input radiation (Fig. 3.20a). Another pattern appeared in the case of strong irradiation ($I_0 = 3.6 \times 10^{11}$ W cm^{-2}) on the surface of GaAs when the mirror was moved towards the CCD camera. At the central part of the beam (i.e., at maximum intensity), the nonlinear growth of reflectance led to the increase of reflected intensity. This growth was caused by the positive adding to the refractive index of GaAs and corresponding increase

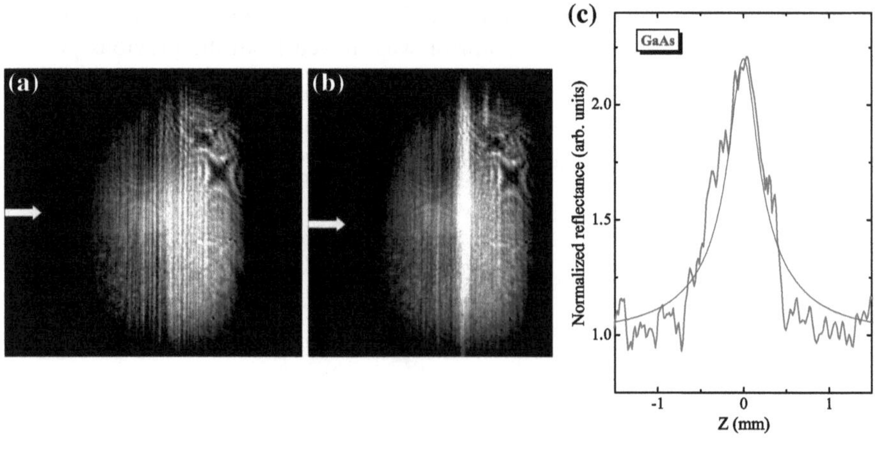

Fig. 3.20 Reflected beam shapes in the cases of **a** weak and **b** strong intensities of the femtosecond pulse focused on the surface of GaAs plate. **c** Normalized SSSR curve of the high-intensity reflected beam registered by a CCD camera. *Thin curve* in Fig. 3.20c is a fit of the second term of Eq. (3.24) with the experimental result (*thick curve*). Adapted from [156] with permission from Springer Science+Business Media

of the Fresnel reflection (Fig. 3.20b). The normalized reflectance $R(z)$ depicted from the shape of reflected beam is presented in Fig. 3.20c. One can see a considerable increase of normalized reflectance at the central part of this SSSR curve. When one returns the mirror to the previous position, a featureless pattern analogous to the Fig. 3.20a appeared again, without the indication of any damage of the surface of GaAs.

To calculate the nonlinear refraction of non-transparent samples, the following expression for the normalized reflectance along the z-axis was used for the fitting with the experimental data [157]:

$$R(z,\Theta) = 1 - \frac{4\frac{R_1(\Theta)}{R_0} I_0 k_2 x}{(x^2+9)(x^2+1)} + \frac{2\frac{R_1(\Theta)}{R_0} I_0 \gamma (x^2+3)}{(x^2+9)(x^2+1)} + \frac{\left(\frac{R_1(\Theta)}{R_0}\right)^2 I_0^2(\gamma^2+k_2^2)}{(x^2+9)(x^2+1)}. \tag{3.24}$$

Here I_0 is the incident radiation intensity at the focal plane; k_2 is the nonlinear extinction coefficient, R_0 is the coefficient of linear reflection; $x = z/z_0$, $z_0 = 0.5\,kw^2$ is the diffraction length of the beam, w is the radius of the beam waist, $k = 2\pi/\lambda$ is the wave number, $R_1(\Theta) = \frac{2n_0^3 \cos\Theta - 4n_0 \cos\Theta \sin^2\Theta}{n_0^4\cos^2\Theta - n_0^2 + \sin^2\Theta}\left[n_0^2 - \sin^2\Theta\right]^{-0.5}$ [158], n_0 is the linear refractive index, and Θ is the angle of radiation incidence. In Eq. (3.24), the first term is responsible for nonlinear absorption, the second term characterizes nonlinear refraction, and the third term characterizes their joint influence.

A characteristic feature of the rz-scan is the use of an open aperture for investigating the nonlinear refraction, while for the conventional transmitting procedure, an open aperture makes it possible to analyze the nonlinear absorption. It should be noted that the scheme without the aperture in the rz-scan allows one to exclude the effect of nonlinear absorption on the results of experiments. Taking into account an insignificant role of the nonlinear absorption with regard to the nonlinear refraction in the case of GaAs, which is seen from the experimental $R(z)$ dependence showing the absence of the valley, one can use the second term of Eq. (3.24) for the fitting with the experimental data. A thin curve in the Fig. 3.20c shows the fit with the experimentally observed normalized reflectance. The corresponding fitting value of the nonlinear refractive index was found to be $\gamma = (1.7 \pm 0.3) \times 10^{-11}\,\text{cm}^2\,\text{W}^{-1}$.

The data of the nonlinear refractive indices of GaAs in the mid-infrared range ($3 \times 10^{-13}\,\text{cm}^2\,\text{W}^{-1}$, $\lambda = 1.7\,\mu\text{m}$) measured using the conventional transmission z-scan are presented in [159]. However, close and below the edge of transmission band, one can expect the increase of the optical Kerr effect. Indeed, the value of γ increases drastically when one approaches toward the band gap energy. Theoretical predictions show that the nonlinear refraction in this region also changes from a positive value (for $E < 0.65E_g$, E and E_g are the photon energy and band gap energy, respectively) to a negative one. Close to the band gap energy of GaAs ($E_g = 1.42\,\text{eV}$), the nonlinear refractive index was reported to be $-2.1 \times 10^{-6}\,\text{cm}^2\text{W}^{-1}$ [160]. In the meantime, previous studies of the γ of GaAs in the spectral range below its band gap (538 nm [154], 532 nm [26]) measured using the rz-scans have shown the positive

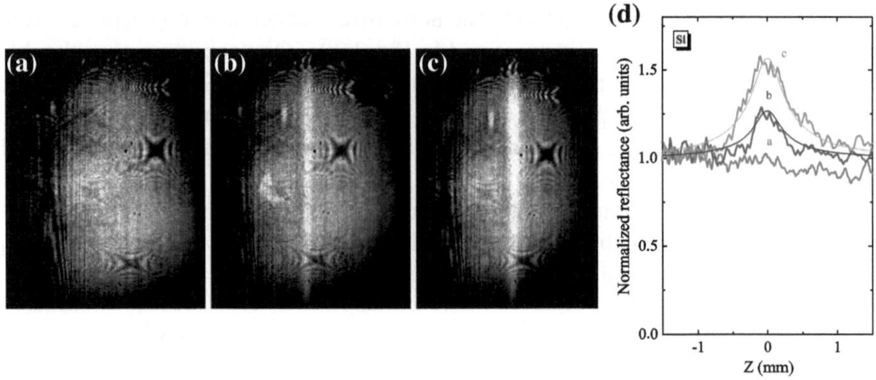

Fig. 3.21 Reflected beam shapes in the cases of **a** weak (9×10^8 W cm^{-2}); **b** moderate (1.2×10^{11} W cm^{-2}), and **c** higher (1.8×10^{11} W cm^{-2}) intensities of the laser beam irradiated the Si plate; **d** normalized SSSR curves of the reflected laser beam in above-presented three cases. *Thin curves* in Fig. 3.21d are the fits of the second term of Eq. (3.24) with the experimental results (*thick curves*). Reproduced from [156] with permission from Springer Science+Business Media

sign of nonlinear refraction. In particular, a value of $\gamma = 7.8 \times 10^{-8}$ cm^2 W^{-1} was reported when the dye laser radiation at 538 nm was used.

An accurate knowledge of the nonlinear optical parameters of silicon is essential for various applications of this material (cross-phase modulation in silicon waveguides, wavelength conversion, Raman lasing and amplification, etc). However, as in the case of GaAs, current knowledge of the nonlinear optical parameters of silicon is limited to the near-infrared region. A steady decrease of the nonlinear refractive index of Si from 3×10^{-14} cm^{2}W^{-1}($\lambda = 1.9$ µm) to 1×10^{-14} cm^{2}W^{-1}($\lambda = 1.2$ µm) in the mid-infrared region was reported by Lin et al. [161].

The SSSR measurements carried out using the silicon plate have shown a drastic increase of the γ for shorter wavelength (790 nm). Figure 3.21a–c presents the shapes of the beam reflected from the surface of Si at different intensities of the femtosecond pulse. The normalized reflectance of sample in these three cases is presented in Fig. 3.21d. Here also shown the fitting procedure using the Eq. (3.24). Thin curves represent the fits for (b) 1.2×10^{11} W cm^{-2} and (c) 1.8×10^{11} W cm^{-2} intensities of laser radiation at the surface of silicon plate. The fitting nonlinear refractive index of silicon was found to be $\gamma = (2.6 \pm 0.5) \times 10^{-11}$ cm^2 W^{-1}.

In rz-scan, the nonlinear refraction is analyzing by the measurement of the wavefront distortion of a laser beam due to photoinduced modification in the reflection coefficient. One can assume considerably smaller modifications of the wavefront compared with the conventional transmission z-scan, since in the latter case, one has an advantage of analyzing the accumulated nonlinear optical process using whole sample's length. In most cases, the modification of refractive index is below the measurement sensitivity, while the increase of laser intensity is undesirable due to optical damage of the sample through the accumulative effect of multiple shots.

One of potential applications of SSSR is the analysis of the nonlinear refraction of semiconductor saturable mirrors (SESAM) currently used for mode-locking as a standard technique for self-starting femtosecond lasers. The manufacture of SESAM is a complicated and expensive process requiring accurate irradiation of this structure to prevent the surface damage. Previously, application of the oblique incidence of a polarized beam allowed increasing the sensitivity of rz-scan technique at the Brewster angle [153]. The tilting at Brewster angle could enhance the sensitivity of the SSSR method, as well. At the same time, one has to examine the applicability of rz-scan relations and used approximations for the case of the tilted samples at these conditions.

3.8.2 Single-Shot y-Scan for Characterization of the Nonlinear Optical Parameters of Transparent Materials

So far, a few single-shot methods based on the analysis of laser beam shape in the far field after propagation of the material were offered. A single-shot technique using a phase object at the entry of a 4f (quarter-phase) coherent imaging system to characterize the value of nonlinear refractive index was reported in [162, 163]. Instead z-scanning the material they analyzed the transverse modification in the image acquired with a CCD camera. Another method is based on the use of two crossed cylindrical lenses to obtain an astigmatic beam [164]. The sample is placed between the focal planes where the beam has a round intensity profile. By evaluating the far-field beam profile it is possible to estimate the nonlinear phase shift. Third method uses the conventional open aperture z-scan and extracts the two-dimensional transverse modifications of the whole far-field pattern by processing all the pixels of the CCD camera [165]. The key element of the fourth single-shot method is a spatial intensity variation of an elliptical Gaussian beam in conjunction with an array detector [166]. In their method, one has to record the changes in the relative intensity of the pump or the probe beams. However, some of these methods did not allow defining the sign of nonlinearity, which is a drawback of above single-shot techniques.

In this subsection, a new approach for analysis of the nonlinear optical parameters of transparent materials using single shot is discussed [167]. The method is based on the analysis of the spatial shape of laser beam after propagation through the tilted sample, which is placed at the focal plane of cylindrical lens. The use of the tilted sample allows different parts of the sample to be simultaneously irradiated at different distances from the focusing lens, thus allowing the standard z-scan-like data to be acquired in a single shot based on the intensity variation in z-axis followed the phase variations along the sample's y-axis. The CS_2 was used to validate the approach in the case of the media demonstrating the nonlinear refraction and absorption.

The 14 mm diameter (FW1/e^{2}M) laser beam was focused by a cylindrical lens ($f = 150$ mm) on a sample tilted at the angle of 45°, as it shown in Fig. 3.22. The light

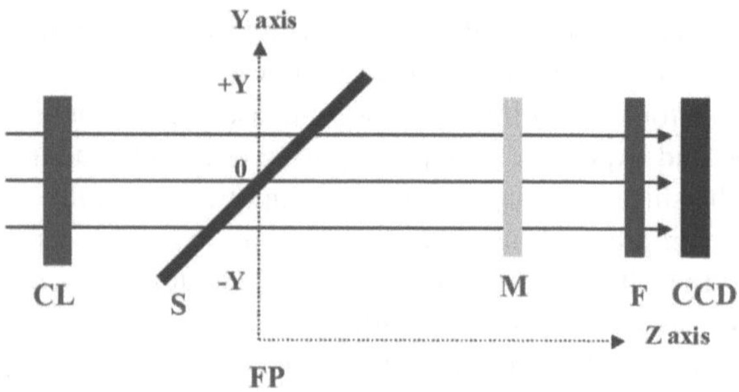

Fig. 3.22 Experimental setup for the single-shot measurements of the nonlinear optical characteristics of transparent materials. *CL* cylindrical lens, *S* sample, *FP* focal plane, *M* 99.7 % reflecting mirror (or dense filter), *F* filter, *CCD* charge-coupled device camera. Reproduced from [167] with permission from IOP Publishing

source was a Ti:sapphire laser ($\lambda = 790$ nm). For these experiments, the two ranges of pulse duration were used (120 fs and 210 ps of compressed and chirped radiation respectively). The beam waist dimensions at the focal plane of cylindrical lens were 14 mm × 25 μm. The laser was operated at a 10 Hz pulse repetition rate. The shutter was used for separation of single shot. The focusing of laser radiation was carried out in such a manner that the size of laser beam in the plane of Fig. 3.22 (i.e. along the y-axis) remained unchanged during propagation through this scheme. A sample was placed at a position of z-axis when the central part of sample coincided with the focal plane of cylindrical lens. A CCD camera registered the shape of propagated radiation in the far field (i.e. at the distance of 600 mm from the samples).

Initially, a dense filter (or 99.7 % reflecting mirror, which can leak a small (~0.3 %) amount of radiation) was placed in front of the cylindrical lens. The obtained image of the laser beam leaked through the dense filter (or mirror) was used to calibrate the spatial shape of radiation along the y-axis. Then a strong filter (or mirror) was moved from previous position toward the CCD camera (as shown in Fig. 3.22). In that case the intensity of focused radiation became strong enough to cause the appearance of the nonlinear optical processes influencing the beam phase modulation along the y-axis, which was further transformed in the spatial shape variation in the far field.

The method is related with the single-shot z-scan of the sample. However, since the changes of laser intensity along the z-axis cause the variation of beam shape along the y-axis, the method can be dubbed as a "Single-Shot Y-Scan of Tilted Sample" (SSYSTS). This method requires only one laser shot to define the changes in the spatial distribution of laser radiation along the y-axis in the far field, which further could be processed to define the nonlinear refractive index and nonlinear absorption coefficient of the sample.

The example of SSYSTS is presented in Fig. 3.23. A 1 mm thick silica glass cell was filled with the carbon disulfide, which is a commonly used reference medium for the z-scans and pump-probe measurements of the nonlinear optical parameters

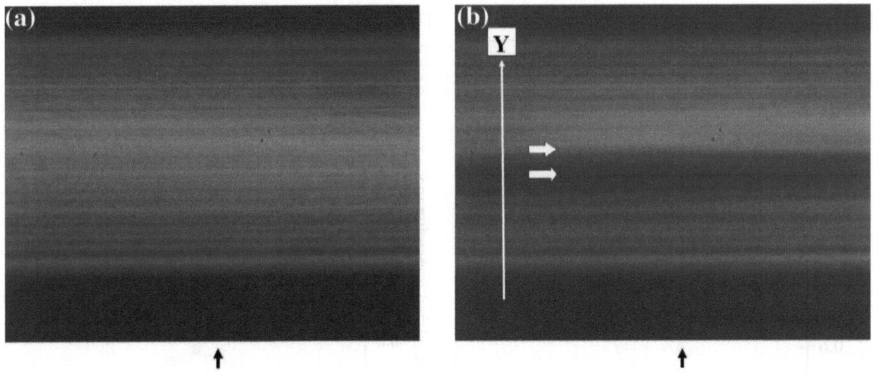

Fig. 3.23 Beam shape images obtained in the cases of **a** weak and **b** strong irradiation using the 120 fs pulse propagating through the tilted CS_2-containing cell. *Black vertical arrows* show the direction at which the analysis of spatial distribution was carried out. These data were further used for preparation of the normalized SSYSTS plots. Reproduced from [167] with permission from IOP Publishing

of media. Its nonlinear refractive index was precisely defined during multiple measurements in a broad spectral range and showed no considerable variation in the femtosecond timescale ($\gamma \sim 3.5 \times 10^{-15}\,\mathrm{cm^2\,W^{-1}}$), while the ten-fold increase of this parameter was reported for the pulses longer than 1.5 ps [33], which was caused by the influence of laser-induced molecular reorientation (see also previous section). The high transparency down to $\lambda = 380\,\mathrm{nm}$ makes CS_2 an excellent choice for the demonstration of the advantages of the proposed method in different spectral ranges.

Figure 3.23a shows the raw image of the beam shape after the propagation of a weak probe radiation ($t = 120\,\mathrm{fs}$, $\lambda = 790\,\mathrm{nm}$) through the cell, when the 99.7 % reflecting mirror was placed in front of the cylindrical lens. This image demonstrates a featureless Gaussian-like intensity distribution along the y-axis. The change in the position of mirror toward the CCD camera led to considerably stronger irradiation of the sample, which in turn caused the phase variations of different parts of the beam propagating through the sample. Most significant changes occurred close to the focal plane, which were further transformed in the intensity variations in the far field (Fig. 3.23b). One can see a considerable change in the spatial distribution along the y-axis, when the bottom part of the beam (close to its central part) was weakened (horizontal bottom arrow), while the upper part of the beam showed some enhancement of its intensity (horizontal upper arrow) compared to the weak irradiation.

The initial processing of these data consisted on the definition of the spatial distribution of the propagated beam along the y-axis for these two cases. Then the spatial distribution curve defined from Fig. 3.23b was normalized by dividing to the curve obtained from the Fig. 3.23a to exclude the influence of the heterogeneities of laser beam shape caused by occasional diffraction patterns originated from the damaged parts of optical elements, as well as the heterogeneities of beam shape itself. This processing resulted in the dependence of the normalized intensity on the

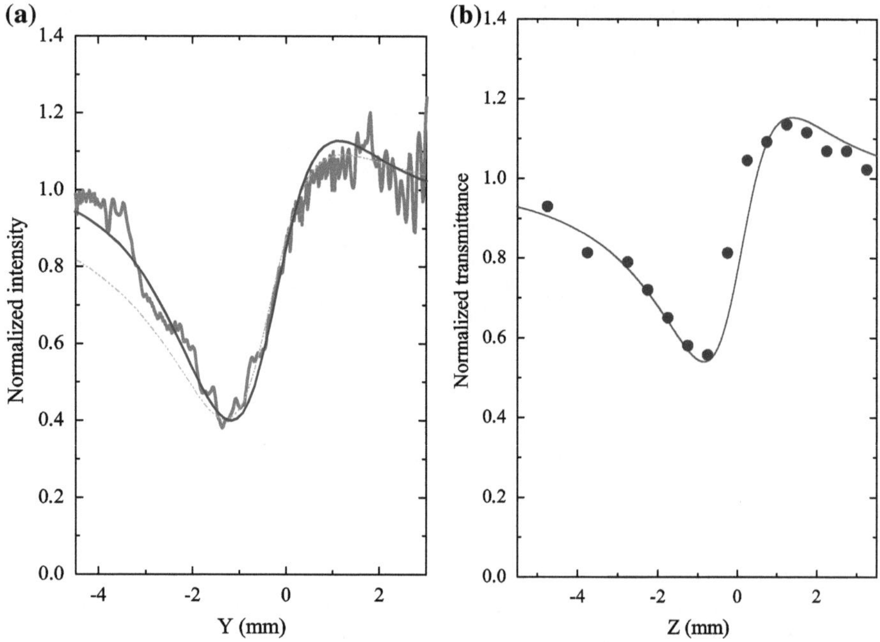

Fig. 3.24 **a** SSYSTS and **b** Standard closed aperture z-scan of the 1 mm thick cell filled with the CS_2 in the case of irradiation by the 120 fs pulses. *Thin dotted line* in **a** and *thin solid line* in **b** show the theoretical fits for the SSYSTS and z-scan. *Thin solid line* in **a** is the fitting curve defined from the [167] in the case of cylindrical focusing. Reproduced from [167] with permission from IOP Publishing

position of laser beam along the y-axis (I(y), Fig. 3.24a, thick line). One can see that this curve resembles the normalized transmittance dependence T(z) of the standard closed aperture z-scan of the medium possessing the positive nonlinear refraction and strong nonlinear absorption (when the valley considerably exceeds the peak). The standard closed aperture z-scan measurements were performed using the spherical lens and observed analogous pattern in the case of 120 fs pulses at approximately identical irradiation conditions (i.e., using the laser intensity at the focal plane in the range of $I_0 = 1 \times 10^{10}$ W cm^{-2}, Fig. 3.24b), when a deep valley followed by a relatively small peak. Different ratios between the valley and peak in these two cases (Fig. 3.24a, b) can be attributed to some difference in the peak intensities at the focal planes of cylindrical and spherical lenses, as well as to different focusing conditions.

The standard z-scan was obtained by measuring the normalized transmittance at 15 positions of the CS_2-containing cell moving along the z-axis. Each point in the Fig. 3.24b is the averaged measurement of normalized transmittance obtained during twenty laser shots. Here one has to take into account the shot-to-shot instability of laser intensity. The conventional technique of the fitting of these results allowed for defining the nonlinear optical parameters of CS_2 ($\gamma = 4 \times 10^{-15}\,\mathrm{cm^2\,W^{-1}}$, $\beta = 3 \times 10^{-13}\,\mathrm{cm\,W^{-1}}$), which were close to previously reported data measured at analogous pulse duration and intensity ($t = 110\,\mathrm{fs}$, $I_0 = 3 \times 10^{10}\,\mathrm{W\,cm^{-2}}$, $\gamma = 3 \times 10^{-15}\,\mathrm{cm^2\,W^{-1}}$, $\beta = 4.2 \times 10^{-13}\,\mathrm{cm\,W^{-1}}$ [33, 55]). One can see a necessity in multiple experimental measurements and cumbersome actions during definition of the sample's nonlinearities in the case of z-scan.

Analogous fitting procedure was applied in the case of SSYSTS and showed lack of coincidence between the experimental data obtained at cylindrical focusing conditions and the theory based on the spherical focusing conditions (thin dotted curve in Fig. 3.24a). The application of the relations [24, 29] commonly used for fitting of the open and closed aperture z-scans seems inappropriate for analysis of the SSYSTS. At the same time the analysis quantifying the beam propagation characteristics of the cylindrical beam interacting with the nonlinear medium compared with the symmetrical beam used in a z-scan will be useful for the definition of the applicability of the proposed technique at different focusing configurations. The fitting using the theory of SSYSTS developed in [167] is presented in Fig. 3.24 as a solid line.

To diminish the role of nonlinear absorption in these measurements one can decrease the pulse energy at the focal plane. Another way is the application of longer pulses, when one can obtain relatively symmetric pattern of both the SSYSTS and z-scan in the case of CS_2. As it was mentioned, for the carbon disulfide, an increase of pulse duration above a few picoseconds leads to appearance of additional nonlinear refractive mechanism (i.e. molecular reorientation), which causes a considerable increase of nonlinear refraction [33]. To examine this approach, the 210 ps pulses of the uncompressed radiation of Ti:sapphire laser ($\lambda = 790\,\mathrm{nm}$) were used for both the SSYSTS and z-scan of CS_2. Figure 3.25 shows these scans, which were obtained at considerably smaller intensity compared to the previous case. One can see that the curves in Fig. 3.25 are relatively symmetric with regard to initial normalized transmittance. The fitted value of z-scan at $t = 210\,\mathrm{ps}$ was $\gamma = 4 \times 10^{-14}\,\mathrm{cm^2\,W^{-1}}$, which is close to the data reported for this range of probe pulse durations. In the case of SSYSTS, the calculations [167] gave the value of nonlinear refractive index of carbon disulfide approximately equal to the γ of CS_2 defined from the z-scan measurements ($\gamma \approx 3.5 \times 10^{-14}\,\mathrm{cm^2\,W^{-1}}$). At these conditions, the SSYSTS curve of carbon disulfide is more suitable for comparison with other samples.

The results of fitting using the relations [167] are presented in Fig. 3.25b by the thin solid line. Here the fitting using relations [24] is also shown taking into account the spherical focusing of the probe radiation (thin dotted line). One can see the discrepancy between the fitting curve and experimental data in the latter case (analogous to Fig. 3.24a) and the reasonable coincidence in the former case.

In described studies, the SSYSTS appear noisy due to some defects on the sample, mirrors, and filters. One can remind that this is a single-shot method. The averaging of the curves of single-shot scan gives relatively smooth pattern. One can also compare this result with the averaged results obtained during standard z-scan procedure. About few hundred shots were used for the results presented in Fig. 3.25a.

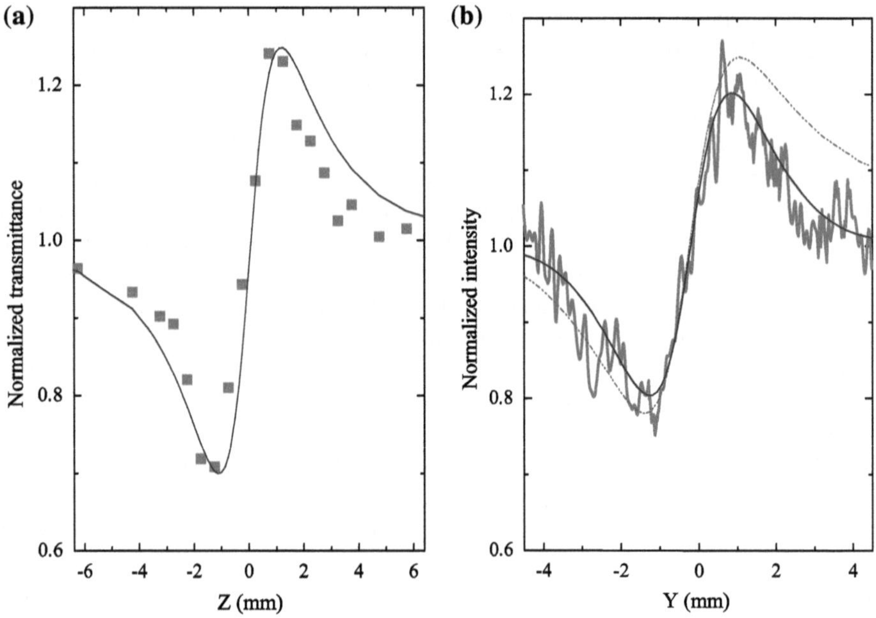

Fig. 3.25 **a** Standard closed aperture z-scan and **b** SSYSTS of the 1 mm thick cell filled with the CS_2 in the case of irradiation by the 210 ps pulses. *Thin solid line* in **a** shows the theoretical fit using the relations of conventional z-scan. *Thin dotted and solid lines* in **b** shows the theoretical fits using the relations of conventional z-scan and relations [167]. Reproduced from [167] with permission from IOP Publishing

3.9 Optical Limiting in Various Media

The study of the optical limiting (OL) of laser radiation in various materials opens the possibility of using these materials as laser shutters for protection against intense laser radiation and is important in investigating the fundamental properties of nonlinear optical media. The OL mechanisms have a different origin. The reverse saturated absorption (RSA) occurring due to large cross sections of the absorption from excited levels is responsible for the OL in solutions of some metal cluster compounds [168, 169], fullerenes and phthalocyanines [17, 99, 170]. Strong nonlinear refraction causes the OL in some inorganic clusters [171]. Nonlinear scattering dominates in limiting processes observed in carbon suspensions [172, 173]. Two- and three-photon absorption is responsible for the OL in semiconductor structures and crystals [24, 174].

Interest in optical limiting in fullerenes was aroused when this effect was demonstrated for the first time in [17] for the C_{60} solution in toluene at a wavelength of $\lambda = 532\,\text{nm}$. The reverse saturated absorption was considered as a mechanism of the OL. Subsequent investigations of the OL in the C_{60} and C_{70} toluene solutions [6, 67, 175–178] showed that the thermal self-focusing observed for nanosecond pulses at

$\lambda = 532$ nm is caused by two-photon absorption via a triplet state. RSA in fullerene solutions was observed up to the near IR spectral range (710–740 nm), where the photon energy is lower than the energy of the lower excited singlet state [79]. In this case, the nonlinear absorption was associated with the excitation of molecules to the lowest triplet state via direct absorption from the ground state. The nonlinear optical characteristics of fullerene-doped polyimide films were studied in [65, 179] in an attempt to optimize the OL.

A series of colloidal metal complexes was studied in [180] using pico- and nanosecond laser pulsed radiation at $\lambda = 532$ nm. The limiting characteristics of such metal clusters were shown to exceed their corresponding values for C_{60} toluene solutions. The dimensional effects of OL were observed in [181] when studying the nonlinear-optical response of gold clusters in aqueous solutions. Gold and silver colloidal aggregates were studied in [98, 182, 183], and the possibility of their usage as optical limiters was shown. The nature of OL in colloidal suspensions of spherical particles was analyzed in [184]. These colloidal structures are characterized by a fast (<360 fs) nonlinear optical response and high nonlinearity ($\chi^{(3)} = 1.5 \times 10^{-7}$ esu for a silver cluster [185, 186]) for femtosecond pulses. Investigations of OL in colloidal metals are limited despite their fairly strong self-action effect and nonlinear absorption during interaction with the nano- and picosecond pulses.

The OL in some semiconductor structures (ZnSe, GaAs, CuCl, and CdTe), caused by two-photon absorption and generation of free charge carriers, was studied in [187–189]. OL processes were also observed in semiconductor-doped optical glasses [190–192]. Nonlinear optical characteristics of PbS nanoparticles were studied in [193] using picosecond radiation of a Nd:YAG laser and its second harmonic, and a saturated absorption was observed at 532 nm and two-photon absorption was observed at 1064 nm. The study [194, 195] of thin films of amorphous chalcogenides (As_2S_3) and doped films of $BaTiO_3$ showed these materials to be promising as optical limiters.

In this section, we discuss the studies of the OL in fullerene solutions, colloidal metals, and semiconductor structures irradiated by the fundamental and second harmonic of pico- and nanosecond pulsed radiation of a Nd:YAG laser [196]. The experimental scheme was described in detail in previous sections. A conventional z-scan technique with an open aperture was applied for these studies. A train of picosecond pulses was generated by a Nd:YAG laser. A single pulse ($t = 35$ ps) separated from the train was amplified to an energy of 2 mJ. The laser radiation was focused by a lens with a focal length of 25 cm. The sample under study was placed in the focal point of the lens. The beam waist diameters were 150 and 100 μm for the fundamental radiation and its second harmonic, respectively. The energy of laser radiation was measured using a calibrated photodiode and a digital voltmeter. In order to vary this energy, the calibrated neutral filters were used. As the source of nanosecond pulses, a Q-switched Nd:YAG laser ($\lambda = 1064$ and 532 nm, $E = 18$ mJ, $t = 28$ ns) was used. The maximum radiation intensity in the beam waist ($\lambda = 1064$ nm) was 8×10^9 W cm^{-2} for the nanosecond pulses and 1.2×10^{11} W cm^{-2} for the picosec-

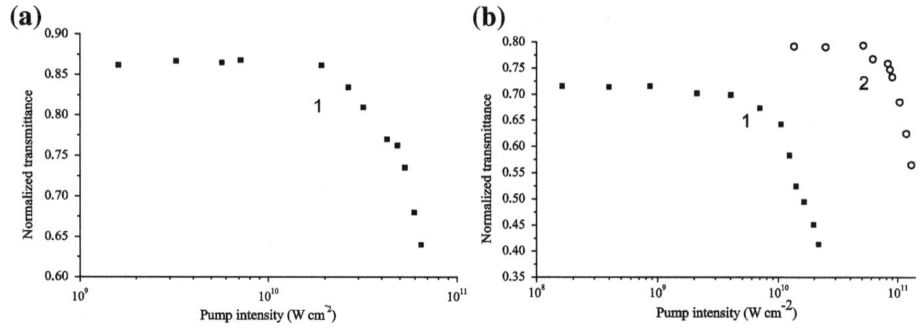

Fig. 3.26 Normalized transmission of (1) C_{60} and (2) C_{70} fullerene solutions in toluene versus intensity of 35 ps laser radiation for the probe wavelengths of **a** 1064 and **b** 532 nm. Reproduced from [196] with permission from Springer Science+Business Media

ond ones. The maximum radiation intensity was chosen to be below the optical breakdown and multiphoton ionization of the materials under study.

3.9.1 Optical Limiting in Fullerenes

The OL was studied in the toluene solutions of C_{60} (0.5 and 0.05 wt%) and C_{70} (0.1 and 1 wt%) placed in 2 mm thick cell. The optical limiting in C_{60} and C_{70} fullerenes has commonly been studied by using nanosecond laser pulses at $\lambda = 532$ nm. We discuss below the experimental results obtained with picosecond pulses at the wavelengths $\lambda = 1064$ and 532 nm. Curve 1 in Fig. 3.26a shows the dependence of a normalized transmission on the pump radiation intensity ($\lambda = 1064$ nm) that is typical for the optical limiting. It should be noted that the OL in that case was well pronounced only for the C_{60} solution, while the C_{70} solution exhibiting no marked change in transmission. Curves 1 and 2 in Fig. 3.26b correspond, respectively, to the transmission of C_{60} and C_{70} solutions at a wavelength of 532 nm. In this case, the OL effect was observed for all the samples studied.

The nonlinear absorption coefficient was determined from the Eq. (3.10). The nonlinear absorption coefficient at $\lambda = 1064$ nm for the 0.5 % C_{60} toluene solution was equal to 1.5×10^{-10} cm W^{-1}. The absence of nonlinear absorption in the C_{70} solutions at $\lambda = 1064$ nm can be explained by the fact that the resonance lines of C_{70} lie in the ultraviolet region. In contrast, the resonance line of C_{60} lies near 565 nm, which suggests the possibility of two-photon absorption. The values of β at $\lambda = 532$ nm were found to be 2.2×10^{-10} cm W^{-1} for C_{60} and 4.5×10^{-11} cm W^{-1} for C_{70}.

The nonlinear absorption of the studied fullerenes can be related to both the RSA due to the aggregate excitation to higher energy levels with larger absorption cross sections than that of the ground state (for $\lambda = 532$ nm) and the two-photon absorption of 1064 nm laser radiation with the subsequent RSA (for C_{60}).

3.9.2 Optical Limiting in Colloidal Solutions

The preparation procedure for colloidal metal solutions is described in [197]. Previously, the OL in colloidal metal structures was studied by using the second harmonic of Nd:YAG laser ($\lambda = 532$ nm). The results of the studies of the OL in colloidal metals both for pico- and nanosecond pulses at a wavelength of 1064 nm are presented below. The optical limiting of picosecond pulses is studied in aggregated solutions of gold and silver.

Figure 3.27 shows the normalized transmission as a function of the laser radiation intensity for aggregated silver and gold in the case of 35 ps pulses. It is noteworthy that the OL was not observed at the first stages of the solution aggregation. The study of nonlinear absorption using the z-scan with an open aperture shows that the two-photon absorption occurs only for aggregated solutions. Indeed, the peak in surface plasmon resonance for the aggregated solutions was observed near 565 nm for gold and at 580 nm for silver, while the absorption peaks for the nonaggregated solutions lie at 525 and 415 nm, respectively.

In the case of nanosecond pulses, the OL was observed both for aggregated and nonaggregated solutions. Figure 3.28 shows typical dependences for platinum and copper. The optical limiting in that case results from the RSA. It should be noted that the OL in a field of nanosecond pulses previously observed in a Pt:ethynyl complex dissolved in dichloroethane [19] was also considered to result from the RSA. Figure 3.29 demonstrates the OL for an aggregated solution of colloidal silver. Comparing the curves for colloidal silver obtained using the nanosecond and picosecond pulses one can see that the OL is somewhat steeper in the case of the picosecond pulses, which is caused by the combined effect of the Kerr self-defocusing and the two-photon absorption. In the case of nanosecond pulses, the two-photon absorption

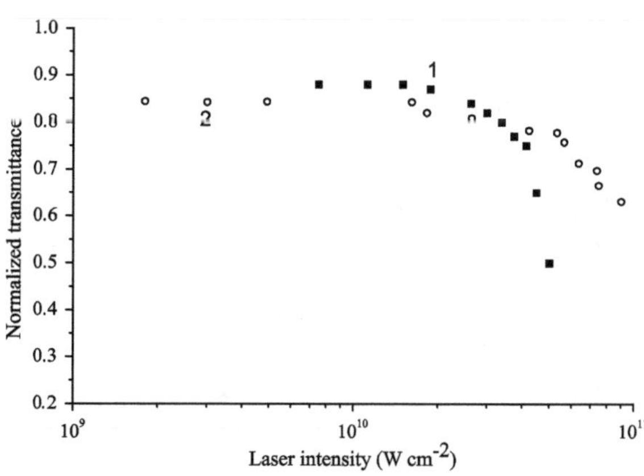

Fig. 3.27 Normalized transmission of colloidal (1) silver and (2) gold versus intensity of pulsed ($t = 35$ ps) laser radiation at a wavelength of 1064 nm. Reproduced from [196] with permission from Springer Science+Business Media

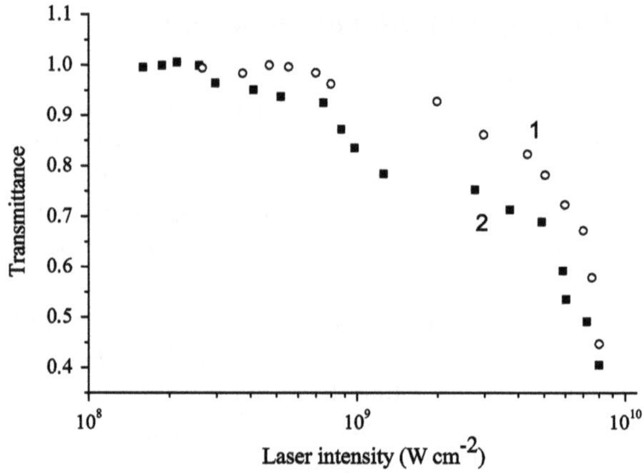

Fig. 3.28 Normalized transmission of colloidal (1) platinum and (2) copper versus intensity of nanosecond ($t = 28$ ns) laser radiation at a wavelength of 1064 nm. Reproduced from [196] with permission from Springer Science+Business Media

is insignificant and the main role belongs to the RSA and thermal self-defocusing. It is also noteworthy that the nonlinear response, which is inertialess in the case of picosecond pulses, provides better limiting in the suggested scheme. The nonlinear absorption coefficient for colloidal gold was found to be 9.8×10^{-13} cm W^{-1} at $\lambda = 1064$ nm and 9.4×10^{-12} cm W^{-1} at $\lambda = 532$ nm for the nanosecond pulsed radiation.

In [198], the OL was studied in PVP metal complexes at 1064 and 532 nm. Despite the absence of nonlinear absorption in the IR region, the PVP metal complexes exhibit distinct defocusing properties. This circumstance suggested the possibility of using

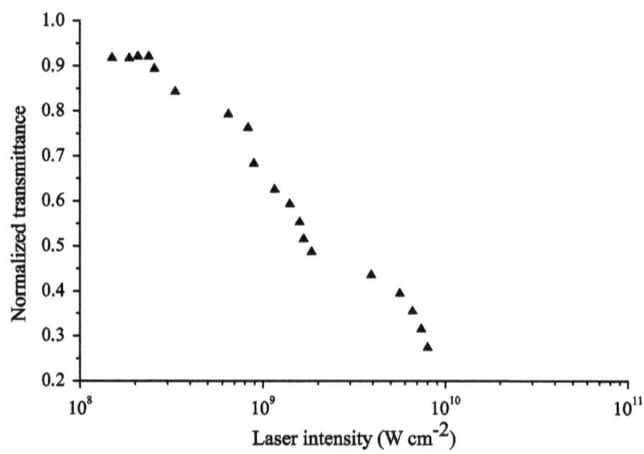

Fig. 3.29 Normalized transmission of colloidal silver versus intensity of laser radiation of nanosecond duration at $\lambda = 1064$ nm. Reproduced from [196] with permission from Springer Science+Business Media

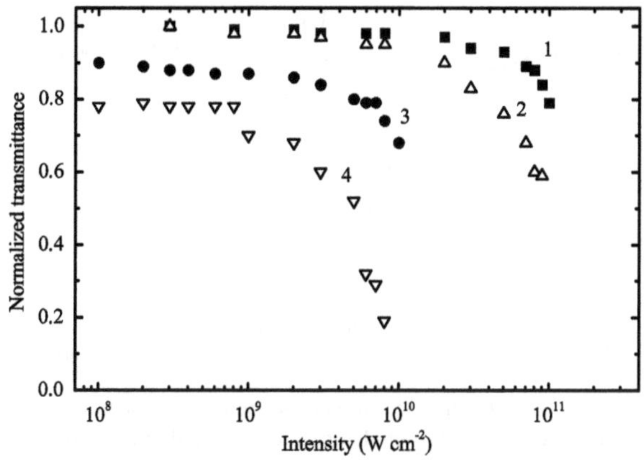

Fig. 3.30 Normalized transmissions as the functions of the radiation intensity at the wavelengths 1064 (1,2) and 532 nm (3,4) for the cobalt-doped PVP at Co concentrations of 2 % (1,3) and 13.5 % (2,4). Adopted from [97] with permission from Springer Science+Business Media

a self-action (in this case, self-defocusing) to obtain the OL in the IR region. Curves 1 and 2 on Fig. 3.30 present the dependences of the normalized transmission of PVP solutions with different cobalt concentrations on the incident radiation intensity at 1064 nm obtained by the closed aperture z-scan. In this case, the cells with solutions were located in the region of minimal transmission, i.e. in the region of the dip in the dependence $T(z)$. A different case was observed for $\lambda = 532$ nm (Fig. 3.30, curves 3 and 4) [97]. In that case, the OL was caused by both self-defocusing (playing a decisive role for IR radiation) and the nonlinear absorption (to a greater extent). The nonlinear absorption could be caused by two-photon absorption and reverse saturated absorption (at high radiation intensities). However, the contributions of these processes cannot be estimated due to the absence of data on the spectral parameters and lifetimes of the excited singlet and triplet states of metal complexes studied in those experiments.

3.9.3 *Optical Limiting in Semiconductors*

Two-photon absorption in semiconductors has been a subject of intense study since the very beginning of the development of nonlinear optics. This process is a powerful tool for the spectroscopic study of various macroscopic characteristics of semiconductors [199]. Below, we discuss the results of the studies of two-photon absorption in some semiconductor structures carried out with the aim to determine the possibility of their usage as optical limiters [196]. We analyze the studies of As_2S_3, $As_{20}S_{80}$, $2As_2S_3/As_2Se_3$, and $3As_2S_3/As_2Se_3$ chalcogenide films. All the film samples were 10 μm thick. The radiation wavelength in these experiments satisfied the condition $E_g/2 < h\omega < E_g$, where E_g is the band gap and $h\omega$ is the radiation quantum energy.

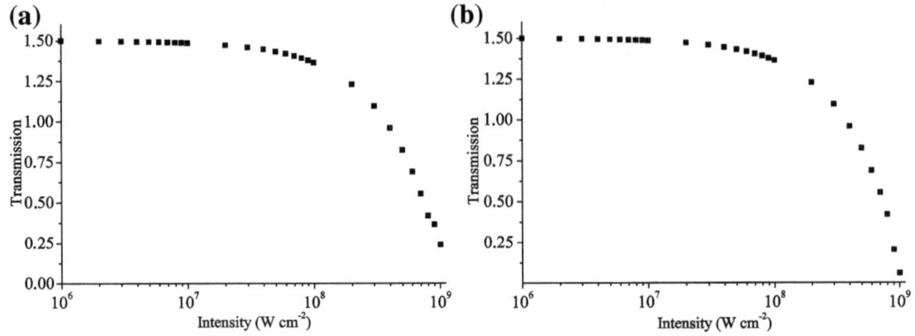

Fig. 3.31 Calculated normalized transmissions versus intensity of laser radiation ($\lambda = 1064$ nm) for **a** $3As_2S_3/As_2Se_3$ and **b** $2As_2S_3/As_2Se_3$ chalcogenide films. Adopted from [196] with permission from Springer Science+Business Media

The measurements of nonlinear optical parameters using the closed aperture scheme show that $\chi^{(3)}$ of some chalcogenide films is fairly high. The highest $\chi^{(3)}$ (-5.78×10^{-9} esu at $\lambda = 1064$ nm) was observed for the $2As_2S_3/As_2Se_3$ film. In connection with this, a theoretical analysis of the OL in chalcogenide films was performed. The calculated normalized transmission dependences for the films $3As_2S_3/As_2Se_3$ (curve a) and $2As_2S_3/As_2Se_3$ (curve b) are shown in Fig. 3.31. The most efficient (12.5-fold) limiting is observed for the $2As_2S_3/As_2Se_3$ film. Among other interesting films are the As_2S_3 and $As_{20}S_{80}$, which attenuated the laser radiation by a factor of 5, and the $3As_2S_3/As_2Se_3$ film, whose attenuation factor was equal to 3. The optical limiting for all the films was observed in the intensity range from 10^7 to 10^9 W cm^{-2}.

In OL experiments with semiconductors, the samples were placed at the point of minimal transmission, i.e., at the focal point of open aperture z-scheme. GaAs thin crystal was also analyzed for comparison with chalcogenide films. Similar dependences were studied for the other films. The most pronounced (25-fold) limiting was observed for the As_2S_3 film; the $As_{20}S_{80}$, $2As_2S_3/As_2Se_3$, and $3As_2S_3/As_2Se_3$ films attenuated the radiation by a factor of 4, 6.6, and 7.5, respectively. A 2.3-fold limiting was observed for the GaAs sample.

As_2S_3 ($Eg = 2.33$ eV) is apparently the most extreme semiconductor structure for which the condition $E_g/2 < h\omega < E_g$ for the Nd:YAG laser radiation is met and, hence, two-photon absorption is still allowed. In this case, the two-photon absorption is responsible for the optical limiting at the wavelength of laser radiation satisfying the above relation, which was also confirmed by the study of the nonlinear absorption using the scheme with an open aperture. The nonlinear absorption coefficients, in particular, for GaAs and As_2S_3, were found to be 31×10^{-9} and 8×10^{-6} cm W^{-1}, respectively.

Efficient OL materials should have a high transmission for low fluxes (or intensities) of radiation and a limited transmission for high fluxes (or intensities). Since the optical limiters are based mainly on two effects, two-photon absorption and RSA, real conditions of their application become the determining factor. The two-photon absorption is a function of the radiation intensity, while the RSA depends on the energy density (flux) of radiation. Thus, if one needs to limit the picosecond radiation, the two-photon absorption is preferable and, conversely, materials exhibiting the RSA will be more efficient if one needs to limit longer (from nanosecond to microsecond) pulses.

References

1. V. Singh, P. Aghamkar, Appl. Opt. **51**, 2288 (2012)
2. T. Godin, R. Moncorge, J.L. Doualan, M. Fromager, K. Ait-Ameur, R.A. Cruz, T. Catunda, J. Opt. Soc. Am. B **29**, 1055 (2012)
3. G.H. Yang, J.H. Han, Z.Y. Huo, Y.Z. Nie, Y.Z. Gu, Phys. Stat. Sol. A **209**, 966 (2012)
4. B.Q. Wang, S.H. Chen, Z.L. Huang, M. Ming, Appl. Surf. Sci. **258**, 5319 (2012)
5. V. Krishnakumar, G. Shanmugam, R. Nagalakshmi, J. Phys. D **45**, 165102 (2012)
6. X. Lu, Q.C. Liu, Z.Y. Liu, S.H. Sun, P.J. Ding, B.W. Ding, B.T. Hu, Appl. Opt. **51**, 2045 (2012)
7. D.G. Kong, Q. Chang, Y.C. Gao, H.A. Ye, L.X. Zhang, G. Shi, X.R. Zhang, Y.X. Wang, K. Yang, Y.L. Song, Phys. B **407**, 1279 (2012)
8. S. Suresh, A. Ramanand, D. Jayaraman, P. Mani, Rev. Adv. Mater. Sci. **30**, 175 (2012)
9. S. Krishnan, C.S.S. Sandeep, R. Philip, N. Kalarikkal, Chem. Phys. Lett. **529**, 59 (2012)
10. M. Fu, K. Wang, H. Long, G. Yang, P.X. Lu, F. Hetsch, A.S. Susha, A.L. Rogach, Appl. Phys. Lett. **100**, 063117 (2012)
11. T.R. Tsai, C.H. Yu, Appl. Phys. Lett. **99**, 241101 (2011)
12. E. Fazio, F. Barreca, S. Spadaro, G. Curro, F. Neri, Mater. Chem. Phys. **130**, 418 (2011)
13. T.S.C. Girisun, S. Dhanuskodi, G. Vinitha, Mater. Chem. Phys. **129**, 9 (2011)
14. T.R. Kumar, R.J. Vijay, R. Jeyasekaran, S. Selvakumar, M.A. Arockiaraj, P. Sagayaraj, Opt. Mater. **33**, 1654 (2011)
15. U. Kurum, R.M. Oksuzoglu, M. Yuksek, H.G. Yaglioglu, H. Cinar, A. Elmali, Appl. Phys. A **104**, 1025 (2011)
16. S. Liu, J.S. Wei, F.X. Gan, J. Appl. Phys. **110**, 033503 (2011)
17. L.W. Tutt, A. Kost, Nature **356**, 225 (1992)
18. W.L.J. Hasi, Z.W. Lu, M.L. Fu, H.H. Lu, S. Gong, D.Y. Lin, W.M. He, Appl. Phys. B **95**, 711 (2009)
19. J. Staromlynska, T.J. McKay, P. Wilson, J. Appl. Phys. **88**, 1726 (2000)
20. R. Adair, L.L. Chase, S.A. Payne, J. Opt. Soc. Am. B **4**, 875 (1987)
21. Y. Xiong, Q. Zhang, X. Sun, W. Tan, X. Xin, W. Ji, Appl. Phys. A **70**, 85 (2000)
22. R.A. Ganeev, J. Opt. A **6**, S3 (2004)
23. S.R. Fridberg, P.W. Smith, IEEE J. Quantum. Electron. **23**, 2089 (1987)
24. M. Sheik-Bahae, A.A. Said, T.H. Wei, D.J. Hagan, E.W. Van Stryland, IEEE J. Quantum Electron. **26**, 760 (1990)
25. D.V. Petrov, A.S.L. Gomes, C.B. de Araujo, Appl. Phys. Lett. **65**, 1067 (1994)
26. R.A. Ganeev, A.I. Ryasnyansky, Phys. Stat. Sol. A **202**, 120 (2005)
27. M. Sheik-Bahae, M. Ebrahimzadeh, Opt. Commun. **142**, 294 (1997)
28. R.A. Ganeev, A.I. Ryasnyansky, A.L. Stepanov, C. Marques, R.C. Da Silva, E. Alves, Opt. Commun. **253**, 205 (2005)
29. X. Liu, S. Guo, H. Wang, L. Hou, Opt. Commun. **197**, 431 (2001)
30. J.F. Reintjes, *Nonlinear Optical Parametrical Processes in Liquids and Gases* (Academic Press, London, 1984)

31. I.A. Kulagin, R.A. Ganeev, R.I. Tugushev, A.I. Ryasnyansky, T. Usmanov, J. Opt. Soc. Am. B **23**, 75 (2006)
32. R. Adair, L.L. Chase, S.A. Payne, Phys. Rev. B **39**, 3337 (1989)
33. R.A. Ganeev, A.I. Ryasnyansky, H. Kuroda, Opt. Spectrosc. **100**, 108 (2006)
34. I.M. Catalano, A. Cingolani, A. Minafra, Phys. Rev. B **5**, 1629 (1972)
35. R. Simon, H. Gerhardt, S. Szatmari, Opt. Lett. **14**, 1207 (1989)
36. I.M. Catalano, A. Cingolani, J. Appl. Phys. **50**, 5638 (1979)
37. M. Sheik-Bahae, P. Mukhredjee, H.S. Kwok, J. Opt. Soc. Am. B **3**, 379 (1986)
38. P. Yeh, IEEE J. Quantum Electron. **25**, 484 (1989)
39. D.D. Nolte, *Photorefractive Effects and Materials* (Kluver Academic Publishers, Boston, 1995)
40. P. Yeh, *Introduction to Photorefractive Nonlinear Optics* (Willey, New York, 1993)
41. B. Edvold, P.E. Andersen, P. Bushhave, P.M. Petersen, IEEE J. Quantum Electron. **30**, 175 (1994)
42. S.A. Podoshvedov, Y.V. Miklyaev, Opt. Commun. **171**, 301 (1999)
43. R.A. Ganeev, A.I. Ryasnyansky, B. Palpant, S. Debrus, J. Appl. Phys. **97**, 104303 (2005)
44. Y. Kawata, T. Tanaka, S. Kawata, Appl. Opt. **35**, 5308 (1996)
45. D. Richter, A. Grunnet-Jepsen, J. Takacs, L. Solumar, IEEE J. Quantum Electron. **30**, 1645 (1994)
46. J.-L. Ferrier, J. Gazengel, X.N. Phu, G. Rivoire, Opt. Commun. **58**, 343 (1986)
47. J.M.C. Jonathan, G. Roosen, P. Roussignol, Opt. Lett. **13**, 234 (1988)
48. M. Sylla, D. Rouede, R. Chevalier, X.N. Phu, G. Rivoire, Opt. Commun. **90**, 391 (1992)
49. B. Taheri, S.A. Holmstrom, R.C. Powell, J.J. Song, A. Munoz, I. Foldvari, A. Peter, Opt. Mater. **3**, 251 (1994)
50. S.P. Aithal, P.P. Kiran, N.D. Rao, J. Nonlin, Opt. Phys. Mater. **9**, 217 (2000)
51. G.C. Duree, J.L. Shultz, G.J. Salamo, M. Segev, A. Yariv, B. Crosignani, P. Di Porto, E.J. Sharp, R.R. Neurgaonkar, Phys. Rev. Lett. **71**, 533 (1993)
52. R.A. Ganeev, A.I. Ryasnyansky, R.I. Tugushev, M.K. Kodirov, F.A. Ahmedjanov, T. Usmanov, Opt. Quantum Electron. **36**, 807 (2004)
53. R.A. Ganeev, I.A. Kulagin, A.I. Ryasnyansky, R.I. Tugushev, T. Usmanov, Opt. Spectrosc. **94**, 561 (2003)
54. G. Lesaux, G. Roosen, A. Brun, Opt. Commun. **56**, 374 (1986)
55. R.A. Ganeev, M. Baba, A.I. Ryasnyansky, M. Suzuki, M. Turu, H. Kuroda, Opt. Commun. **231**, 431 (2004)
56. C.Y. Chien, G. Korn, J.S. Coe, J. Squier, G. Mourou, R.S. Craxton, Opt. Lett. **20**, 353 (1995)
57. M. Sheik-Bahae, D.C. Hutchings, D.J. Hagan, E.W. Van Stryland, IEEE J. Quantum Electron. **27**, 1296 (1991)
58. R.K. Choubey, R. Trivedi, M. Das, P.K. Sen, P. Sen, S. Kar, K.S. Bartwal, R.A. Ganeev, J. Cryst. Growth **311**, 2597 (2009)
59. D. Milam, M.J. Weber, A.J. Glass, Appl. Phys. Lett. **31**, 822 (1977)
60. I.A. Kulagin, R.A. Ganeev, R.I. Tugushev, A.I. Ryasnyanskii, T. Usmanov, Quantum Electron. **34**, 657 (2004)
61. R.A. Ganeev, I.A. Kulagin, A.I. Ryasnyansky, R.I. Tugushev, T. Usmanov, Opt. Commun. **229**, 403 (2004)
62. N.M. Bityurin, V.I. Bredikhin, V.N. Genkin, Quantum Electron. **8**, 1377 (1978)
63. A. Seilmeter, W. Kaiser, Appl. Phys. **23**, 113 (1980)
64. M.P. Joshi, S.R. Mishra, H.S. Rawat, S.C. Mehendale, K.C. Rustagi, Appl. Phys. Lett. **62**, 1763 (1993)
65. R.A. Ganeev, A.I. Ryasnyansky, M.K. Kodirov, T. Usmanov, Opt. Commun. **185**, 473 (2000)
66. D. Vincent, J. Cruickshank, Appl. Opt. **36**, 7794 (1997)
67. S. Couris, E. Koudoumas, A.A. Ruth, D. Leach, J. Phys. B **28**, 4537 (1995)
68. N.V. Kamanina, L.N. Kaporskii, B.V. Kotov, Opt. Commun. **152**, 280 (1998)
69. R.A. Ganeev, A.I. Ryasnyansky, V.I. Redkorechev, K. Fostiropoulos, G. Priebe, T. Usmanov, Opt. Commun. **225**, 131 (2003)

70. R.A. Ganeev, M. Baba, H. Kuroda, G.S. Boltaev, R.I. Tugushev, T. Usmanov, Eur. Phys. J. D **64**, 109 (2011)
71. W.J. Blau, H.J. Byrne, D.J. Cardin, T.J. Dennis, J.P. Hare, H.W. Kroto, R. Taylor, D.R.M. Walton, Phys. Rev. Lett. **67**, 1423 (1991)
72. H. Hoshi, N. Nakamura, Y. Mauyama, T. Nakagawa, S. Suzuki, H. Shiromaru, Y. Achiba, Jpn. J. Appl. Phys. **30**, L1397 (1991)
73. X.K. Wang, T.G. Zhang, W.P. Lin, S.Z. Liu, G.K. Wong, M.M. Kappes, R.P.N. Chang, J.B. Ketterson, Appl. Phys. Lett. **60**, 810 (1992)
74. G. Wang, J. Wen, Q. Houng, S. Qian, X. Lu, J. Phys. D Appl. Phys. **36**, 84 (1999)
75. Y. Wang, L.T. Cheng, J. Phys. Chem. **96**, 1530 (1992)
76. H. Hoshi, T. Manaka, K. Ishikawa, H. Takezoe, Jpn. J. Appl. Phys. **36**, 6403 (1997)
77. D. Neher, G.I. Stegeman, F.A. Tinker, N. Peyghambarian, Opt. Lett. **17**, 1491 (1992)
78. R.A. Ganeev, A.I. Ryasnyansky, N.V. Kamanina, I.A. Kulagin, M.K. Kodirov, T. Usmanov, J. Opt. B **3**, 88 (2001)
79. S.R. Mishra, H.S. Rawat, S.C. Mehendale, Appl. Phys. Lett. **71**, 46 (1997)
80. L.W. Tutt, T.F. Bogess, Prog. Quantum Electron. **17**, 299 (1993)
81. J.W. Perry, K. Mansour, I.Y. Lee, X.L. Wu, P.W. Bedworth, C.T. Chen, D. Ng, R. Marder, P. Miles, T. Wada, M. Tian, H. Sasabe, Science **273**, 1533 (1996)
82. R.A. Ganeev, A.I. Ryasnyansky, V.I. Redkorechev, K. Fostiropoulos, G. Priebe, T. Usmanov, Fuller. Nanotubes Carbon Nanostruct. **12**, 327 (2004)
83. S.R. Mishra, H.S. Rawat, M. Langhate, Opt. Commun. **147**, 328 (1998)
84. D.Y. Kim, W.E. Torruellas, J. Kang, C. Bosshard, G. Stegeman, P. Vidakovic, J. Zyss, W.E. Moerner, R. Twieg, G. Bjorklund, Opt. Lett. **19**, 868 (1994)
85. A.I. Ryasnyansky, R.A. Ganeev, G. Priebe, V.I. Redkorechev, K. Fostiropoulos, T. Usmanov, Fuller. Nanotubes Carbon Nanostruct. **13**, 131 (2005)
86. R.A. Ganeev, M. Baba, M. Morita, A.I. Ryasnyansky, M. Suzuki, M. Turu, H. Kuroda, J. Opt. A **6**, 282 (2004)
87. R.A. Ganeev, M. Baba, M. Morita, A.I. Ryasnyansky, M. Suzuki, H. Kuroda, J. Opt. A **6**, 1076 (2004)
88. R.A. Ganeev, S.R. Kamalov, M.K. Kodirov, M.R. Malikov, A.I. Ryasnyansky, R.I. Tugushev, S.U. Umidullaev, T. Usmanov, Opt. Commun. **184**, 305 (2000)
89. V.I. Bezrodnyi, A.A. Ishchenko, Appl. Phys. B **73**, 283 (2001)
90. R.A. Ganeev, R.I. Tugushev, A.A. Ishchenko, N.A. Derevyanko, A.I. Ryasnyansky, T. Usmanov, Appl. Phys. B **76**, 683 (2003)
91. R.A. Ganeev, A.I. Ryasnyansky, M.K. Kodirov, T. Usmanov, Opt. Spectrosc. **91**, 878 (2001)
92. R.A. Ganeev, A.I. Ryasnyansky, R.I. Tugushev, H. Kuroda, A.A. Ishchenko, N.A. Derevyanko, T. Usmanov, Nonlinear Opt. Quantum Opt. **32**, 187 (2004)
93. G.A. Swartzlander, H. Yin, A.E. Kaplan, J. Opt. Soc. Am. B **6**, 1317 (1989)
94. M. Samoc, A. Samoc, D. Luther-Devies, H. Reisch, U. Scherf, Opt. Lett. **23**, 1295 (1998)
95. G.S. Pan, R. Kesavamoorthy, S.A. Asher, J. Am. Chem. Soc. **120**, 6525 (1998)
96. F.M. Qureshi, S.J. Martin, X. Long, D.D.C. Bradley, F.Z. Henari, W.J. Blau, E.C. Smith, C.H. Wang, A.K. Kar, H.L. Anderson, Chem. Phys. **231**, 87 (1998)
97. R.A. Ganeev, A.I. Ryasnyansky, M.K. Kodirov, S.R. Kamalov, V.A. Li, R.I. Tugushev, T. Usmanov, Appl. Phys. B **74**, 47 (2002)
98. Y.P. Sun, J.E. Riggs, H.W. Rollings, R. Gudaru, J. Phys. Chem. B **103**, 77 (1999)
99. J.W. Perry, K. Mansour, S.R. Marder, D. Alvarez, J.K. Perry, P. Choong, Opt. Lett. **19**, 625 (1994)
100. S.V. Rao, N.K.M. Naga Srinivas, D.N. Rao, L. Giribabu, B.G. Maiya, R. Philip, G.R. Kumar, Opt. Commun. **182**, 255 (2000)
101. G. Fang, Y. Song, Y. Wang, X. Zhang, C. Li, L.C. Song, P.C. Liu, Opt. Commun. **183**, 523 (2000)
102. S. Guha, K. Kang, P. Porter, J.F. Poach, D.E. Remy, F.J. Aranda, N. Rao, Opt. Lett. **17**, 264 (1992)
103. F. Hache, D. Ricard, C. Flytzanis, U. Kreibig, Appl. Phys. A **47**, 347 (1988)

104. L. Yang, K. Becker, F.M. Smith, R.H. Marguder, R.F. Haglund, L. Yang, R. Dorsinville, R.R. Alfano, R.A. Zuhr, J. Opt. Soc. Am. B **11**, 457 (1994)
105. D. Ricard, P. Roussignol, C. Flytzanis, Opt. Lett. **10**, 511 (1995)
106. D.A. Weitz, M. Oliveria, Phys. Rev. Lett. **52**, 1433 (1984)
107. K. Uchida, S. Kaneko, S. Omi, C. Hata, H. Tanji, Y. Asahara, A.J. Ikushima, J. Opt. Soc. Am. B **11**, 1236 (1994)
108. R.A. Ganeev, M. Baba, A.I. Ryasnyansky, M. Suzuki, H. Kuroda, Opt. Commun. **240**, 437 (2004)
109. M.J. Bloemer, J.W. Haus, P.R. Ashley, J. Opt. Soc. Am. B **7**, 790 (1990)
110. R.A. Ganeev, A.I. Ryasnyansky, M.K. Kodirov, S.R. Kamalov, T. Usmanov, J. Phys. D **34**, 1602 (2001)
111. S.C. Mehendale, S.R. Mishra, K.S. Bindra, M. Laghate, T.S. Dhami, K.S. Rustagi, Opt. Commun. **133**, 273 (1997)
112. V.V. Slabko, S.V. Karpov, V.I. Zaitsev, A.P. Popov, J. Phys. Condens. Matter. **5**, 7231 (1993)
113. A.I. Ryasnyansky, P. Palpant, S. Debrus, R.A. Ganeev, A.L. Stepanov, N. Gan, C. Buchal, S. Uysal, Appl. Opt. **44**, 2839 (2005)
114. T. Karali, N. Can, L. Valberg, A.L. Stepanov, P.D. Townsend, C. Buchal, R.A. Ganeev, A.I. Ryasnyansky, H.G. Belik, M.L. Jessett, C. Ong, Phys. B **363**, 88 (2005)
115. M. Baba, M. Ichihara, R.A. Ganeev, M. Suzuki, H. Kuroda, M. Morita, D. Rau, T. Ishii, M. Iwamura, Appl. Phys. Lett. **84**, 2394 (2004)
116. A.L. Stepanov, R.A. Ganeev, A.I. Ryasnyansky, T. Usmanov, Nucl. Instrum. Methods B **206**, 624 (2003)
117. R.A. Ganeev, A.I. Ryasnyansky, A.L. Stepanov, T. Usmanov, Phys. State Solid B **241**, 935 (2004)
118. R.A. Ganeev, A.I. Ryasnyansky, A.L. Stepanov, T. Usmanov, Phys. State Solid B **241**, R1 (2004)
119. R.A. Ganeev, A.I. Ryasnyansky, A.l. Stepanov, T. Usmanov, Opt. Quantum Electron. **36**, 949 (2004)
120. R.A. Ganeev, A.I. Ryasnyansky, A.L. Stepanov, T. Usmanov, Phys. State Solid B **238**, R5 (2003)
121. Y. Akiyama, K. Midorikawa, Y. Matsunawa, Y. Nagata, M. Obara, H. Tashiro, K. Toyoda, Phys. Rev. Lett. **69**, 2176 (1992)
122. S. Kubodera, Y. Nagata, Y. Akiyama, K. Midorikawa, M. Obara, H. Tashiro, K. Toyoda, Phys. Rev. A **48**, 4576 (1993)
123. C.-G. Wahlström, S. Borgström, J. Larsson, S.-G. Pettersson, Phys. Rev. A **51**, 585 (1995)
124. W. Theobald, C. Wülker, F.R. Schäfer, B.N. Chichkov, Opt. Commun. **120**, 177 (1995)
125. M. Bellini, C. Corsi, M.C. Gambino, Phys. Rev. A **64**, 023411 (2001)
126. J.H. Kim, C.H. Nam, Phys. Rev. A **65**, 033801 (2002)
127. V. Tosa, H.T. Kim, I.J. Kim, C.H. Nam, Phys. Rev. A **71**, 063807 (2005)
128. R.A. Ganeev, M. Suzuki, M. Baba, H. Kuroda, J. Opt. Soc. Am. B **23**, 1332 (2006)
129. S.R. Friberg, P.W. Smith, IEEE J. Quantum Electron. **23**, 2089 (1987)
130. M.J. Morgan, C.Y. She, R.L. Carman, IEEE J. Quantum Electron. **11**, 259 (1975)
131. D.N. Nikogosyan, *Properties of Optical and Laser Related Materials: A Handbook* (Wiley, Chichester, 1997)
132. M. Falconieri, G. Salvetti, Appl. Phys. B **69**, 133 (1999)
133. H.-S. Albrecht, P. Heist, J. Kleinschmidt, D.V. Lap, Appl. Phys. B **57**, 193 (1993)
134. S. Couris, M. Renard, O. Faucher, B. Lavorel, R. Chaux, E. Koudoumas, X. Michaut, Chem. Phys. Lett. **369**, 318 (2003)
135. S. Couris, E. Koudoumas, F. Dong, S. Leach, J. Phys. B **29**, 5033 (1996)
136. T. Kawazoe, H. Kawaguchi, J. Inoue, O. Haba, M. Ueda, Opt. Commun. **160**, 125 (1999)
137. M.-T. Zhao, B. Singh, P.N. Prasad, J. Chem. Phys. **89**, 5535 (1988)
138. R.A. Ganeev, M. Baba, A.I. Ryasnyansky, M. Suzuki, M. Turu, H. Kuroda, Appl. Phys. B **78**, 433 (2004)

139. R. DeSalvo, A.A. Said, D.J. Hagan, E.W. Van Stryland, M. Sheik-Bahae, IEEE J. Quantum Electron. **32**, 1324 (1996)
140. D. von der Linde, H. Schüler, J. Opt. Soc. Am. B **13**, 216 (1996)
141. H. Du, G.Q. Xu, W.C. Chin, L. Huang, W. Li, Chem. Mater. **14**, 4473 (2002)
142. H.P. Li, C.H. Kam, Y.L. Lam, W. Ji, Opt. Commun. **190**, 351 (2001)
143. A.A. Said, M. Sheik-Bahae, D.J. Hagan, T.-H. Wei, J. Wang, J. Young, E.W. Van Stryland, J. Opt. Soc. Am. B **9**, 405 (1992)
144. J. Etchepare, G. Grillon, J.P. Chambaret, G. Hamoniaux, A. Orszag, Opt. Commun. **63**, 329 (1987)
145. M. Sheik-Bahae, A.A. Said, D.J. Hagan, E.W. Van Stryland, Opt. Eng. **30**, 1228 (1991)
146. J.-G. Tian, W.-P. Zang, C.-Z. Zhang, G. Zhang, Appl. Opt. **34**, 4331 (1995)
147. M. Bass, H.H. Barrett, IEEE J. Quantum Electron. **8**, 338 (1972)
148. X. Liu, S. Guo, H. Wang, N. Ming, J. Nonlinear Opt. Phys. Mater. **10**, 431 (2001)
149. Y. Morel, A. Irimia, P. Najechalski, Y. Kervella, O. Stephan, P.L. Baldeck, C. Andraud, J. Chem. Phys. **114**, 5391 (2001)
150. C.L. Zhan, D.H. Li, D.Q. Zhang, Y.J. Li, D.Y. Wang, T.X. Wang, Z.Z. Lu, L.Z. Zhao, Y.X. Nie, D.B. Zhu, Chem. Phys. Lett. **353**, 138 (2002)
151. G.Y. Zhou, X.M. Wang, D. Wang, Z.S. Shao, M.H. Jiang, Appl. Opt. **41**, 1120 (2002)
152. D.Y. Wang, C.L. Zhan, Y. Chen, Y.J. Li, Z.Z. Lu, Y.X. Nie, Chem. Phys. Lett. **369**, 621 (2003)
153. M. Martinelli, S. Bian, J.R. Leite, R.J. Horowicz, Appl. Phys. Lett. **72**, 1427 (1998)
154. M. Martinelli, L. Gomes, R.J. Horowicz, Appl. Opt. **39**, 6193 (2000)
155. K. Wang, Q.R. Xing, H.Y. Li, J.P. Li, Z.G. Zhang, N. Zhang, L. Chai, Q.Y. Wang, Opt. Commun. **265**, 369 (2006)
156. R.A. Ganeev, Appl. Phys. B **91**, 273 (2008)
157. A.I. Ryasnyanskii, J. Appl. Spectrosc. **72**, 712 (2005)
158. A.D. Bristow, N. Rotenberg, H.M. van Driel, Appl. Phys. Lett. **90**, 191104 (2007)
159. W.C. Hurlbut, Y.-S. Le, K.L. Vodopyanov, P.S. Kuo, M.M. Fejer, Opt. Lett. **32**, 668 (2007)
160. J.W. Fang, S.Y. Zhang, Appl. Phys. B **67**, 633 (1998)
161. Q. Lin, J. Zhang, G. Piredda, R.W. Boyd, P.M. Fauchet, G.P. Agrawal, Appl. Phys. Lett. **91**, 021111 (2007)
162. G. Boudebs, S. Cherukulappurath, Phys. Rev. A. **69**, 053813 (2004)
163. S. Cherukulappurath, G. Boudebs, A. Monteil, J. Opt. Soc. Am B **21**, 273 (2004)
164. J. Hein, M. Helbig, S. Rentsch, Appl. Opt. **36**, 1173 (1997)
165. A. Marcano, H. Maillotte, D. Gindre, D. Métin, Opt. Lett. **21**, 101 (1996)
166. J. Jayabalan, A. Singh, S.M. Oak, Appl. Opt. **45**, 3852 (2006)
167. R.A. Ganeev, I.A. Kulagin, J. Opt. A **11**, 085001 (2009)
168. W. Ji, H.J. Du, S.H. Tang, S. Shi, J. Opt. Soc. Am. B **12**, 876 (1995)
169. W. Ji, S. Shi, H.J. Du, P. Ge, S.H. Tang, X.Q. Xin, J. Phys. Chem. **99**, 17297 (1995)
170. D.G. McLean, R.L. Sautherland, M.C. Brant, D.M. Brandelik, P.A. Fleitz, T. Pottenger, Opt. Lett. **18**, 858 (1993)
171. Z.R. Chen, H.W. Hou, X.Q. Xin, B. Yu, S. Shi, J. Phys. Chem. **99**, 8717 (1995)
172. T. Xia, A. Dogariu, K. Mansour, D.J. Hagan, A.A. Said, E.W. Van Stryland, S. Shi, J. Opt. Soc. Am. B **15**, 1497 (1998)
173. O. Durant, V.V. Grolier-Mazza, R. Frey, Opt. Lett. **23**, 1471 (1998)
174. R.A. Ganeev, A.I. Ryasnyanskii, M.K. Kodirov, F.R. Akhmedjanov, T. Usmanov, Opt. Spectrosc. **103**, 986 (2007)
175. C.V. Bindhu, S.S. Harilal, V.P.N. Nampoori, C.P.G. Vallabhan, Appl. Phys. B **70**, 429 (2000)
176. S.S. Harilal, C.V. Bindhu, V.P.N. Nampoori, C.P.G. Vallabhan, J. Appl. Phys. **86**, 1388 (1999)
177. V.V. Golovlev, W.R. Garrett, C.H. Chen, J. Opt. Soc. Am. B **13**, 2801 (1996)
178. S.R. Mishra, R.S. Rawat, M.P. Joshi, S.C. Mehendale, J. Phys. B **27**, L157 (1994)
179. N.V. Kamanina, E.L. Aleksandrova, L.N. Kaporski, Tech. Phys. Lett. **26**, 400 (2000)
180. Y.L. Song, C. Zhang, X.L. Zhao, Y.X. Wang, G.Y. Fang, G.C. Jin, S.L. Qu, S.P. Wu, X.Q. Xin, H.Y. Ye, Chem. Lett. **9**, 1076 (2000)

181. L. Francois, M. Mostafavi, J. Belloni, J.F. Delouis, J. Delaire, P. Feneyrou, J. Phys. Chem. B **104**, 6133 (2000)
182. N.N. Lepeshkin, W. Kim, V.P. Safonov, J.G. Zhu, R.L. Armstrong, C.W. White, R.A. Zuhr, V.M. Shalaev, J. Nonlinear Opt. Phys. Mater. **8**, 191 (1999)
183. M.R.V. Sahyun, S.E. Hill, N. Serpone, R. Danesh, D.K. Sharma, J. Appl. Phys. **79**, 8030 (1996)
184. V. Joudrier, P. Bourdon, F. Hache, C. Flytzanis, Appl. Phys. B **67**, 627 (1998)
185. H. Inouye, K. Tanaka, J. Tanahashi, K. Hirao, Phys. Rev. B **57**, 11334 (1998)
186. H. Inouye, K. Tanaka, J. Tanahashi, T. Hattori, H. Nakatsuka, Jpn. J. Appl. Phys. **39**, 5132 (2000)
187. A.A. Said, T. Xia, D.J. Hagan, E.W. Van Stryland, M. Sheik Bahae, J. Opt. Soc. Am. B **14**, 824 (1997)
188. A.Y. Hamad, J.P. Wicksted, J. Appl. Phys. **79**, 3816 (1996)
189. I. Androsch, P. Glas, Opt. Commun. **105**, 125 (1994)
190. B.L. Justus, A.J. Campillo, D.G. Hendershot, D.K. Gaskill, Opt. Commun. **103**, 405 (1993)
191. K.S. Bindra, S.M. Oak, K.C. Rustagi, Opt. Commun. **124**, 452 (1996)
192. S.M. Oak, K.S. Bindra, R. Chari, K.C. Rustagi, J. Opt. Soc. Am. B **10**, 613 (1993)
193. B.L. Yu, G.S. Yin, C.S. Zhu, F.X. Gan, Opt. Mater. **11**, 17 (1998)
194. C.H. Kwak, Y.L. Lee, S.G. Kim, J. Opt. Soc. Am. B **16**, 600 (1999)
195. W.F. Zhang, Y.B. Huang, M.S. Zhang, Z.G. Liu, Appl. Phys. Lett. **76**, 1003 (2000)
196. R.A. Ganeev, A.I. Ryasnyansky, M.K. Kodirov, S.R. Kamalov, T. Usmanov, Opt. Spectrosc. **93**, 789 (2002)
197. R.A. Ganeev, A.I. Ryasnyansky, M.K. Kodirov, S.R. Kamalov, T. Usmanov, Opt. Spectrosc. **90**, 568 (2001)
198. R.A. Ganeev, A.I. Ryasnyansky, S.R. Kamalov, M.K. Kodirov, T. Usmanov, Tech. Phys. **47**, 991 (2002)
199. R.A. Ganeev, A.I. Ryasnyansky, M.K. Kodirov, T. Usmanov, J. Opt. A **4**, 446 (2002)

Chapter 4
Laser Ablation Induced Cluster Formation

4.1 Methods of Laser-Induced Nanoparticle Formation

The structural, optical, and nonlinear optical properties of nanoparticles took much attention due to their potential applications in optoelectronics, mode-locking technologies, optical limiting, etc. Currently, nanoparticle formation during laser ablation of solid-state targets using the ultrashort laser pulses is a widely accepted technique. Together with formation of the ripples of wavelength range sizes, this technique gives the opportunity of creating the exotic structures with variable physical and chemical properties. The increase of surface area of the nanostructured materials allows for enhancement of the velocity of catalytic reaction, provides the opportunity of application of such structures for the information writing, manufacturing of lubricants, semiconductor technologies, etc.

The structural, optical, and nonlinear optical parameters of nanoparticles are known to differ from those of the bulk materials due to the quantum confinement effect. Silver [1], copper [2], and gold [3] are among the most useful metals suited for the nanoparticles preparation for optoelectronics and nonlinear optics. Further search of prospective materials in clustered form, their preparation and application are of considerable importance. In particular, the formation of cluster-containing laser plumes using relatively long (nanosecond and picosecond) laser pulses seems to be an attractive area of studies due to availability of such lasers in many laboratories. The knowledge of the conditions when clusters are presented in the plasma plumes opens the door for the study of various properties of these media, in particular their

We analyze the methods for nanoparticles formation using the laser ablation technique and show the results of the morphological and optical characterization of these structures. In particular, the methods of preparation of the indium nanoparticles by laser ablation of bulk indium and GaAs in liquids will be shown. We discuss the experimental studies of the structural modifications during laser ablation of nanoparticle-containing targets and analyze the dynamics of nanoparticle formation during laser ablation of silver at different pressures of surrounding noble gases.

R. A. Ganeev, *Nonlinear Optical Properties of Materials*,
Springer Series in Optical Sciences 174, DOI: 10.1007/978-94-007-6022-6_4,

high-order nonlinear optical characteristics, which allow the creation of the coherent sources of efficient high-order harmonic generation of laser radiation [4].

The nanoparticle formation during laser ablation of targets has previously been described as a process of short excitation of electronic gas and transfer of this energy to the atomic cell with further aggregation processes, which continue during evaporation of the material [5, 6]. In the case of bulk target ablation, the attention is taken for creation of the conditions when laser energy is accumulated for a short period at a small area to maintain the conditions of non-equilibrium heating. In that case, the extremely heterogeneous conditions help creating the nanoparticles in the small areas of heated samples. The ablation-induced nanoparticle formation in laser plumes was carefully documented in multiple experiments using femtosecond laser pulses [7, 8]. One can maintain the conditions when the aggregated atoms do not disintegrate during evaporation from the surface. The analysis of the aggregation state of evaporated particles was carried out by different techniques. Among them the time resolved emission spectroscopy, CCD camera imaging of plasma plume, Rayleigh scattering, laser-induced fluorescence, etc, has shown the ability of defining the presence of the nanoparticles in the laser plumes.

Currently, the nanoparticles, fullerene clusters, and carbon nanotubes with different sizes and shapes are commercially available from various manufacturers and can be attached to the surfaces and then evaporated using the laser ablation technique. In that case one has to carefully define the optimum laser intensity, pulse duration, and focusing conditions for heating of the nanoparticles until they evaporate from the targets. The history of ablated nanoparticles in that case can be difficult to analyze using the above techniques due to some restrictions in identification of the clusters in plasma plumes at moderate heating of the targets. In that case the comparison of the size characteristics of initial nanoparticles and deposited debris becomes a versatile approach for definition of the changes of nanoparticle morphology during laser ablation. Another indirect method is the analysis of the harmonic spectra generating in the plasmas containing nanoparticles and atoms/ions of the same origin. Time-of-flight mass spectrometry can also reveal the plasma ingredients.

It is well accepted that, when a solid target is ablated by the laser radiation, the ablating material consists on atoms and clusters. These atoms and clusters tend to aggregate during laser pulse or soon afterwards, leading to the formation of larger clusters. The reported results (see for example [9]) also indicate that the ablation processes in the picosecond and femtosecond time scales considerably differ compared to the nanosecond one. In addition to earlier experimental observations, several theoretical studies have suggested that rapid expansion and cooling of the solid-density matter heated by short laser pulse may result in nanoparticle synthesis via different mechanisms. Heterogeneous decomposition, liquid phase ejection and fragmentation, homogeneous nucleation and decomposition, and photomechanical ejection are among those processes that can lead to nanoparticles aggregation and disintegration [10–12]. Short pulses, contrary to the nanosecond pulses, do not interact with the ejected material thus avoiding complicated secondary laser interactions. Further, short pulses heat a solid to higher temperature and pressure than do longer

pulses of comparable fluence, since the energy is delivered before significant thermal conduction can take place.

Pulsed laser deposition techniques in rare gas ambiences have been used for nanometer-size particle preparation, multi-component thin film deposition, and carbon nanotube syntheses [13]. As a result of frequent collisions of ablated particles with gas atoms in plume, the particles cool down and form nanoparticles [14]. The development of the nanoparticle synthesis at gas conditions and the characterization techniques that make possible the control of nanoparticle sizes within a few nm have attracted renewed interest to the production of metal nanoparticles, as this opens the possibility of taking advantage of their special properties for the development of applications such as new catalysts, tunnel resonance resistors, or optical devices [15]. The search of optimal gas pressure for nanoparticles formation during laser ablation of solids thus remains an open issue.

It is known that metal ablation in air is less efficient than that in vacuum due to re-deposition of the ablated material. The influence of surrounding gas on the cluster formation during laser ablation of the metals by short laser pulses has previously been analyzed only at two conditions, when the target was placed in vacuum or ambient air. The ablation rates in vacuum can be calculated using a thermal model, as well as from basic optical and thermal properties [16]. It is of interest to analyze the influence of the concentration of surrounding gas on cluster formation. One can study this process using noble gases of different atomic numbers at the pressures varying between 10^{-2} torr (i.e. vacuum conditions) and 760 torr (i.e. atmospheric pressure). It is important to discuss whether there is a threshold in gas pressure scale above which the conditions of nanoparticle formation get spoiled.

In this Chapter, we present and analyze the methods for nanoparticles formation using the laser ablation technique and show the results of the morphological and optical characterization of these structures.

4.2 Characterization of the Nanoparticles Synthesized During Laser Ablation of Indium and GaAs in Various Liquids

4.2.1 Structural and Optical Properties of Indium Nanoparticles Prepared by Laser Ablation in Liquids

Further search of prospective nanoparticle-containing materials and their preparation and application are of considerable importance nowadays. For example, metal nanoparticles will be potential additives of lubricating oils in the near future; indium nanoparticles are expected to play a part in future nanometer lubricants. The indium nanoparticles could also be used for single electron transistors, as tags for detection of DNA hybridization, as printing building blocks in nanoxerography, and as starting material for convenient synthesis of InP using phosphide ions.

The "cost" of laser ablation in liquids as a simpler preparation method in the terms of the properties of nanoparticles is a relatively short-term stability of colloidal nanoparticles. This is a main obstacle of laser ablation in liquids [17]. It would be problematic for technological applications of the material in the case when optical properties of nonlinear device containing such structures appear to be unstable under long-term laser irradiation. One of possible proposals to overcome this obstacle is a preparation of the nanoparticles solution using low viscosity liquid and further dissolution of prepared solution in the matrix with high viscosity.

To examine the generality of the method and its potential use for preparing nanoparticles, let us consider the indium, which being a low melting point metal is a good candidate for such studies and can be of various practical interests. Indium nanoparticles have been prepared by different methods [18–23]. In [20], small indium particles were prepared by metal vapor deposition. Indium nanoparticles were also prepared by the evaporation of indium into a polymerized monomer in order to study the thermal stability of the resulting composite material [21]. The description of the synthesis of monodispersed indium nanoparticles by an organometallic route and their arrangement in two- and three-dimensional networks is presented in [22]. The method for indium nanoparticles preparation from bulk indium that involves surface oxidation and dispersion of indium droplets is described in [20].

In this subsection, we present the methods of preparation of the indium nanoparticles by laser ablation of bulk indium in liquids. The structural and optical properties of the solutions of these nanoparticles were discussed in [24]. A nanosecond Nd:YLF laser (pulse duration $t = 20\,\text{ns}$, wavelength $\lambda = 1054\,\text{nm}$, pulse energy $E = 36\,\text{mJ}$, 10 Hz pulse repetition rate) was used for the ablation of indium target in liquids. The laser radiation was focused by a 6 cm focal length lens inside a 5 cm long cell containing ethanol or water. The bulk indium to be ablated was kept close to the back window of the cell, to prevent any optical damage of the input window. The ablation was performed during 15 min (9000 laser shots), after which a yellow-colored suspension was obtained. It may be noted that an increase in the ablation time resulted in a decrease in the efficiency of nanoparticles formation and also led to sedimentation of large nano-aggregates. The absorption spectrum of the prepared nanoparticle-containing suspensions was recorded using a fiber optical spectrometer. The structural analysis of In nanoparticle sizes was carried out using the total reflection x-ray fluorescence spectrometry (TXRF).

The absorption spectrum of the ethanol suspension of ablated indium is presented in Fig. 4.1. The observed absorption maximum at 350 nm can be associated with the surface plasmon resonance of indium nanoparticles. A broad SPR peak indicates the presence of the nanoparticles with different sizes. The absorption spectra of indium nanoparticles dispersed in chloroform exhibit the characteristic peak of the surface plasmon band at 240 nm, which reveals the formation of indium particles, were reported in [20]. They have mentioned that the surface plasmon band at 240 nm is a clear evidence for the existence of the individual particles less than 50 nm in size. In addition, UV absorption measurements at room temperature presented in [23] have shown surface plasmon bands characteristic for the specific nanoparticle size. Surface plasmon bands were observed by them at 281 nm (for the 1.3 nm average

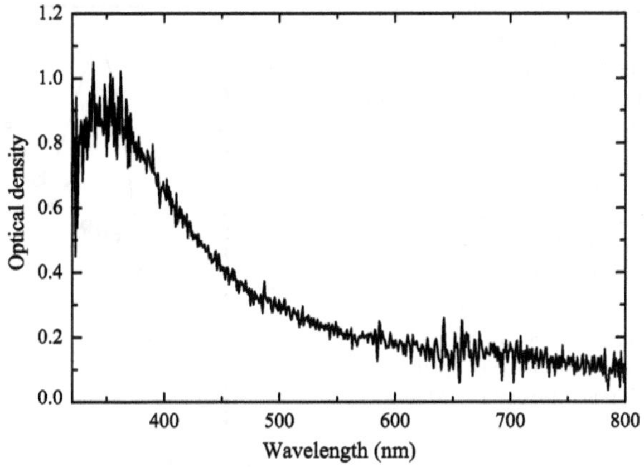

Fig. 4.1 Absorption spectrum of ablated indium suspension in ethanol. Reproduced from [24] with permission from Springer Science+Business Media

particle size), 291 nm (for the 12 nm average particle size) and 319, 340 and 350 nm (for the 107 nm average particle size) respectively.

The surface plasmon properties of nanostructures have been explained by Kreibig and Genzel [25]. As the mean free path of the metal electrons exceeds the particle size, the free-electron contribution to the dielectric function undergoes a modification, thereby causing the surface plasmon resonance frequency peak to be directly proportional to the particle size and the resonance spectral width to be inversely proportional to the particle size [25, 26].

For the sizes of measured In nanoparticles (43 nm) the quantum effects do not take place. The role of higher sized nanoparticles could lead to the red-shift of the absorption peak. At the same time, previous studies of the SPR of indium nanoparticles have shown that it shifts toward the longer wavelength range both in the case of different solvents [27] and morphologies [28] of In clusters, and for different techniques for the preparation of nanoparticles. When the In clusters are embedded in the matrix with a high refractive index, the Mie resonances can also be red-shifted to the visible range. At the same time, the absorption phenomenon in indium nanoparticles due to SPR is found solvent dependent [27]. Shift of about 50 nm for this absorption for In in dichloromethane (290 nm) probably indicates the formation of larger sized indium particles thus corroborating with the early reports (240 nm, [20, 29]).

The indirect confirmation of the variation of the SPR is also presented in the case of reflectance anisotropy spectroscopy of the In islands on GaAs surface, wherein, for higher indium coverage, the shift of the reflectance anisotropy spectra from 250 to 400 nm was observed [29]. The reason of the shift of the resonance from 240 to 350 nm range in described experiments was caused by different morphology of the particles (e.g., varied from spherical to elliptical shape), as it was discussed in literature in the case of silver nanoparticles.

Total reflection x-ray fluorescence measurements were performed using TXRF spectrometer [30]. Figure 4.2 shows the x-ray fluorescence trace recorded in the case of indium nanoparticles deposited after evaporation from the liquid suspension on a

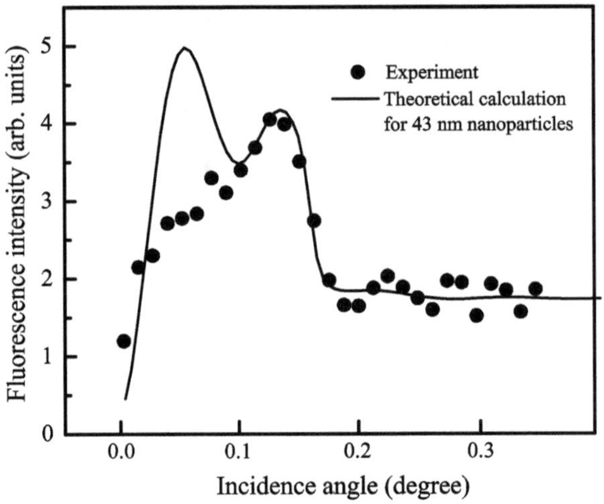

Fig. 4.2 Recorded fluorescence profile of the indium nanoparticles prepared by laser ablation in ethanol and deposited on the plexi glass substrate. The *scattered points* show experimental data, while the *solid line* shows a fitted profile. The critical angle of plexi glass is ~0.17° for 8.50 keV x-ray energy. Reproduced from [24] with permission from Springer Science+Business Media

plexi glass substrate. The average size of the indium nanoparticles was determined by fitting the recorded fluorescence profile using CATGIXRF program. Solid line presents the best theoretical fit, from which the average size of the nanoparticles was estimated to be 43 nm. The transmission electron microscope (TEM) measurements also confirmed the presence of indium nanoparticles in the prepared suspensions, with effective size distribution lying between 20 and 100 nm.

4.2.2 Laser Ablation of GaAs in Liquids: Structural and Optical Characteristics of Colloidal Suspensions

The semiconductor quantum dots, semiconductor nanoparticles-doped glasses, and colloidal microcrystallites have attracted extensive attention over the past two decades. In particular, the interest in GaAs nanoparticles was attributed to their potential applications in nonlinear optics. The chemical techniques for the preparation of GaAs nanoparticles were discussed by Malik et al. [31]. The laser ablation was also widely used for the preparation of GaAs nanoparticles on the surfaces of various materials. Laser ablation of GaAs was reported in various publications (see [32, 33] and references therein). Most of these studies were performed in vacuum conditions for the preparation of thin GaAs films. The application of laser ablation of GaAs in liquids and analysis of structural, optical, and nonlinear optical properties of colloidal GaAs is another approach for the formation and study of the clusters of this semiconductor. This technique has a number of advantages in comparison with the ion implantation in solids (simplicity, cost, variability, etc). Meanwhile, this technique can be applied for the preparation of colloidal semiconductor suspensions [34, 35].

High nonlinear optical susceptibilities ($\sim 10^{-8}$ esu) of such nanoparticles were related with quantum confinement effects. Quantum confinement is attributed to the small-size effect, which alters the electronic structures of nanocrystals from those of the bulk materials. This effect has been extensively studied due to scientific interest and technical importance. The third-order optical nonlinearity of confined GaAs quantum dots was predicted to be significantly higher with comparing to bulk GaAs [36]. An interest in the synthesis, characterization, and application of colloidal semiconductor "quantum dot" materials has grown markedly since these studies are aimed to bridge the gap between solids and atoms. This increased attention reflects both strong size dependence of optical and electronic properties of these materials, and the fact that the materials are relatively easy to make with well-controlled particle sizes.

In this subsection, we discuss the studies of the optical and structural properties of GaAs nanoparticles prepared by laser ablation in various liquids [37]. Laser ablation of GaAs in liquids was carried out using second harmonic radiation of Nd:YAG laser. Laser radiation (wavelength $\lambda = 532$ nm, pulse duration $t = 9$ ns, pulse energy $E = 30$ mJ, 10 Hz pulse repetition rate) was focused by a 50 mm focal length lens at normal incidence onto the surface of GaAs wafer placed inside the 10 mm thick cell filled with liquid of different viscosity (water, ethanol, ethylene glycol, or silicon oil). The fluence of 532 nm radiation at the target surface was measured to be 20 J cm^{-2}. The solution was constantly stirred during the interaction of laser radiation with GaAs target to prevent the growth of concentration of GaAs nanoparticles in front of ablated area. The prepared suspensions were analyzed by TEM and spectroscopic techniques to confirm the appearance of GaAs nanoparticles. A volume ratio of GaAs in the liquid suspensions was estimated to be $\sim 2 \times 10^{-4}$.

The appearance of stable GaAs nanoparticles was confirmed by the TEM studies of these solutions. Figure 4.3 presents the TEM image of GaAs nanoparticle in ethylene glycol (GaAs:EG). The TEM measurements conducted just after laser ablation have shown the appearance of nanoparticles with size distribution ranging from 5 to 200 nm. This distribution was considerably narrowed after the sedimentation of long-sized nanoparticles with only small nanoparticles (ranging from 5 to 15 nm) dominated in the GaAs suspension. The spacing of 0.32 nm was exhibited in the high-resolution TEM micrographs, which corresponds to the d (111) of a GaAs zinc-blend structure, analogous to those reported in [33]. An elemental analysis of GaAs nanoparticles using energy dispersive X-ray (EDX) spectroscopy also confirmed the presence of GaAs in these suspensions (Fig. 4.4; the copper and carbon lines are originated from the grid material). Those studies have shown that the particles tend to be gallium-rich (the accounted weight ratio Ga:As $\sim$ 1.4:1), which was higher with comparing to other reported values (1.2:1 [31, 38]).

The absorption spectra of suspensions were similar to those of quantum-confined GaAs nanocrystals embedded in glasses [39], exhibiting a broad absorption through the visible and UV ranges. Curve 1 of Fig. 4.5 shows the absorption spectrum of GaAs deposited in ethylene glycol. The absorption curve was considerably blue-shifted with comparing to the bulk GaAs (curve 5). The linear absorption coefficients of GaAs nanoparticles-contained ethylene glycol at 397 and 795 nm were measured

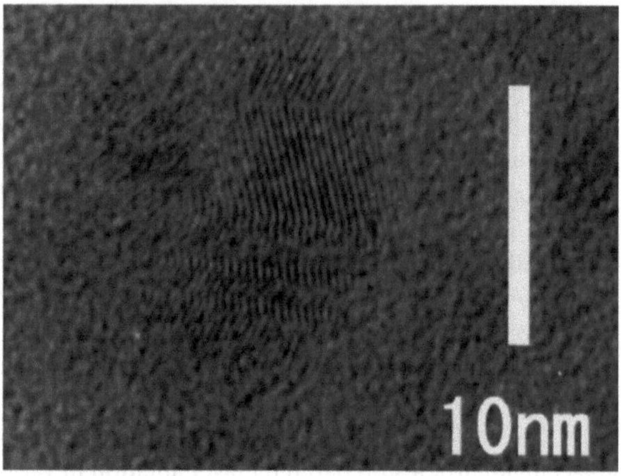

Fig. 4.3 The TEM micrograph of single GaAs nanoparticle from the GaAs:EG solution. Reproduced from [37] with permission from Springer Science+Business Media

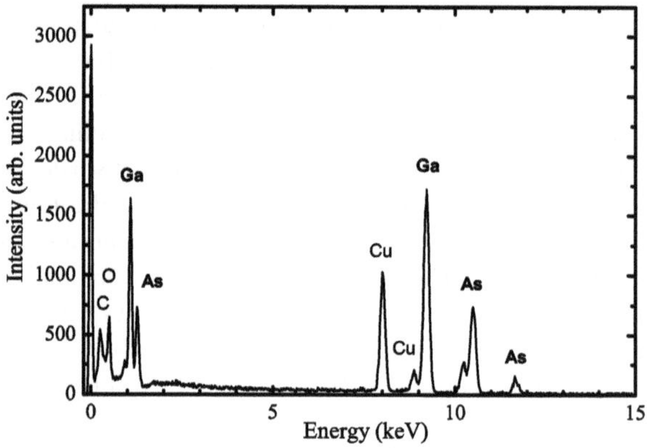

Fig. 4.4 The energy dispersive X-ray spectrum of GaAs:ethanol solution. Reproduced from [37] with permission from Springer Science+Business Media

to be 1.6 and 0.35 cm^{-1}, respectively. The variations of absorption spectra of GaAs nanoparticles in water, ethylene glycol, and ethanol caused by sedimentation of long-sized nanoparticles were observed during a period of three weeks (Fig. 4.5, curves 1 and 2). In the case of GaAs ablated in silicon oil (GaAs:SO) the nanoparticle size distribution and absorption spectra remained constant due to high viscosity of the host material. In that case the inhomogeneous distribution of GaAs particles ranging from big blocks (0.2–10 μm) to nanoparticle-sized structures was observed.

There were some previous reports that the absorption spectra of large semiconductor nanoparticles appear to be the same as the ones of bulk semiconductors. In particular, the laser ablation was applied for the preparation of CdS and ZnSe colloidal solutions and it was shown that there were no considerable distinctions in absorption spectra of nanoparticle solutions and bulk crystals [34]. However, for nanoparticles with sizes smaller than Bohr radius of exciton, a considerable blue shift in absorption spectra caused by quantum confinement effect is a common feature. The blue shift in absorption spectra of small GaAs nanoparticles was reported in [31].

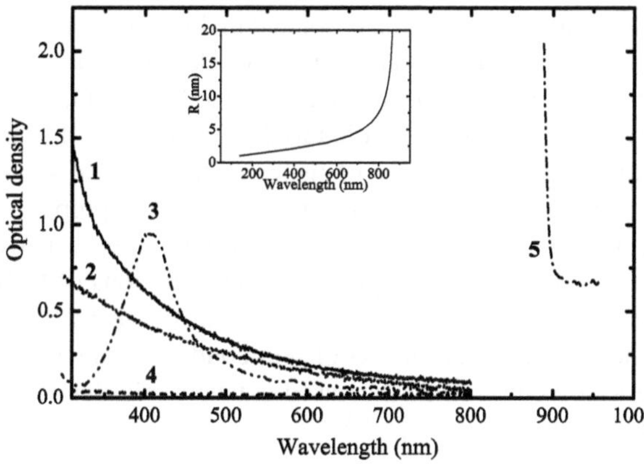

Fig. 4.5 The absorption spectra of 10 mm thick GaAs:EG solution *1* one and *2* three weeks after ablation, *3* silver:EG solution produced at analogous conditions, *4* pure ethylene glycol, and *5* 500 μm thick bulk GaAs wafer. *Inset* the calculated dependence of band gap wavelength on the radius of GaAs nanoparticles. Reproduced from [37] with permission from Springer Science+Business Media

In particular, the absorption edges of nanoparticles with the sizes of 4.47 and 5.45 nm were located at 494 and 580 nm, respectively. In present studies no fixed absorption edge for GaAs solutions was observed due to the presence of nanoparticles of different sizes. The band gap shift in that case was undefined, that was appeared as a featureless absorption curve.

When the radius of semiconductor nanoparticles becomes close or smaller than Bohr radius of exciton, then their optical properties can be altered. The radius of Bohr exciton can be calculated by

$$R_b = \frac{4\pi\varepsilon_0\varepsilon\hbar^2}{e^2}\left(\frac{1}{m_e} + \frac{1}{m_h}\right) \tag{4.1}$$

here ε_o and ε are the dielectric constants of free space and semiconductor ($\varepsilon = 12.8$ for GaAs), $\hbar$ is the Planck constant, m_e and m_h are the effective masses of electron and hole ($m_e = 0.067\, m_o$, and $m_h = 0.2\, m_o$ for GaAs, m_o is the electron mass), and e is the charge of electron. The value of R_b for GaAs was calculated to be 13.5 nm.

The change of band gap energy of nanoparticles can be written as:

$$\Delta E_g = \frac{\pi e R_b}{8\varepsilon_0\varepsilon R^2} - \frac{1.786e^2}{4\pi\varepsilon_0\varepsilon R}, \tag{4.2}$$

where R is the radius of nanoparticles.

The dependence of band gap wavelength on the radius of GaAs nanoparticles is presented in the inset of Fig. 4.5. Using this dependence and results of spectral measurements the average sizes of GaAs nanoparticles were estimated to be in the range of $R = 1.7 - 8$nm that was close to the TEM data.

The initial absorption spectra of GaAs ablated in different liquids (water, ethylene glycol, ethanol) were similar. However, after two weeks the absorption of GaAs:water and GaAs:ethanol solutions was weakened relative to GaAs:EG, probably due to the stronger viscosity in the latter case that led to the smaller rate of sedimentation.

It is worth noting the opportunity of using of the organic solvents instead of inorganic ones for various reasons. One of the advantages of preparing nanoparticles in organic solvents is that the distribution of particle sizes is narrower than that in water, as it was shown for polyvinylpyrrolydone solutions of CdS nanoparticles [40]. The stabilization of metal nanoparticle nonlinearities using PVP stabilizers was shown in [41]. The nonlinear parameters of semiconductor nanoparticle solutions in the case of organic solvents can be increased considerably due to the formation of clusters with controlled sizes.

4.3 Synthesis and Analysis of Nanostructured Thin Films Prepared by Laser Ablation of Metals in Vacuum

The studies on nanoparticles prepared using metal vapor deposition [42], reduction of some salts by alkalides [19], and solution dispersion method [20] have revealed many interesting structural and optical properties of these materials. The laser ablation of metals in vacuum is among the perspective techniques that can also be successfully applied to the preparation of nanoparticle-containing media.

A comparison of the morphology of the deposition by nanosecond and picosecond ablation shows unequivocally the advantages of short-pulse ablation for the preparation of nanoparticles during laser ablation in vacuum [5]. To prove the generality of the vacuum deposition method and its potential use for preparing nanoparticles, various metals were analyzed at different focusing conditions of the laser radiation. The most interesting features of laser ablation and nanoparticle formation during irradiation of the solid targets have been commonly observed in the case of short laser pulses (100 fs–1 ps). In this section, we discuss the results of studies of the formation of nanoparticles of silver, chromium, stainless steel, and indium in vacuum using much longer (subnanosecond) laser pulses [43]. Those studies showed that the nanoparticles could be efficiently formed using long laser pulses under tight focusing conditions.

The ablation of various materials was carried out in vacuum using uncompressed pulses (of 300 ps duration) from a chirped-pulse amplification Ti:sapphire laser system. The samples were placed inside a vacuum chamber. Uncompressed pulses from the Ti:sapphire laser ($\lambda = 793\,\text{nm}$, $\tau = 300\,\text{ps}$, $E = 30\,\text{mJ}$, 10 Hz pulse repetition rate) were focused on a bulk target at two regimes of focusing. In the first case (referred to as tight focusing), the intensity of laser radiation was in the range of $2 \times 10^{12}\,\text{W}\,\text{cm}^{-2}$, and in the second case (referred to as weak focusing), the intensity was considerably lower ($4 \times 10^{10}\,\text{W}\,\text{cm}^{-2}$). Float glass, silicon wafer, and various metal strips (silver, copper, and aluminum) were used as the substrates and were

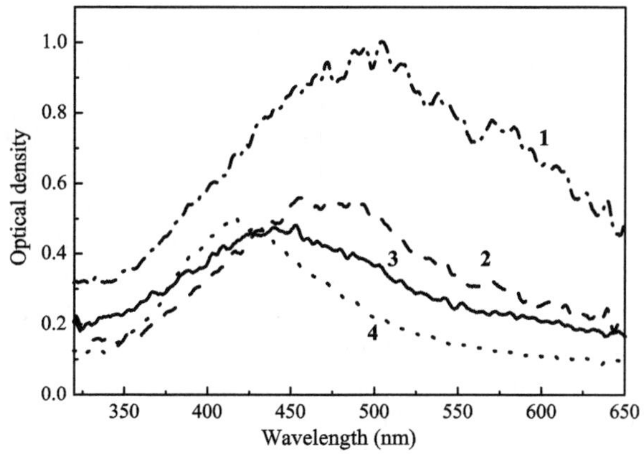

Fig. 4.6 Absorption spectra (curves 1–3) of the silver films deposited at different focusing conditions and the absorption spectrum of silver nanoparticles (curve 4) implanted inside the silica glass plate by ion bombardment. Reproduced from [43] with permission from Optical Society of America

placed at a distance of 50 mm from the targets. The structure of the deposited films of ablated metals was analyzed using different techniques. For this purpose, the nanoparticle-containing films were deposited on different substrates. The presence of nanoparticles was inferred by analyzing the spatial characteristics and spectral absorption of the deposited material.

The absorption spectra of the materials deposited on transparent substrates (float glass) were used to determine the presence or absence of nanoparticles. The presence of nanoparticles was inferred from the appearance of strong absorption bands associated with surface plasmon resonances. Figure 4.6 presents the absorption spectra of silver films deposited on float glass substrates. Earlier work [44] has shown that the SPR of spherical silver nanoparticles induces a strong absorption in the range of 410–480 nm depending on the preparation technique. In the discussed study, a variation in the position of the absorption of Ag deposition was observed, which depended on the conditions of excitation and evaporation of bulk target by the 300 ps pulses. However, in all these cases, the peaks of SPR were centered in the range of 440–490 nm (Fig. 4.6, curves 1–3). In the case of the deposition of a silver film at weak focusing conditions, no absorption peaks were observed in this region, indicating the absence of nanoparticles. Curve 4 of Fig. 4.6 shows analogous measurements made on a sample of Ag nanoparticles embedded in silica glasses using the ion bombardment, reported in [45]. In that work, the thickness of the implanted layer was 60 nm, and the size of the silver nanoparticles was reported to vary from 4 to 8 nm. It is observed that the absorption curve of this sample is quite similar to those of silver deposited on the glass surfaces (Fig. 4.6, curves 1–3). The only difference is that the position of the peak of SPR (415 nm) was on the shorter wavelength side (Fig. 4.6, curve 4).

Much attention has been devoted during the past few years to precisely determine the spatial arrangement in two- and three-dimensional structures of metals. However, ordering and using the nanomaterials necessitate synthesis of monodispersed individual nanoparticles, for which no general method is presently available. We describe, later in the text, the analysis of synthesized nanoparticles by laser ablation of bulk targets at two different conditions of ablation. The structure of the ablated silver was analyzed by studying the films deposited on the silicon substrates. One of the aims of this study was to investigate whether the plasma plumes contain nanoclusters. The presence of the latter could be responsible for the enhancement of the nonlinear optical characteristics of plasma media. In particular, high-order harmonic conversion efficiency may be affected because of the quantum confinement effect during propagation of the femtosecond laser pulse through the nanoparticle-containing plasma. Harmonic generation using single atoms and multiparticle aggregates has been reported in [46] for argon atoms and clusters, which showed the advantages of using the clusters for harmonic generation. The same has been shown in the case of harmonic generation in the plasma plumes containing nanoparticles [47]. More details of this phenomenon will be discussed in Chap. 6.

TXRF measurements were performed using a fluorescence technique [30] for the analysis of the structural properties of the deposited material. The angular dependence of the fluorescence intensity in the total reflection region [48, 49] could be successfully used to identify the presence of nanoparticles on a flat surface. This was done for the deposition in the tight focusing condition of the laser. In the case of weak focusing, it showed a thick film-like deposition of metal, without any inclusion of nanoparticles (see the TXRF image in the case of the ablation of indium at weak focusing conditions, Fig. 4.7). It can be seen from this figure that the fluorescence intensity of In-Lα decreased abruptly below the critical angle for indium (0.36° at 8.5 keV). For incident angles larger than the critical angle, In-Lα fluorescence intensity increased monotonically. It showed a behavior similar to that for a thick film of atomic indium deposited on the substrate.

Figure 4.8 shows the X-ray fluorescence trace recorded in the case of silver nanoparticles deposited on a glass substrate at tight focusing conditions. It can be seen from this figure that the angle-dependent fluorescence profile of the silver film shows the presence of nanostructure on the flat surface, as well as indicates a monoatomic

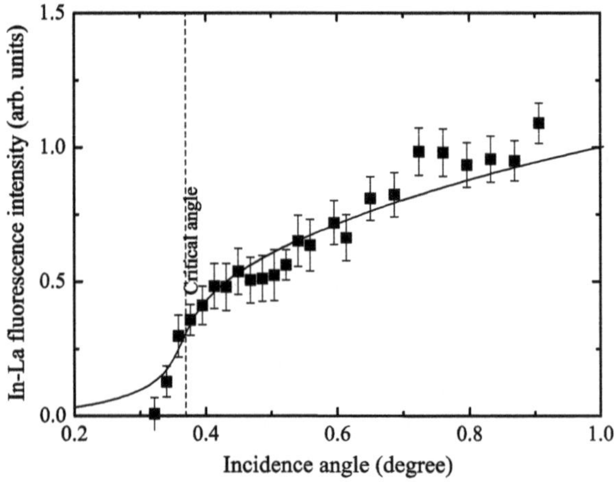

Fig. 4.7 Recorded x-ray fluorescence profile for an indium deposition prepared by ablation at the weak focusing conditions. The profile shows that the indium is deposited in the form of a continuous mono-atomic film instead of nanoparticles. *Solid line* shows the fitted profile. The critical angle of indium is ~0.36° for 8.50 keV x-ray energy. Reproduced from [43] with permission from Optical Society of America

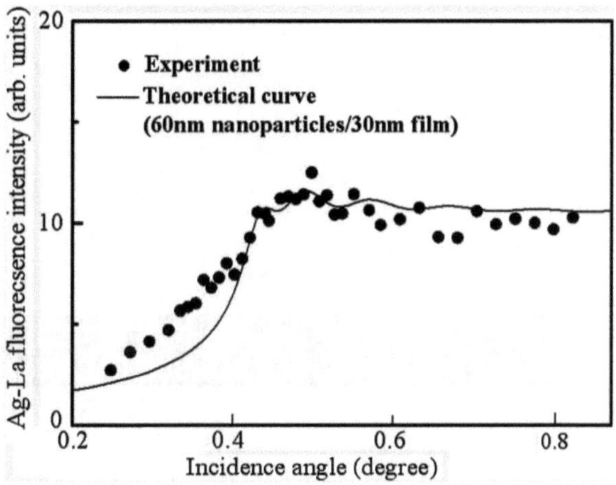

Fig. 4.8 Recorded x-ray fluorescence profile of the silver nanoparticles prepared by the laser ablation in vacuum and deposited on a float glass substrate. The *dots* show experimental data, while the *solid line* shows a fitted profile. Reproduced from [43] with permission from Optical Society of America

layer. The average size of the silver nanoparticles was determined by fitting the recorded fluorescence profile using the CATGIXRF program [47, 50]. The solid line presents the best theoretical fit, from which the average size of the nanoparticles was estimated to be 60 nm, while the thickness of the layer of monoatomic silver particles was estimated to be 30 nm. The TEM measurements also confirmed the presence of silver nanoparticles in these deposited films.

SEM studies of the structural properties of the deposited films showed that, in the case of tight focusing condition, these films contained a lot of nanoparticles with variable sizes. In the weak focusing condition, the concentration of nanoparticles was considerably smaller compared to that in the tight focusing condition. Figure 4.9a shows the SEM images of the deposited chromium nanoparticles on the surface of a silicon wafer. In the case of weak focusing, the deposited film was almost homogeneous with a few nanoparticles appearing in the SEM images (see Fig. 4.9a showing the deposition of chromium), while at tight focusing conditions, plenty of nanoparticles ranging from 30 to 100 nm appeared in the SEM images (see Fig. 4.9b showing the deposition of stainless steel). The average size of those nanostructures was estimated to be 60 nm. An enlarged SEM image of the silver nanoparticles prepared in tight focusing conditions is presented in Fig. 4.9c. The average size of these spherical clusters was also measured to be 60 nm. The same behavior was observed in the case of other targets. Those studies revealed that the material of the target does not play a significant role in the formation of nanoparticles in the case of laser ablation using 300 ps laser pulses in tight focusing conditions. Hence, further studies on the influence of substrate material on the nanoparticle deposition were carried out using the silver target.

Figure 4.10 shows the SEM images of silver deposited on a copper substrate. One can see a considerable difference in the concentrations of nanoparticles in the cases of weak and tight focusing. There is a special interest in silver nanoparticles because of their potential applications. In general, highly dispersed metals have a much higher

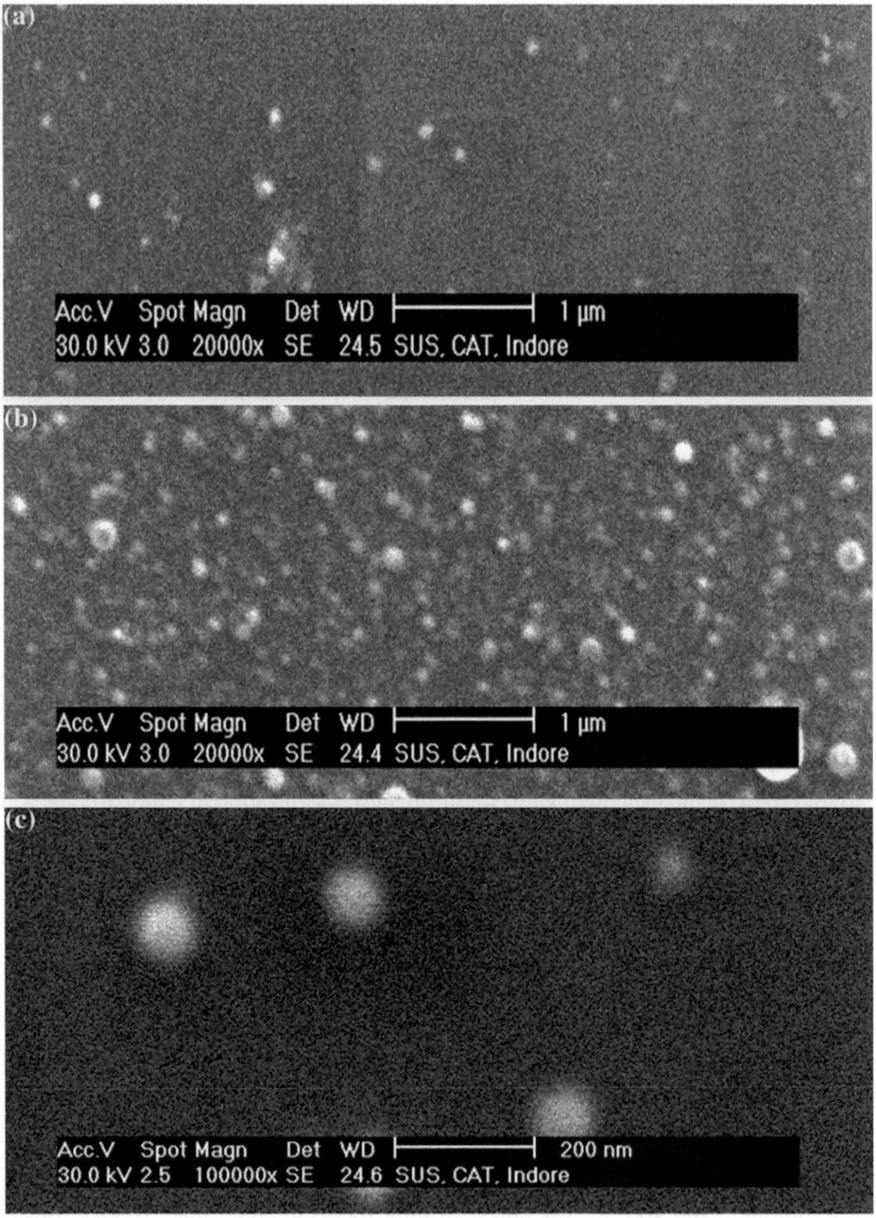

Fig. 4.9 SEM images of the chromium, stainless steel, and silver nanoparticles deposited on the silicon wafer. These images were taken at **a** weak focusing conditions (chromium deposition), **b** tight focusing conditions (stainless steel deposition), and **c** tight focusing conditions (silver deposition). The average size of the spherical clusters in all the three cases was measured to be 60 nm. Reproduced from [43] with permission from Optical Society of America

Fig. 4.10 SEM images of the silver nanoparticles deposited on a copper substrate at **a** weak focusing conditions, and **b** tight focusing conditions. Reproduced from [43] with permission from Optical Society of America

surface area for a given volume, and hence they can be useful for efficient catalytic conversion. Silver nanoparticles are widely used for surface-enhanced Raman scattering. Silver nanoparticles have an advantage over other metal nanoparticles (e.g., gold and copper) from the point of view of the position of the SPR of silver, which is far from the interband transitions. This enables one to investigate the optical and nonlinear optical effects in the silver nanoparticles by focusing on the surface plasmon contribution.

Further studies on the characteristics of nanosized structures of the deposited materials were carried out using the atomic force microscope (AFM). Figure 4.11a shows the AFM image of the silver nanoparticles deposited on the surface of copper strip. The average size of silver nanoparticles was 65 nm. In contrast to this, the AFM images obtained from the deposited films prepared under the weak focusing condition showed considerably smaller number of nanoparticles. Figure 4.11b shows an AFM picture of silver deposition prepared under these conditions. This image indicates the presence of very few nanoparticles. The same difference in AFM pictures was observed in the case of indium deposition under the two focusing conditions.

To characterize the ablation process, the temporal and spectral characteristics of the radiation emitted by the plume were also studied. The oscilloscope traces showed a considerable increase in the duration of the plasma emission in the case of tight focusing that could be expected considering the excitation conditions. A combined analysis of the spectra and oscilloscope traces in the cases of two different regimes

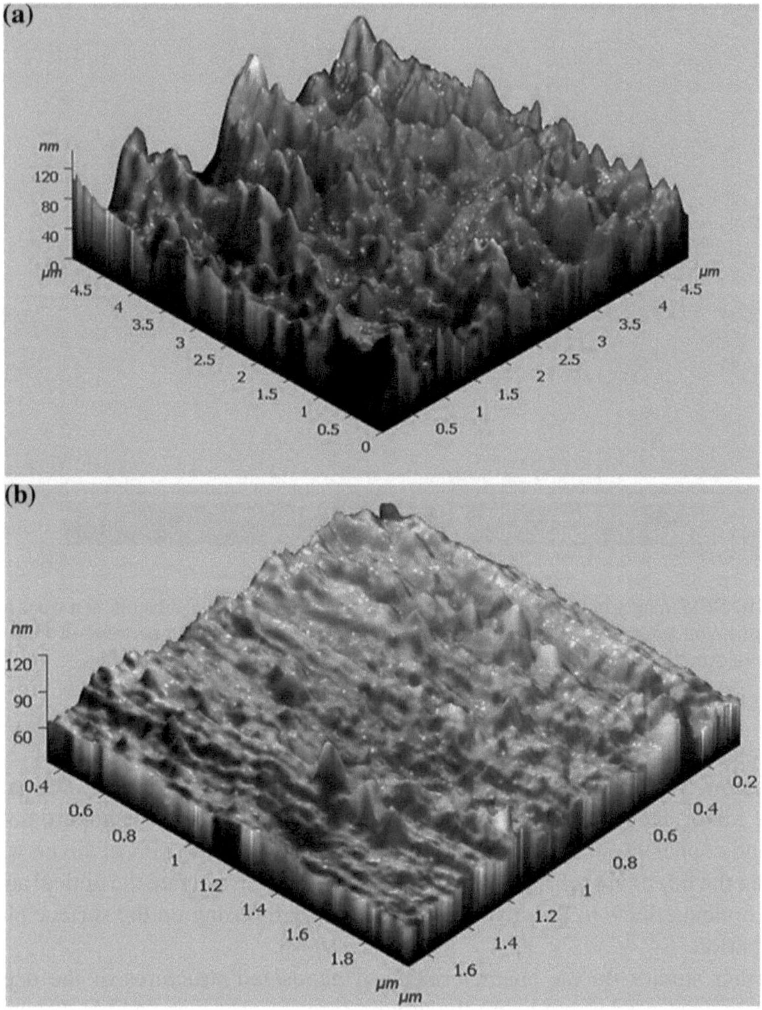

Fig. 4.11 AFM images of **a** the silver nanoparticles deposited on a copper strip at the tight focusing conditions, and **b** the silver nanoparticles deposited on an aluminum strip at the weak focusing conditions. Very few nanoparticles are seen in the case of **b** weak focusing conditions compared to the case of **a** tight focusing conditions. Reproduced from [43] with permission from Optical Society of America

of the excitation of plasma revealed that the structureless continuum appearing in the spectrum in the case of tight focusing is due to the emission from hot nanoparticles produced during laser ablation. Such hot nanoparticles behave like black body radiators emitting for a longer time till they get cooled down.

The model developed in [5] for aluminum predicts that, for short laser pulses at intensities in the range 10^{12}–$10^{13}\,\mathrm{W\,cm^{-2}}$, the adiabatic cooling drives the system into a metastable region of its phase diagram, resulting in the production of a relatively large fraction of nanoparticles. At larger intensities ($\geq 10^{14}\,\mathrm{W\,cm^{-2}}$), the system can never reach the metastable region, resulting in an almost atomized plume.

Pulsed laser deposition using short pulses has gained some interest because of a number of advantages over other processes, such as the possibility of producing materials with a complex stoichiometry and a narrowed distribution of nanoparticle sizes with reduced porosity. Typically, laser deposition is carried out in an ambient gas, which quenches the ablated plume, thus controlling the mean particle size [51]. However, some previously reported studies [5, 52], as well as this study, suggest that nanoparticles are generated as a result of some relaxation processes of the extreme material state reached by the irradiated target surface.

4.4 Characterization of Nanoparticles During Laser Ablation of Nanoparticle-Containing Targets

The ablation-induced nanoparticle formation in laser plumes has carefully been analyzed in multiple experiments [5, 7, 8, 53–59]. In the meantime, only few studies were reported when nanoparticles already exist at the surface of ablated targets. Currently, the nanoparticles with different sizes and shapes are commercially available from various manufacturers and can be attached to the surfaces. In that case one has to carefully define the optimum laser intensity, pulse duration, and focusing conditions for heating of the nanoparticles until they evaporate from the targets. The comparison of the size characteristics of initial nanoparticles and deposited debris becomes a versatile approach for definition of the changes of nanoparticle morphology during laser ablation. The maintenance of the original properties of nanoparticles allows one to analyze the optical and nonlinear optical properties of nanoparticle-containing laser plasma at well-defined conditions.

In this section, we discuss the experimental studies of the structural modifications during laser ablation of nanoparticle-containing targets [60]. The experiments were carried out using the Ti:sapphire laser. To create the ablation, part of uncompressed radiation from the Ti:sapphire laser ($\lambda = 800\,\mathrm{nm}$, $t = 210\,\mathrm{ps}$) was focused on a target placed in the vacuum chamber. The spot size of this radiation on the target surface was maintained in the range of 0.5–0.8 mm. The intensity of this radiation on the target surface was varied between 2×10^9 and $5 \times 10^{10}\,\mathrm{W\,cm^{-2}}$. The laser fluence during ablation was in the range of 0.4–1 $\mathrm{J\,cm^{-2}}$. The chamber was maintained at the vacuum pressure of 8×10^{-4} Pa. The debris from plasma plume was deposited on nearby placed Si wafer and Al foil. The distance between the target and substrates was 40–70 mm. The structure of deposited material was analyzed by the scanning electron microscopy. The commercially available nanoparticles (Cr_2O_3, In_2O_3, Ag, Sn, Au, Cu powder) were glued on the glass or silver substrates by mixing with the drop of superglue. The bulk materials of the same origin as nanoparticle powder were also used to compare the ablation from monoparticle-containing targets.

The sizes of original nanoparticles were varied in the range of 30–300 nm. Figure 4.12 presents the SEM images of some nanoparticles (silver, tin, and gold) before the ablation. The presence of nanoparticles in the plumes was confirmed by analyzing the morphology of the ablated material deposited on the substrates. It was shown that, at optimal excitation conditions, the nanoparticles remain almost intact in the plasma plume. The laser ablation was carried out at different laser intensities on the surface of targets. One has to maintain the intensity when the size characteristics of deposited material remained intact with regard to the initial morphology of nanoparticles. The intensity of subnanosecond pulse at which these conditions were satisfied was in the range of 3×10^9–1×10^{10} W cm^{-2}. At these conditions, the sizes of the nanoparticles deposited on the substrates during laser ablation were close to those glued on the surface of targets. One can note that the duration of deposition plays important role in the aggregation of the debris on the substrates. The deposition of silver nanoparticles over 5 min at $I = 1 \times 10^{10}$ W cm^{-2} and 10 Hz pulse repetition rate led to the aggregation of deposited debris and creation of a film containing large grains (Fig. 4.13a). Furthermore, the increase of laser intensity above certain level ($I \sim 3 \times 10^{10}$ W cm^{-2}, laser fluence 4 J cm^{-2}) led to considerable disintegration or aggregation of nanoparticles on the target surface. This followed by the appearance of chaotic drops of silver aggregates on the substrate surface (Fig. 4.13b). Another pattern was observed in the case of ablation of the bulk targets. The ablation of bulk silver led to the appearance of deposited nanoparticles (~15 nm) on the substrates (Fig. 4.13c). This process was observed at considerably higher laser intensity ($I \sim 6 \times 10^{10}$ W cm^{-2}). No nanoparticle formation was observed on the substrates at moderate intensities of laser radiation on the bulk target surface ($5 \times 10^9 - 2 \times 10^{10}$ W cm^{-2}).

The SEM images of deposited nanoparticles in most cases, when the optimal excitation of nanoparticle-containing targets was maintained, also revealed that they remain approximately the same as the initial powders. The sizes of deposited Cr_2O_3 clusters were in the range of 30–150 nm. The same can be said about the Au (40–180 nm), Cu (60–200 nm) and other deposited clusters. The SEM images of initial and deposited In_2O_3 and Cr_2O_3 nanoparticle debris are presented in Fig. 4.14. The images of deposited Sn and Au nanoparticles are shown in Fig. 4.15. The intensity of laser pulse at the nanoparticle-containing targets was kept during those studies in the range of 3×10^9–8×10^9 W cm^{-2}. The elemental consistence of ablated material and debris on the surface of deposited substrates was confirmed using the EDX spectroscopy (Fig. 4.16).

The size distribution of tin nanoparticles on the target and after the ablation and deposition on the Si substrate at the optimal conditions of ablation is presented in Fig. 4.17. One can see that, at optimal conditions of laser ablation ($I = 8 \times 10^9$ W cm^{-2}), the morphology of ablated powder basically retained the same as before the ablation. At the same time broader wings of size distribution point out the appearance of both small and large nanoparticles due to partial aggregation and disintegration of some nanoparticles (Fig. 4.17b). Note that the mean sizes of nanoparticles before and after irradiation remained almost unchanged (130 and 150 nm respectively).

Fig. 4.12 SEM images of various nanoparticle powders taken before the ablation. **a** Ag, **b** Sn, **c** Au. Reproduced from [60] with permission from Springer Science + Business Media

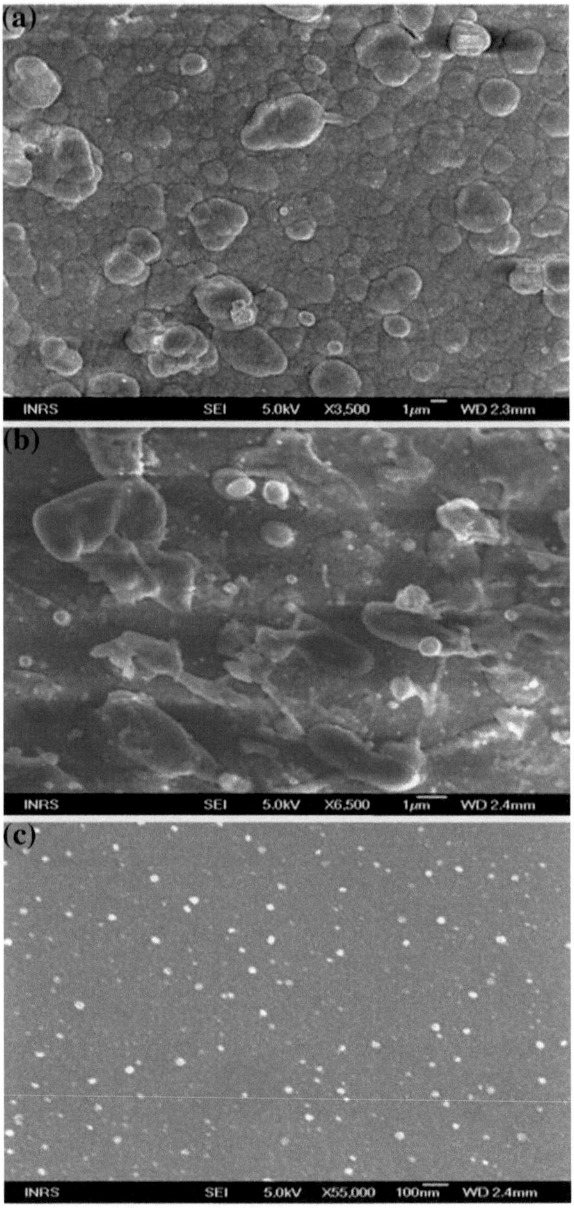

Fig. 4.13 SEM images of the **a** appearance of the grained silver film created on the Si substrate during ablation ($I \sim 1 \times 10^{10}\,\mathrm{W\,cm^{-2}}$) of the nanoparticle-containing target, **b** chaotic deposition of silver debris deposited on the Si substrate at the intensities exceeding threshold of disintegration of nanoparticles on the target surface ($I \geq 2 \times 10^{10}\,\mathrm{W\,cm^{-2}}$), and **c** silver nanoparticles produced during laser ablation of silver bulk target at $I \sim 6 \times 10^{10}\,\mathrm{W\,cm^{-2}}$. Reproduced from [60] with permission from Springer Science+Business Media

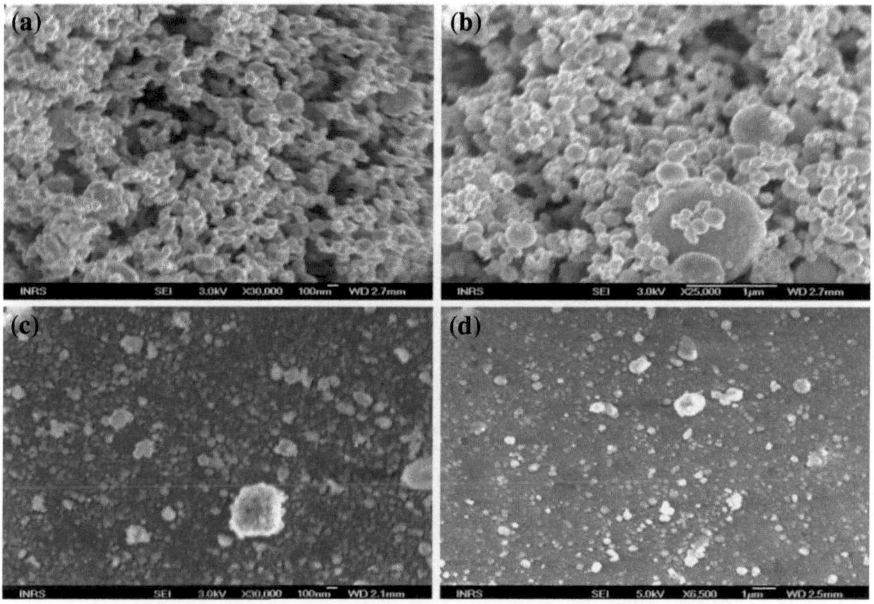

Fig. 4.14 SEM images of the initial **a** In_2O_3 and **b** Cr_2O_3 nanopowders and of the debris of **c** In_2O_3 and **d** Cr_2O_3 nanoparticles deposited on the Al substrate during laser ablation of nanopowder-containing targets. Reproduced from [60] with permission from Springer Science+Business Media

4.5 Nanoparticle Formation During Laser Ablation of Metals at Different Pressures of Surrounding Noble Gases

Pulsed laser deposition in vacuum using short pulses has gained the interest because of a number of advantages over other processes, such as the possibility of producing materials with a complex stoichiometry and a narrowed distribution of nanoparticle sizes with reduced porosity. When laser deposition is carried out in an ambient gas, the latter quenches the ablated plume. Some previously reported studies [5, 52] suggest that nanoparticles are generated as a result of some relaxation processes of the extreme material state reached by the irradiated target surface. The influence of surrounding gas on the conditions of cluster formation during laser ablation of the metals by short laser pulses has previously been analyzed only at two conditions, when the target was placed in vacuum or ambient air. It is of interest to analyze the influence of the concentration of surrounding gas on the cluster formation. One can study this process using noble gases of different atomic numbers (Z) at the pressures varying between 10^{-2} torr (i.e vacuum conditions) and 760 torr (i.e. atmospheric pressure). It would be interesting to analyze whether there is a threshold in gas pressure scale above which the conditions of nanoparticle formation get spoiled.

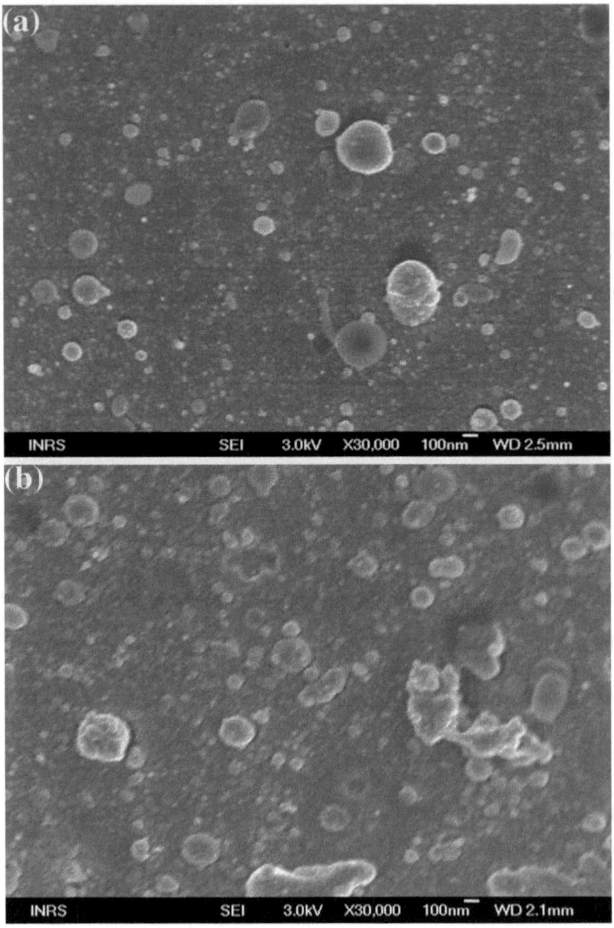

Fig. 4.15 SEM images of deposited debris from the **a** Au and **b** Sn nanoparticle-containing targets. Reproduced from [60] with permission from Springer Science+Business Media

In this section, we analyze the dynamics of nanoparticle formation during laser ablation of silver at different pressures of surrounding noble gases (helium and xenon) and discuss the formation of silver nanoparticles at these conditions, while comparing with those in the case of vacuum and atmospheric air [61]. The experiments were carried out using the mode locked Nd:YAG laser. To create the ablation, a single pulse from the pulse train of oscillator was amplified ($\lambda = 1064$ nm, $t = 38$ ps, $E = 10$ mJ, 2 Hz pulse repetition rate) and focused on a target placed in the vacuum chamber. The laser pulses were focused on a bulk target (Ag or Cu plates) at two regimes of focusing. In the first case (tight focusing), the intensity of laser radiation was in the range of (0.5-2) $\times$ 10^{12} W cm^{-2}, and in the second case (weak focusing), the intensity was considerably smaller ($\sim$4 $\times$ 10^{10} W cm^{-2}). Note that in the latter

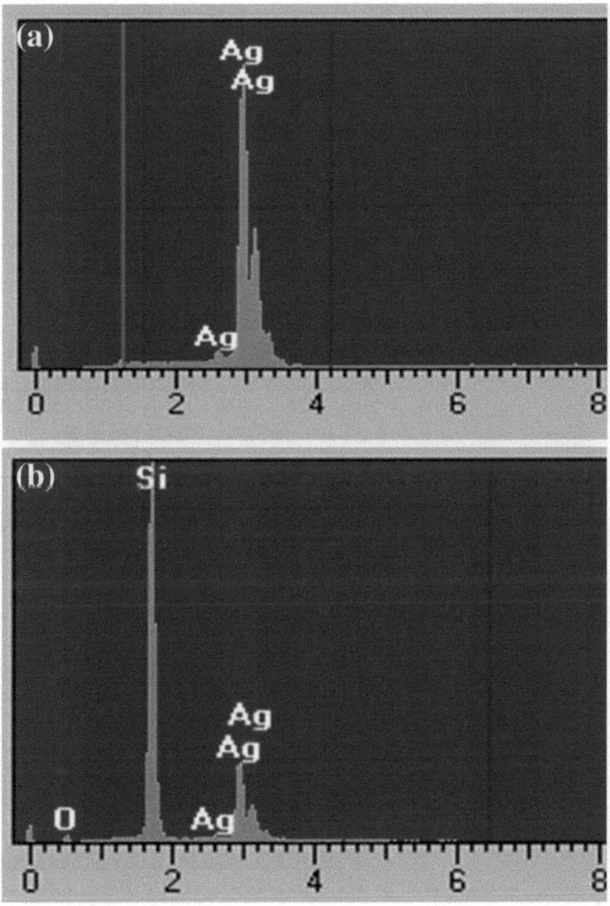

Fig. 4.16 EDX spectra of the **a** silver powder and **b** debris from the silver nanoparticle-containing target deposited on the Si substrate

case no nanoparticles were observed on the deposition of the debris. The laser fluence during ablation at tight focusing conditions was in the range of few tens of $\mathrm{J\,cm^{-2}}$. The chamber was maintained at the vacuum pressure of 8×10^{-4} torr. The pressure was varied by adding different noble gases up to $p = 300$ torr. The debris from the plasma plume was deposited on the BK7 glass plates. The distance between the target and substrates was 40 mm. The presence of nanoparticles was inferred by analyzing the spatial characteristics of the deposited material and the spectral absorption of the deposited material. The structure of deposited material was analyzed by an atomic force microscope. The absorption spectra of the deposited films were analyzed by a fiber spectrometer. The targets were ablated during 30 min using the picosecond pulses at 2 Hz pulse repetition rate. In the case of vacuum deposition, relatively thick ($\sim$0.6–0.9 μm) Ag and Cu films were produced. The deposition in air conditions at analogous parameters of experiments showed considerably thinner layers

Fig. 4.17 Histograms of the tin nanoparticle size distribution **a** *before* and **b** *after* ablation of the nanopowder-containing target. Reproduced from [60] with permission from Springer Science+Business Media

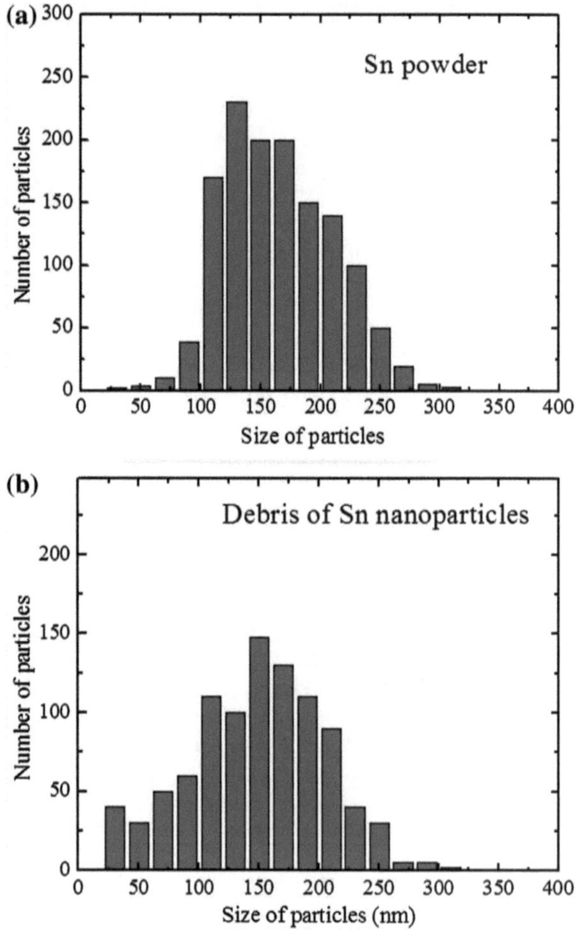

of deposited material (~0.03–0.06 μ m). The nanoparticle formation was mostly described as a process of short excitation of electronic gas and transfer of this energy to the atomic cell with further aggregation processes, which continue during evaporation of the material [6].

The analysis of deposited films has shown a considerable difference in their morphology. Figure 4.18 shows the AFM images of the Ag and Cu films deposited in vacuum at tight focusing conditions. Here the histograms of nanoparticle size distribution corresponding to the AFM images of these films are also shown. The mean sizes of nanoparticles were 20 nm (in the case of Ag ablation) and 60 nm (in the case of Cu ablation). The sharp images of nanoparticles indicate that concentration of the deposited atomic layer containing single particles is insignificant. To define the relative concentrations of nano- and monoparticles, one has to use the time-of-flight mass spectroscopy, which was not available in those studies. However, analysis of AFM images has clearly indicated a difference in the "sharpness" of nanoparticle images in

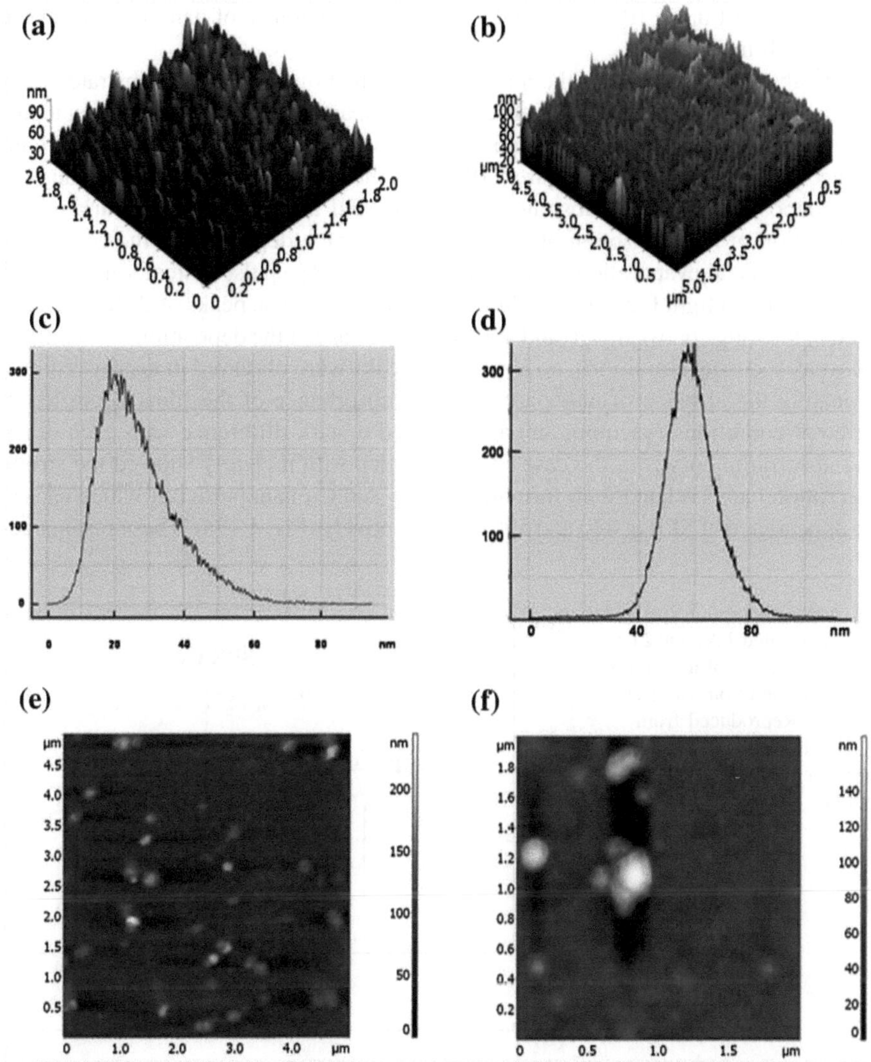

Fig. 4.18 AFM images of the **a** Ag and **b** Cu films deposited during laser ablation in vacuum and corresponding histograms of **c** Ag and **d** Cu nanoparticle size distribution. AFM images of the **e** Ag and **f** Cu films deposited during laser ablation in air. Reproduced from [61] with permission from Springer Science+Business Media

the case of vacuum ($p < 10^{-2}$ torr) and air deposition. Figure 4.18e shows the AFM image of the deposited Ag film produced during laser ablation at air conditions. One can see a considerable decrease of nanoparticle concentration on this image. The same feature was observed in the case of the Cu film deposited in air (Fig. 4.18f). The analysis of nanoparticle formation at different pressures of surrounding gas describing below showed a decrease of the "sharpness" of nanoparticle images at the pressures of up to 10–30 torr, with further disappearance of nanoparticles in the deposited films.

The absorption spectra of the materials deposited on transparent substrates were used to determine the presence or absence of nanoparticles. The presence of nanoparticles was inferred from the appearance of strong absorption bands associated with surface plasmon resonance. Figure 4.19 shows the absorption spectra of Ag and Cu films deposited on glass substrates at vacuum and air conditions. A variation of the position of the absorption of Ag deposition, which depended on the conditions of excitation and evaporation of bulk targets by the 38 ps pulses interacting with the metal surface at tight focusing conditions was observed. The peaks of SPR were centered in the range between 470 and 490 nm. In the case of the deposition of Ag film at the weak focusing conditions, no absorption peaks were observed in nearby regions, indicating the absence of nanoparticles. The absorbance of the films deposited at different conditions (vacuum, air) also showed a stark difference with each other. The absorption curves in the case of vacuum deposition clearly showed the strong absorption bands related with formation of Ag and Cu nanoparticles, which caused the appearance of SPR at 470 and 590 nm respectively (Fig. 4.19a). These resonances

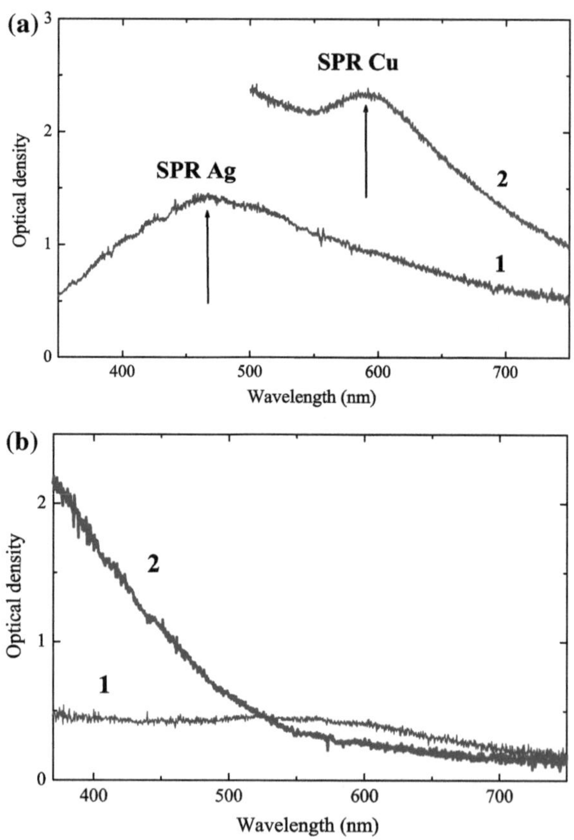

Fig. 4.19 Absorption spectra of the deposited *1* Ag and *2* Cu films in the cases of **a** vacuum and **b** air conditions of laser ablation. Reproduced from [61] with permission from Springer Science+Business Media

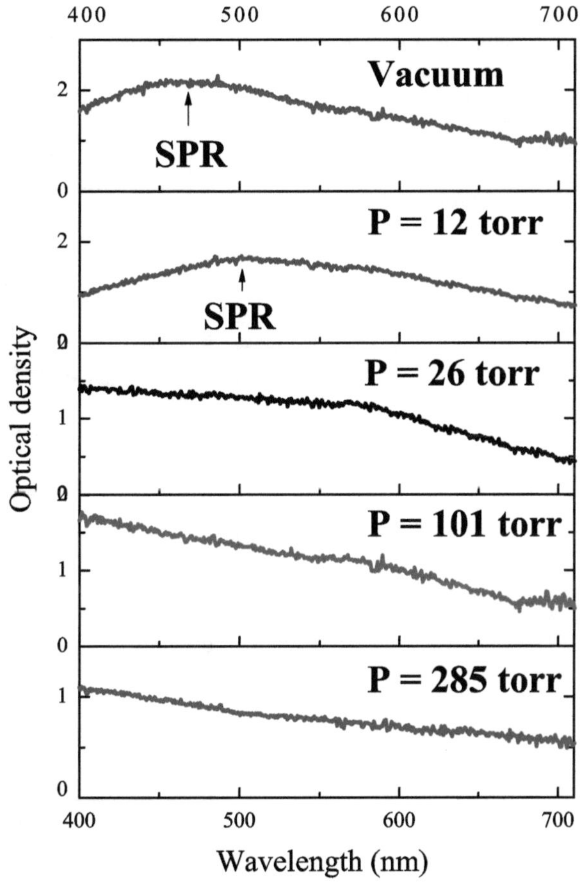

Fig. 4.20 Absorption spectra of the silver films deposited at vacuum conditions and different pressures (12, 26, 101, and 285 torr) of xenon. Reproduced from [61] with permission from Springer Science+Business Media

are a direct consequence of dielectric confinement and can be interpreted in terms of a collective motion of the electrons in nanoparticles. In contrast, in the case of laser ablation at air conditions, no resonance absorption bands appeared in the absorption spectra of deposited materials (see Fig. 4.19b).

The correlation between the morphology and absorption studies, showing nanoparticles and SPR only in the case of vacuum deposition, allowed concluding about the absorption measurements as a versatile method for definition of nanoparticle formation. Below we analyze the studies of nanoparticle formation at different pressures of noble gases. As it was mentioned, the ablation of bulk silver in vacuum led to the appearance of deposited nanoparticles (~20 nm) on the substrates. This process was carried out at relatively high laser intensity ($I > 5 \times 10^{11}$ W cm^{-2}). No nanoparticles were observed on the substrates at moderate intensities of laser radiation on the bulk target surface (5×10^9–7×10^{10} W cm^{-2}). The light (He, $Z = 4$) and heavy (Xe, $Z = 131$) noble gases were used to define the influence of weight characteristics of the surrounding particles on the nanoparticle formation during laser ablation of silver target at different gas pressures.

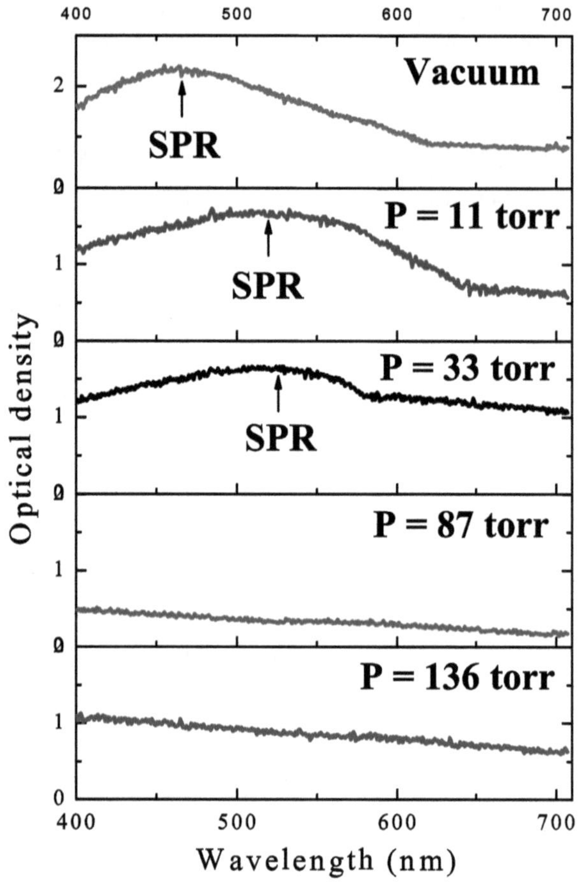

Fig. 4.21 Absorption spectra of the silver films deposited at vacuum conditions and different pressures (11, 33, 87, and 136 torr) of helium. Reproduced from [61] with permission from Springer Science+Business Media

Figure 4.20 shows the absorption spectra of deposited Ag films in the case of different pressures of xenon. One can see that SPR appears up to 12 torr. Above this pressure, only monotonic growth of absorption towards blue side was observed. The same measurements in the case of helium (Fig. 4.21) showed analogous tendency, when the SPR of silver nanoparticles disappeared from the absorption spectra of the deposited films obtained at gas pressure above 33 torr. Note a shift of SPR towards the longer wavelengths with increase of gas pressure. The difference in threshold pressures for these two gases can be explained as follows. With the growth of surrounding particles, the formation of nanoparticles becomes suspended due to some impeding processes. One can assume that, above the threshold pressure, these particles start to play negative role in the aggregation of silver particles during the heating of ablated targets by short pulses. Depending on atomic weight, they can diminish the probability of accumulation of Ag particles in the clusters at different pressures of surrounding gas. Heavy xenon particles more strongly affect the cluster formation compared with lighter helium particles. This follows with smaller threshold pressure in the former case, above which the nanoparticle formation almost stops.

References

1. R.A. Ganeev, A.I. Ryasnyansky, A.L. Stepanov, C. Marques, R.C. da Silva, E. Alves, Opt. Commun. **253**, 205 (2005)
2. M. Falconieri, G. Salvetti, E. Cattaruza, F. Gonella, G. Mattei, P. Mazzoldi, M. Piovesan, G. Battaglin, R. Polloni, Appl. Phys. Lett. **73**, 288 (1998)
3. S. Debrus, J. Lafait, M. May, N. Pinçon, D. Prot, C. Sella, J. Venturini, J. Appl. Phys. **88**, 4469 (2000)
4. R.A. Ganeev, J. Phys. B: At. Mol. Opt. Phys. **40**, R213 (2007)
5. S. Amoruso, G. Ausanio, R. Bruzzese, M. Vitiello, X. Wang. Phys. Rev. B **71**, 033406 (2005)
6. D. Perez, L.J. Lewis, Phys. Rev. B **67**, 184102 (2003)
7. D. Scuderi, O. Albert, D. Moreau, P.P. Pronko, J. Etchepare, Appl. Phys. Lett. **86**, 071502 (2005)
8. J. Perrière, C. Boulmer-Leborgne, R. Benzerga, S. Tricot, J. Phys. D: Appl. Phys. **40**, 7069 (2007)
9. R. Teghil, L. D.'Alessio, A. Santagata, M. Zaccagnino, D. Ferro, D. J. Sordelet. Appl. Surf. Sci. **210**, 307 (2003)
10. T.E. Glover, J. Opt. Soc. Am. B **20**, 125 (2003)
11. V. Kabashin, M. Meunier, J. Appl. Phys. **94**, 7941 (2003)
12. H.O. Jeschke, M.E. Garsia, K.H. Bennemann, Phys. Rev. Lett. **87**, 015003 (2001)
13. Y. Suda, T. Nishimura, T. Ono, M. Akazawa, Y. Sakai, N. Homma, Thin Solid Films **374**, 287 (2000)
14. T. Makimura, Y. Kunii, K. Murakami, Jpn. J. Appl. Phys. **35**, 4780 (1996)
15. J. Gonzalo, A. Perea, D. Babonneau, C.N. Afonso, N. Beer, J.-P. Barnes, A.K. Petford-Long, D.E. Hole, P.D. Townsend, Phys. Rev. B **71**, 125420 (2005)
16. S.I. Anisimov, Y.A. Imas, G.S. Romanov, Y.V. Khodyko, *High Power Radiation Effect in Metals* (Nauka, Moscow, 1970)
17. R.A. Ganeev, A.I. Ryasnyansky, Opt. Commun. **246**, 163 (2005)
18. Q. Chen, M. Tanaka, K. Furuya, J. Surf. Anal. **5**, 348 (1999)
19. K.-L. Tsai, J.L. Dye, J. Am. Chem. Soc. **113**, 1650 (1991)
20. Y. Zhao, Zh Zhang, H. Dang, J. Phys. Chem. B. **107**, 7574 (2003)
21. G.T. Cardenas, E.C. Salgado, J. Morales, H.Z. Soto, J. Appl. Polym. Sci. **73**, 1239 (1999)
22. K. Soulantica, A. Maisonnat, M.-C. Fromen, M.-J. Casanove, P. Lecante, B. Chaudret, Angew. Chem. Int. Ed. **40**, 448 (2001)
23. Meliorum Technologies Proprietary, http://www.meliorum.com/
24. R.A. Ganeev, A.I. Ryasnyanskiy, U. Chakravarty, P.A. Naik, H. Srivastava, M.K. Tiwari, P.D. Gupta, Appl. Phys. B **86**, 337 (2007)
25. U. Kreibig, L. Genzel, Surf. Sci. **156**, 678 (1985)
26. C.F. Bohren, D.R. Huffman, *Absorption and Scattering of Light by Small Particles* (Wiley, New York, 1983)
27. P.K. Khanna, K.-W. Jun, K.B. Hong, J.-O. Baeg, R.C. Chikate, B.K. Das, Mater. Lett. **59**, 1032 (2005)
28. T.V. Shubina, D.S. Plotnikov, A. Vasson, J. Leymarie, M. Larsson, P.O. Holtz, B. Monemar, H. Lu, W.J. Schaff, P.S. Kop'ev, J. Cryst. Growth **288**, 230 (2006)
29. N. Esser, A.M. Frisch, A. Röseler, S. Schintke, C. Goletti, B.O. Fimland, Phys. Rev. B **67**, 125306 (2003)
30. M.K. Tiwari, B. Gowrishankar, V.K. Raghuvanshi, R.V. Nandedkar, K.J.S. Sawhney, Bull. Mater. Sci. **25**, 435 (2002)
31. M.A. Malik, P. O'Brien, S. Norager, J. Smith, J. Mater. Chem. **13**, 2591 (2003)
32. J. Perriere, E. Millon, M. Chamarro, M. Morcrette, C. Andreazza, Appl. Phys. Lett. **78**, 2949 (2001)
33. A. Borowiec, M. Mackenzie, G.C. Weatherly, H.K. Haugen, Appl. Phys. A **77**, 411 (2003)
34. K.V. Anikin, N.N. Melnik, A.V. Simakin, G.A. Shafeev, V.V. Voronov, A.G. Vitukhnovsky, Chem. Phys. Lett. **336**, 357 (2002)

35. R.A. Ganeev, A.I. Ryasnyansky, T. Usmanov, Opt. Quantum Electron. **35**, 211 (2003)
36. L. Banyai, M. Lindberg, S.W. Koch, Opt. Lett. **13**, 212 (1988)
37. R.A. Ganeev, M. Baba, A.I. Ryasnyansky, M. Suzuki, H. Kuroda, Appl. Phys. B **80**, 595 (2005)
38. S.S. Kher, R.L. Wells, Chem. Mater. **6**, 2056 (1994)
39. B.L. Justus, R.J. Tonucci, A.D. Berry, Appl. Phys. Lett. **61**, 3151 (1992)
40. H. Yao, S. Takahara, H. Mizuma, T. Kosegi, T. Hayashi, Jpn. J. Appl. Phys. **35**, 4633 (1996)
41. R.A. Ganeev, A.I. Ryasnyansky, M.K. Kodirov, S.R. Kamalov, V.A. Li, R.A. Tugushev, T. Usmanov, Appl. Phys. B **74**, 47 (2002)
42. X.M. Yan, J. Ni, M. Robbins, H.J. Park, W. Zhao, J.M. White, J. Nanopart. Res. **4**, 525 (2002)
43. R.A. Ganeev, U. Chakravarty, P.A. Naik, H. Srivastava, C. Mukherjee, M.K. Tiwari, R.V. Nandedkar, P.D. Gupta, Appl. Opt. **46**, 1205 (2007)
44. R.A. Ganeev, M. Baba, A.I. Ryasnyansky, M. Suzuki, H. Kuroda, Opt. Commun. **240**, 437 (2004)
45. R.A. Ganeev, A.I. Ryasnyansky, A.L. Stepanov, T. Usmanov, Phys. Status Solidi B **241**, R1 (2004)
46. T.D. Donnelly, T. Ditmire, K. Neuman, M.D. Perry, R.W. Falcone, Phys. Rev. Lett. **76**, 2472 (1996)
47. R. A. Ganeev M. Suzuki, M. Baba, M. Ichihara, H. Kuroda, J. Appl. Phys. **103**, 063102 (2008)
48. M.J. Bedzyk, G.M. Bommarito, J.S. Schildkraut, Phys. Rev. Lett. **62**, 1376 (1989)
49. D.K.G. de Boer, Phys Rev. B **44**, 498 (1991)
50. M.K. Tiwari, K.J.S. Sawhney, B. Gowri Sankar, V.K. Raghuvanshi, R.V. Nandedkar, Spectrochim. Acta B **59**, 1141 (2004)
51. K. Sturm, S. Fäahler, H.U. Krebs, Appl. Surf. Sci. **154–155**, 462 (2003)
52. S. Amoruso, R. Bruzzese, N. Spinelli, R. Velotta, M. Vitiello, X. Wang, G. Ausanio, V. Iannotti, L. Lanotte, Appl. Phys. Lett. **84**, 4502 (2004)
53. T.E. Glove, J. Opt. Soc. Am. B **20**, 125 (2003)
54. D. Perez, L.J. Lewis, Phys. Rev. B **67**, 184102 (2003a)
55. W. Marine, L. Patrone, B. Lukyanchuk, M. Sentis, Appl. Surf. Sci. **154–155**, 345 (2000)
56. S. Amoruso, G. Ausanio, A.C. Barone, R. Bruzzese, L. Gragnaniello, M. Vitiello, X. Wang, J. Phys. B: At. Mol. Opt. Phys. **38**, L329 (2005)
57. S. Noël, J. Hermann, I. Itina, Appl. Surf. Sci. **253**, 6310 (2007)
58. S. Amoruso, R. Bruzzese, X. Wang, G. Ausanio, L. Lanotte, J. Phys. B: At. Mol. Opt. Phys. **40**, 1253 (2007)
59. M. Kaku, S. Suetake, Y. Senba, S. Kubodera, M. Katto, T. Higashiguchi, Appl. Phys. Lett. **92**, 181503 (2008)
60. R.A. Ganeev, L.B. Elouga Bom, T. Ozaki, Appl. Phys. B **96**, 491 (2009)
61. R.A. Ganeev, G.S. Boltaev, R.I. Tugushev, T. Usmanov, Appl. Phys. A **100**, 119 (2010)

Chapter 5
Low-Order Nonlinear Optical Characterization of Clusters

Nowadays, low-order nonlinearities of nanoparticles attract an attention and are among the mostly studied characteristics of these structures due to various potential applications. Below we present a retrospective of the most recent studies of the nanoparticles' nonlinearities.

Metal nanoparticles are among the most studied ones. Nonlinear and thermo optical properties of the gold and silver nanocolloids were investigated in [1]. The signs of the nonlinear refraction coefficients of these colloids were negative. The linear and nonlinear absorption coefficients of gold nanocolloid and silver nanocolloid are determined to be $0.9\,\mathrm{cm}^{-1}$, $2.3\times10^{-3}\,\mathrm{cm\,W}^{-1}$ and $0.22\,\mathrm{cm}^{-1}$, $1.9\times10^{-4}\,\mathrm{cm\,W}^{-1}$, respectively. Nanostructured silver and linear carbon chain (LCC) particle water colloids were prepared in [2] by a pulsed laser ablation procedure. The Ag nanoparticles induce a limiting threshold reduction, an increased nonlinear absorption coefficient and a marked asymmetrical peak/valley profile of the (Ag:LCC) when compared to the LCC. All these nonlinear contributions determine the increase of the third-order susceptibility, while maintaining a significantly high linear transmission (75 %) at 532 nm and high photostability. The magnitude of the nonlinear optical response of these nanohybrids makes them promising candidates for potential optoelectronic applications.

Metal nanoclusters composite glasses synthesized by ion implantation have been shown as promising nonlinear photonic material. In [3], the nonlinear absorption measurements of gold nanoparticles implanted in four structurally different types of silicate glasses are reported. All targets containing gold nanoparticles in a layer 500 nm under the surface of the glass have been prepared by ion implantation with subsequent annealing. The targets were characterized by UV-VIS absorption spectroscopy, transmission electron microscopy, and by the z-scan technique. The resulting nanoparticles differed in size and shape, as well as depth distribution characteristic

We present a retrospective of the most recent studies of the nanoparticles' nonlinearities. We discuss some of studies of the nonlinearities of various nanoparticles (indium, silver, semiconductors, nanocrystallite photorefractive media) prepared by different methods.

R. A. Ganeev, *Nonlinear Optical Properties of Materials*,
Springer Series in Optical Sciences 174, DOI: 10.1007/978-94-007-6022-6_5,

for glasses with different chemical compositions. With the z-scan technique, it was shown that the nanoparticles produced in silicate glasses exhibit substantial two-photon absorption (TPA). The TPA coefficient differed depending on size, shape, and depth distribution of the metal nanoclusters and the structure and composition of the glass substrates. The highest TPA coefficient ($16.25\,\mathrm{cm\,GW^{-1}}$) was found for the glass BK7 in which the largest non-spherical nanoparticles have been observed in the thinnest layer.

The nonlinear optical properties of Au-Ag nanoplanets produced by ion implantation and irradiation in silica, experimentally investigated by means of the single beam z-scan technique were studied in [4]. The measurements provided experimental evidence of the intense local-field enhancement theoretically demonstrated for these plasmonic nanosystems. In particular, this has a dramatic impact on their nonlinear absorption behavior and results in a tunable changeover from reverse saturated absorption to saturated absorption by slightly varying the pump intensity and in the possibility to activate and observe nonlinear phenomena of the electron dynamics otherwise inaccessible in the intensity range that can be employed to study these materials. Finally, for the nanoplanet configuration, a dramatic decrease of the intensity-dependent absorption coefficient was found, which could be very promising for obtaining optical gain materials.

Silver nanocolloid was prepared in [5] by chemical reduction method and its nonlinear absorption properties were investigated by using open aperture z-scan with nanosecond laser pulses operating at 532 nm. A switch over from saturated absorption to reverse saturated absorption was observed when the input intensity is increased from 28.1 to $175.8\,\mathrm{MW\,cm^{-2}}$. The underlying mechanism responsible for the observed switching behavior is the interplay between ground state plasmon band bleaching and excited state absorption.

A simple and universal method for fabricating various kinds of metal and semiconductor (Si, Ge, Bi, and Cu) nanoparticle-glass composites by using metallic Al as a reducing agent in the raw materials of the glass batches was reported in [6]. The fabricated nanoparticle glass composites exhibit large third-order optical nonlinearities ($\chi^{(3)}$ up to 10^{-8} esu) and an ultrafast response time (within picoseconds), which makes them possible to manufacture ultrafast all-optical switches.

In [7], the nonlinear optical absorption of gold nanoparticles dispersed in polyvinyl alcohol (Au:PVA) is investigated using open aperture z-scan technique. It was reported that the nature of the nonlinear absorption depends on the excitation wavelength. The optical limiting capability of the sample is demonstrated at a wavelength of 532 nm.

Metal nanocluster composite glass was formed by Cu ion implantation into silica using metal vapor vacuum arc ion source and analyzed in [8]. Third-order nonlinear optical properties of the nanoclusters were measured at 1064 nm excitations using z-scan technique. The mechanisms responsible for the nonlinear response were discussed. Third-order nonlinear susceptibility $\chi^{(3)}$ of this kind of sample was determined to be 4.2×10^{-8} esu.

In [9], ellipsoidal gold nanoparticles-embedded bismuthate glasses have been prepared via a facile melt-annealing approach. Femtosecond z-scan measurement shows that the nanocomposites exhibit a maximum third-order nonlinear susceptibility of 4.88×10^{-10} esu at 800 nm, which is two orders higher than that of the host glass. Optical Kerr shutter measurement demonstrates ultrafast response time (in scale of sub-picosecond) of the intraband transition enhanced third-order nonlinearities.

An integrated plasmonic nanocomposite showing a nonlinear optical response changing its sign with wavelength, depending on its position with respect to the surface plasmon resonance of the nanocomposite was analyzed in [10]. This nanocomposite is a SiO_2 matrix containing both, embedded quasi-spherical Ag nanoparticles and silicon quantum dots. The wavelengths used for the picosecond z-scan study were 355 and 532 nm, which are localized at both sides of the surface plasmon resonance of the Ag nanoparticles (395 nm), and 1064 nm, which is localized well far away of it. The integrated plasmonic system shows a positive nonlinear refraction below the plasmon resonance, changing to a negative value for wavelengths above resonance. On the other hand, below resonance the nonlinear absorption is cancelled due to opposite responses from the individual nanosystems, while above resonance only saturated absorption is observed.

Tetrahedral, icosahedral, and cubic gold nanoparticles have been synthesized in the experiments described in [11] by reducing chloroauric acid with ethylene glycol in the presence of polyvinyl pyrrolidone. The nonlinear properties of the gold nanoparticles have been investigated by using the open aperture z-scan technique with the nanosecond pulse at a wavelength of 532 nm. It shows out that all the gold nanoparticles possess strong optical nonlinearities when being excited at their surface plasmon resonances. It is identified that the nonlinear properties of the gold nanoparticles are dependent on their shapes, and the icosahedra gold nanoparticles show much better nonlinear optical properties than the tetrahedral and cubic ones.

Semiconductor nanoparticles also attracted the attention during last time due to their strong nonlinear optical properties. In particular, in [12], CdTe nanocrystals were synthesized by wet chemical route. The z-scan studies were carried out to measure the sign and magnitude of the optical nonlinearity of CdTe nanoparticles. The calculated values of n_2, β, and $\chi^{(3)}$ from z-scan measurements were $-5.45\times10^{-9}\,\mathrm{cm^2\,W^{-1}}$, $-1.048\times10^{-4}\,\mathrm{cm\,W^{-1}}$ and 24.64×10^{-6} esu, respectively.

In [13], ZnS nanoparticles were prepared by a simple chemical method and using polyvinyl pyrrolidone as capping agent. The nonlinear optical properties of ZnS nanoparticles in aqueous solution were studied by z-scan technique using CW He-Ne laser at 632.8 nm. The nonlinear absorption coefficient was estimated to be as high as $3.2\times10^{-3}\,\mathrm{cm\,W^{-1}}$ and the nonlinear refractive index was in order of $10^{-8}\,\mathrm{cm^2\,W^{-1}}$. The sign of the nonlinear refractive index was obtained negative that indicated this material exhibits self-defocusing optical nonlinearity.

In [14], colloidal CdSe quantum dots, whose surface traps are controlled with the purification process, are measured with a z-scan technique using a 80 fs pulse duration laser source. Their two-photon absorption coefficient and refractive index are decreased more than one order of magnitude with the increase of the density of surface traps, indicating that the surface states are involved in the two-photon process.

The sodium borosilicate glass containing Bi_2S_3 nanocrystals was prepared by employing both sol-gel and atmosphere control methods in [15]. The third-order nonlinear optical absorption of the glass is investigated in detail using the open aperture z-scan technique at the wavelength of 770 nm with a pulse width of 200 fs. The results show that the transformation from saturated absorption to reverse saturated absorption in the glass is observed with the increase of the input light intensity of the laser used.

There is also increased interest in formation of various ***films, nanowires, etc.*** containing nanoparticles possessing high nonlinearities. Thin films made from a composite of the polymer polyvinyl-alcohol and cobalt oxide (Co_3O_4) nanoparticles were fabricated by spin coating in [16]. Linear and nonlinear optical properties of thin films with thicknesses of hundreds of nanometers were investigated. Reverse saturated absorption and saturated absorption were observed when the films were illuminated with the different fluences. Optical nonlinearities corresponding to reverse saturated absorption were measured by the z-scan technique. A nonlinear refractive index of $10^{-10}\,cm^2\,W^{-1}$ and nonlinear absorption coefficient of $10^3\,cm\,GW^{-1}$ have been measured from 425 to 675 nm. The experimental results show that the Co_3O_4 nanoparticle/PVA composite is a promising material for nonlinear optical devices in the visible range, since it takes advantages of the high optical nonlinearities of transition metal oxides and the superior mechanical properties and convenient fabrication properties of polymers.

The enhanced linear absorption and modified nonlinear absorption of TiO_2 nanowires coated with Ag nanoparticles were reported in [17]. Experimental results indicated that the coated Ag nanoparticles significantly increased the linear absorption of the nanostructures in the wavelength range of visible light. Z-scan experiments showed that when the excitation energy increased, the nonlinear absorption of the TiO_2 nanowires changed from reverse saturated absorption to saturated absorption. When Ag nanoparticles were coated on the TiO_2 nanowires, the reverse saturated absorption was significantly inhibited. Authors suggested that the as-prepared nanostructures may find potential applications in the field of solar cells and all-optical switching.

In this Chapter, we discuss some of studies of the nonlinearities of various nanoparticles prepared by different methods.

5.1 Nonlinear Optical Properties of Indium and Gallium Arsenide Nanoparticles Prepared by Laser Ablation in Liquids

The details of indium nanoparticle suspension preparation are presented in the previous chapter. The investigation of the nonlinear optical characteristics of indium nanoparticles suspensions was carried out in [18] using the second harmonic of the picosecond Nd:YAG laser radiation (532 nm) and standard z-scan technique. The description of the experimental procedure for the measurements of the nonlinearities of such structures can be found elsewhere [19], as well as in some previous sections. The care was taken to prevent any optical breakdown in the indium-containing suspensions. The breakdown intensity of such structures was measured to be $2\times10^{10}\,\mathrm{W\,cm^{-2}}$, while the laser intensities used in discussed nonlinear optical studies did not exceed $4\times10^{9}\,\mathrm{W\,cm^{-2}}$ [20].

The results of the measurements of the nonlinear optical parameters of indium nanoparticles-containing aqueous suspension are presented in Fig. 5.1. The closed aperture z-scan shows a minimum of a normalized transmittance $T(z)$ before the focal position with further maximum after the focus, which characterizes a medium having a positive nonlinear refractive index. At the same time, the minimum peak considerably exceeded the maximum peak, which indicates a nonlinear absorption in the prepared suspension. The influence of thermal nonlinearity was insignificant, since the Nd:YAG laser ($t = 50$ ps) was operated at low-repetition rate (2 Hz). The best theoretical fit of the Eq. (3.5) with the experimental measurements (Fig. 5.1, solid line) gave the nonlinear optical parameters of the solution ($\gamma = 1.24 \times 10^{-14}\,\mathrm{cm^2\,W^{-1}}$ and $\beta = 1.02 \times 10^{-9}\,\mathrm{cm\,W^{-1}}$). The values of the real and imaginary parts of the third-order nonlinear susceptibilities were calculated to be $9.9\times10^{-18}\,\mathrm{cm^2\,V^{-2}}$ and $3.43\times10^{-18}\,\mathrm{cm^2\,V^{-2}}$, respectively. Using these data, the absolute value of $\chi^{(3)}$ was calculated to be $1.05\times10^{-17}\,\mathrm{cm^2V^{-2}}$. One can see that the value of $|\chi^{(3)}|$ is mostly dependent on the nonlinear refractive properties of indium nanoparticles-containing solution. The relatively small value of optical nonlinearities was defined by the small volume part of the indium nanoparticles in the solution ($\sim10^{-5}$). Taking into account this volume part, the nonlinear susceptibility of indium nanoparticles can be estimated to be $1.05\times10^{-12}\,\mathrm{cm^2\,V^{-2}}$.

The analysis of the nonlinear optical characteristics of bulk GaAs was reported in numerous publications [21–27]. Most of these studies were carried out in the spectral

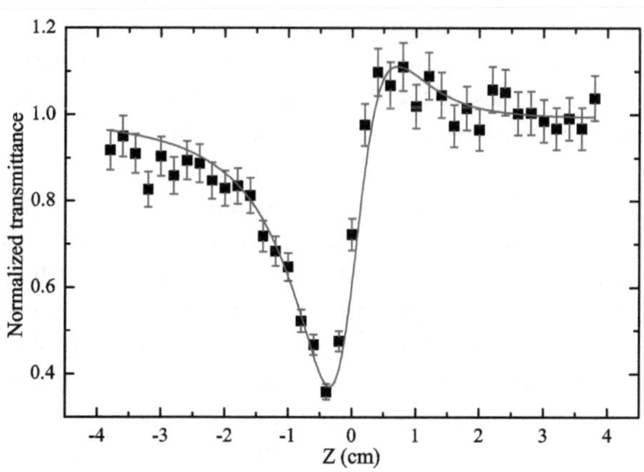

Fig. 5.1 Closed aperture z-scan of indium nanoparticle-containing aqueous suspension. *Solid line* is the best theoretical fit. Reproduced from [20] with permission from Springer Science + Business Media

range of 1 μm, above the band gap energy (E_g= 1.42 eV). Negative nonlinear refraction and strong nonlinear absorption were revealed in those studies. It was reported also about the observation of confinement effects in the case of nanocluster-shaped GaAs and thin GaAs films [28–30]. The values of nonlinear refractive index as high as $\gamma = -2.1 \times 10^{-6}\,\mathrm{cm^2\,W^{-1}}$ were measured close to the band gap energy of GaAs [31]. The nonlinear absorption in a heavily doped *n*-GaAs near the fundamental absorption edge was calculated in [32] on the basis of the model used in [33]. The transmission z-scan technique commonly used for the determination of nonlinear optical parameters cannot be applied in the absorption range of the spectrum for a bulk GaAs crystal. The absorption in visible range restricts measurement of distortions in the transmitted beam, which will require a sample thickness that is near the absorption length. The reflection z-scan technique [34, 35] can overcome some difficulties in the measurements of nonlinear optical characteristics of semiconductors at the photon energy $\hbar\omega > E_g$. This technique measures the nonlinear terms of susceptibility through the consequent changes in the reflection coefficient of the sample.

Quantum confinement is attributed to the small-size effect, which alters the electronic structures of nanocrystals from those of the bulk materials. This effect has been extensively studied due to scientific interest and technical importance. The third-order optical nonlinearity of confined GaAs quantum dots was predicted to be significantly higher with comparing to the bulk GaAs [36]. It was shown that the nonlinear refractive index of GaAs-doped glasses at 1060 nm ($\gamma = -5.6 \times 10^{-12}\,\mathrm{cm^2\,W^{-1}}$) is more than one order of magnitude greater than that of bulk GaAs ($\gamma = -4.1 \times 10^{-13}\,\mathrm{cm^2\,W^{-1}}$). Justus et al have shown that two-photon absorption coefficient of GaAs-doped glass composite is similar to that of bulk GaAs ($\beta = 26 \times 10^{-9}\,\mathrm{cm\,W^{-1}}$), and reported that the nonlinear absorption coefficient is enhancing in the quantum-confined sample [28]. An interest in the synthesis, characterization, and application of colloidal semiconductor "quantum dot" materials has grown markedly since these studies are aimed to bridge the gap between solids and atoms [37]. This increased attention reflects both strong size dependence of optical and electronic properties of these materials, and the fact that the materials are relatively easy to make with well-controlled particle sizes.

Here we describe the studies of the optical, structural, and nonlinear optical properties of GaAs nanoparticles prepared by laser ablation in various liquids, in particular ethylene glycol (EG) and silicon oil (SO) [38]. The nonlinear refraction and nonlinear absorption of these media were analyzed using the radiation of different pulse repetition rate, pulse duration, and wavelength. The method of GaAs nanoparticle solution preparation was presented in previous Chapter. A volume ratio of GaAs in the liquid solutions was estimated to be $\sim 2 \times 10^{-4}$. The GaAs solutions were then studied using the z-scan technique to measure their nonlinear optical characteristics. The 1, 2, 5, and 10 mm thick silica glass cells filled with investigated solutions were used for these studies. The detailed description of experimental setup of z-scan was described in Chap. 3. Various lasers operating at different lasing conditions were used in those z-scan studies. The first one was the Ti:sapphire laser delivering a femtosecond radiation operated at 10 Hz pulse repetition rate ($t = 110$ fs, $\lambda = 795$ nm, $E = 10$ mJ).

The output radiation and its second harmonic with variable pulse duration were also available from this laser. The laser also operated at a Q-switched regime delivering 8 ns pulses. The 795 nm, 80 MHz, 100 fs, 300 mW pulses from the seeding oscillator (Tsunami) were used for the investigation of thermal effect-induced self-interaction in solutions. The 1054 nm, 100 MHz, 280 fs, 100 mW radiation was also used for the investigation of thermal effect-induced nonlinear optical processes caused by IR radiation. The aggregation and sedimentation of GaAs nanoparticles caused variations of their optical and nonlinear optical properties. The nonlinear optical properties of GaAs-containing solutions were analyzed mostly after the stabilization of their structural characteristics. The fresh colloids possessed by higher values of nonlinear optical parameters, however, those measurements have shown poor reproducibility.

The first set of those nonlinear optical studies was carried out using the laser radiation of high pulse repetition rate. The radiation of Ti:sapphire oscillator (λ = 795 nm, W = 300 mW, t = 100 fs, 80 MHz pulse repetition rate) was used to analyze the influence of thermal effect-induced nonlinearities on the propagation of laser radiation through the GaAs-containing solutions. There are different mechanisms of losses of 795 nm radiation in these solutions. Among them the single-photon scattering and absorption and two-photon absorption seem to be most important. However, the peak intensity of laser radiation ($I = 3 \times 10^9$ W cm^{-2}) and small concentration were insufficient to cause the efficient nonlinear absorption in reported experimental conditions. The z-scan studies using CCD images in the case of Kerr-induced nonlinearities were previously reported by Marcano et al [39]. Here we present the scheme for the analysis of the thermal effect-induced self-defocusing using the CCD images (Fig. 5.2). The z_1 and z_2 positions correspond to the peak and valley of normalized transmittance in the case of closed aperture measurements. The analysis of CCD images at different positions of 2 mm thick cell filled with GaAs:EG solution allowed defining the thermal mechanisms of this nonlinear optical effect. More details of these measurements will be presented below.

The difference in the absorption mechanisms leading to the negative nonlinear refraction was clearly seen from the z-scans of GaAs:EG and pure ethylene glycol. Figure 5.3a presents two z-scan curves measured using the 1 mm thick cells filled with GaAs:EG (filled squares) and pure ethylene glycol (open circles). The

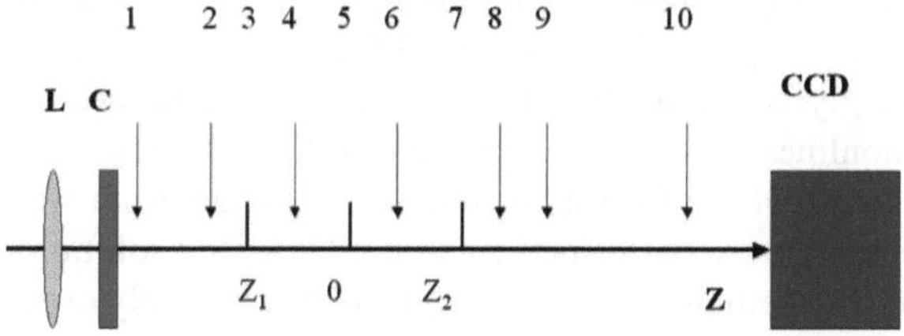

Fig. 5.2 Scheme for the measurements of the thermal-induced self-defocusing. *L* focusing lens, *C* silica glass cell filled with GaAs:EG solution, *CCD* charge coupled device. Positions 1–10 correspond to different positions of the cell. Adapted from [38] with permission from Springer Science + Business Media

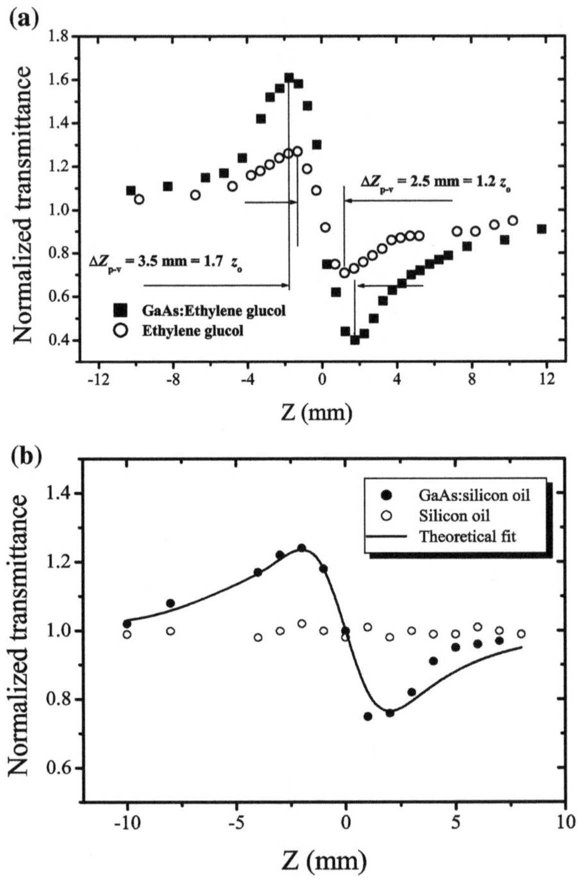

Fig. 5.3 **a** The z-scans measured using the 1 mm thick cells filled with GaAs:EG (*filled squares*) and pure EG (*open circles*) at different average power of 795 nm radiation, and **b** analogous measurements using GaAs:SO and pure silicon oil. Reproduced from [38] with permission from Springer Science + Business Media

linear absorption is responsible for the observed nonlinear refraction in the case of GaAs:EG solution ($\Delta z_{pv} = |z_1 - z_2| = 3.5\,\text{mm} \approx 1.7 z_0$), whereas in the case of pure ethylene glycol the nonlinear absorption is a main mechanism of thermal effect-induced nonlinearity ($\Delta z_{pv} = 2.5\,\text{mm} \approx 1.2 z_0$). Here $z_0 = \pi w_0^2/\lambda = 2\,\text{mm}$ is the diffraction length of the focused beam, and w_0 is the beam waist radius. Analogous measurements were performed using the GaAs:SO solution (Fig. 5.3b). The nonlinear refractive coefficient of ethylene glycol can be easily extracted from those measurements (Fig. 5.3a) taking into account a standard technique of calculations reported in [18]. This value was found to be $\gamma = -6.3 \times 10^{-14}\ \text{cm}^2\,\text{W}^{-1}$. The thermal effect-induced nonlinear refractive index of GaAs:EG solution extracted from the same data presented in Fig. 5.3a was measured to be $\gamma = -4 \times 10^{-13}\,\text{cm}^2\,\text{W}^{-1}$

taking into account experimental conditions of these experiments. Both these data represent the influence of slow nonlinear processes. The similar variations of the beam shape in the case of 1054 nm, 280 fs, 100 MHz, 100 mW radiation propagating through the GaAs-containing solutions confirmed a decisive role of linear absorption (Fig. 5.4). No absorption and beam shape variation were observed at $\lambda = 1054$ nm in the case of pure ethylene glycol, ethanol, and water.

In the next set of nonlinear optical studies the radiation operated at 10 Hz pulse repetition rate was used in order to exclude the influence of slow thermal accumulative processes that were observed at high pulse repetition rates. In the case of nanosecond pulses ($\lambda = 795$ nm, $t = 8$ ns, 10 Hz pulse repetition rate) no fast Kerr-induced nonlinear optical processes were observed in GaAs-containing solutions up to maximum intensities leading to the optical breakdown of the surface of silica glass cells at this pulse duration ($I = 8 \times 10^9$ W cm^{-2}). These studies were carried out using the closed aperture z-scans and time-integrated photodiodes. However, the self-defocusing was observed when the time waveform of propagated pulses was analyzed using fast p-i-n diodes. A detailed description of this technique is presented in [40]. The idea of this technique is based on the analysis of the oscilloscope traces of the pulse propagated through the nonlinear medium in closed aperture z-scan scheme.

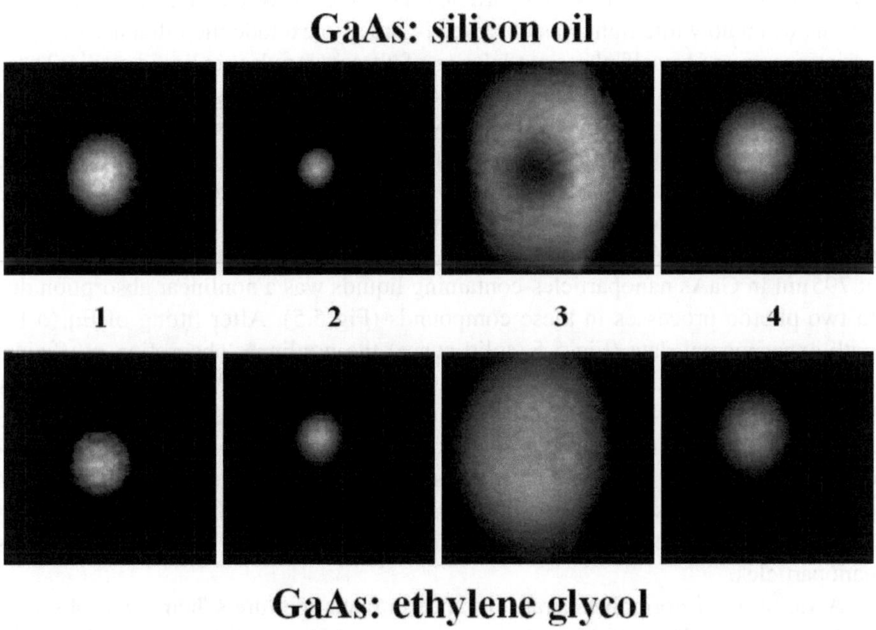

Fig. 5.4 The CCD images of the beam shapes of 1054 nm radiation operated at high pulse repetition rate and propagated through the GaAs:SO and GaAs:EG solutions at different positions of cells in z-scan-like configuration. Positions 1–4 correspond to Nos. 1,3,7 and 10 positions of cells presented in Fig. 5.2. Reproduced from [38] with permission from Springer Science + Business Media

Below we discuss the thermal effect-induced self-interaction process observed in the case of nanosecond pulses. The thermal lens appears with energy absorption growth. This lens effect is more effective for the trailing part of pulse because of the time delay of this process due to the transfer of energy to the solvent and temperature rise of solution. The heating of host dielectric has a rise-time consisting on the time t_1 required for energy transfer from nanoparticles to solvent and time t_2 required for temperature rise of solvent to be transformed into the change of refractive index via density reduction. $t_1$is a function of nanoparticles' sizes. For 10 nm particles this time is of order of 25 ps. t_2 is equal to w_o/υ, where υ is the velocity of acoustic wave in liquid. For $w_o = 21\,\mu$m and $\upsilon = 1500$ m/s, this parameter is equal to $t_2 = 14$ ns. It means that, at least for the central part of focused beam, there are some conditions of self-defocusing in GaAs solution during propagation of nanosecond radiation through the sample. It takes between 4 and 6 ns to achieve the density variations in the central part of focal volume. These estimations of self-defocusing start-up time are qualitatively coinciding with experimental observations of the variations of the trailing part of 8 ns pulses in closed aperture scheme.

The application of short laser pulses (in picosecond and femtosecond ranges) at low pulse repetition rate allows excluding the influence of slow thermal-related nonlinear optical processes and analysing the self-interaction mechanisms caused by the electronic response of GaAs nanoparticles. The studies were carried out at the conditions when no white light was observed in order to exclude the influence of phase variations caused by self-phase modulation of ultrashort pulses in liquid media. These studies using the 795 and 397 nm femtosecond radiation were performed at the laser intensities up to $I = 1 \times 10^{11}$ W cm^{-2}. No influence of GaAs nanoparticles on the nonlinear refractive properties of GaAs:liquid solutions was registered. The influence of the positive refractive nonlinearities of the silica glass and water was observed at the intensities exceeding 4×10^{10} W cm^{-2} ($\lambda = 795$ nm), that eclipsed the influence of those of GaAs nanoparticles. The only nonlinear optical process observed at 795 nm in GaAs nanoparticles-containing liquids was a nonlinear absorption due to two-photon processes in these compounds (Fig. 5.5). After fitting of Eq. (3.10) with experimental data (Fig. 5.5, solid curve) the nonlinear absorption coefficient of GaAs:water solution was calculated to be 0.7×10^{-9} cm W^{-1}. This value only 30 to 40 times below one measured in bulk GaAs in IR range (26×10^{-9} cm W^{-1}) in spite of small volume ratio of GaAs in liquid solutions ($\sim2\times10^{-4}$), thus indicating the influence of quantum confinement on the enhancement of the β of GaAs nanoparticles solution. The third-order susceptibility of GaAs nanoparticles at 795 nm was calculated to be $|\chi^{(3)}| \sim 2\times10^{-9}$ esu assuming the volume ratio of GaAs nanoparticles.

A variation of normalized transmittance in open aperture scheme was observed in the case of 795 nm, 400 fs pulses operated at 10 Hz repetition rate. This regime of laser-colloids interaction at high irradiation leads to the fragmentation of GaAs nanoparticles. As a result, a nonlinear absorption presented by a normalized T(z) dependence was less pronounced in the case of 10 Hz pulse repetition rate in comparison with single-shot interaction. These measurements were repeated at several

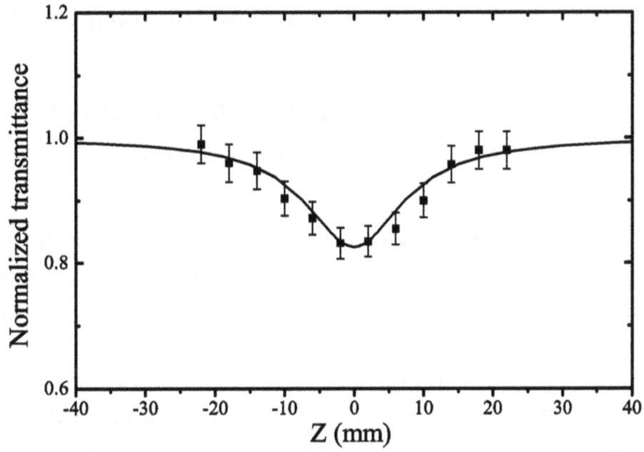

Fig. 5.5 The normalized transmittance of GaAs:water solution using open aperture z-scan at $\lambda = 795\,\text{nm}$. *Solid line* shows the theoretical fit. Reproduced from [38] with permission from Springer Science + Business Media

times to ensure reproducibility, and were also performed in pure liquid-containing cells, which showed no measurable nonlinear response under the same conditions.

Finally, we briefly discuss the application of GaAs nanoparticle-containing liquids as optical limiters. The search of optical limiting in various materials opens new opportunities of their application as laser switching systems for protection from intense laser radiation. Some details on optical limiting studies are presented in Sect. 3.10. The optical limiting processes in GaAs nanoparticle-containing liquids were studied using the subpicosecond radiation ($t = 0.15\,\text{ps}$). Laser radiation was focused by a 100 mm focal length lens inside the 10 mm thick cells containing colloidal GaAs. The transmitted radiation was measured by a calibrated photodiode. Figure 5.6 presents the normalized transmittance dependences as functions of 795 nm pulse energy in the case of GaAs:EG and GaAs:water suspensions and pure water. The optical limiting in suspensions was observed at the pulse energies exceeding $8 \times 10^{-6}\,\text{J}$ and was caused by two-photon absorption that was confirmed by the nonlinear absorption investigations using open aperture z-scan scheme.

The studies of silver nanoparticles in various liquids prepared by the same technique of laser ablation [41] have revealed identical properties with comparing to those reported in this section (see the next section). The viscosity of liquids plays an important role in preserving of the optical and nonlinear optical characteristics of nanoparticles solutions in both cases. The intensity thresholds of femtosecond continuum generation both in Ag- and GaAs-containing suspensions were higher than those for pure liquids. The thermal lens-induced self-defocusing dominated in these suspensions in the case of high pulse repetition rate. However, the appearance of strong absorption band in silver nanoparticles-containing suspensions at ~400 nm associated with surface plasmon resonance was responsible for the enhancement of fast nonlinear optical processes (two-photon absorption, saturable absorption, and Kerr-induced self-defocusing) in the vicinity of this spectral range. None of such features were registered in the case of GaAs suspensions.

5.2 Low-Order Optical Nonlinearities of Silver Clusters

5.2.1 *Low-Order Nonlinearities of Silver Nanoparticles Embedded in Solid Transparent Matrices*

Nonlinear optical characteristics of nanoparticle suspensions (NS) have been under active investigation during last two decades. These materials demonstrate high nonlinearities and fast response [42–47], especially in the frequency region of surface plasmon resonance. Strong nonlinear optical response was observed in copper- and gold-doped transparent media [48–50]. In [43, 51], it has been shown that the nonlinear refractive index of the copper colloidal solution has a considerable value. The same results have been shown for silver NS [52, 53] and gold NS [54, 55]. Optical limiting characteristics of colloidal silver aggregates were studied in [56] at the wavelength of $\lambda = 532$ nm of picosecond and nanosecond lasers. It was shown that at certain conditions the OL in these media can surpass that of C_{60}. The size effects of OL were observed for gold clusters [57]. These colloidal structures possess by fast (<360 fs) nonlinear optical response and high nonlinearity in the femtosecond range ($\chi^{(3)} = 1.5 \times 10^{-7}$ esu for Ag clusters) [58, 59]. At the same time, one should note that there have been few OL studies of NS despite their high self-action and nonlinear absorption properties in picosecond and nanosecond ranges.

Silver nanoparticles have attracted interest due to their potential applications [60]. The analysis of the optical studies of Ag nanoparticles prepared by laser ablation was presented by Brause et al [61]. The nonlinear optical studies of colloidal Ag nanoparticles in various aggregated states were reported in [62]. These colloidal structures demonstrate fast nonlinear optical response and high nonlinear optical susceptibility ($\chi^{(3)} = 1.5\times10^{-7}$ esu for Ag clusters [59]). The dynamics of the spatial parame-

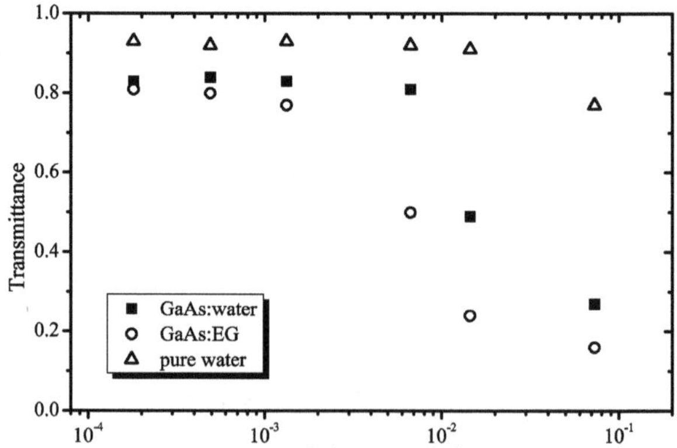

Fig. 5.6 The normalized transmittance dependences as functions of laser radiation energy in the case of GaAs:EG and GaAs:water solutions and pure water. Reproduced from [38] with permission from Springer Science + Business Media

ters of aggregated Ag nanoparticles leads to the variation of their nonlinear optical characteristics. At the same time the mechanisms of these variations have yet to be determined. The location of SPR of silver nanoparticles (400–430 nm) depends on preparation conditions [61]. The enhancement of the nonlinear susceptibility responsible for nonlinear absorption and refraction in this spectral range was reported in [63]. The quantum confinement in the case of small-sized Ag nanoparticles can also lead to the growth of the nonlinear optical properties of these structures.

5.2.1.1 Characterization of Nonlinear Optical Parameters of Silver-Doped Silica Glasses at $\lambda = 1064$ nm

The composite materials based on the insulator matrices doped with metal nanoparticles took the attention more than two decades ago [64]. These investigations are of great interest [65, 66] due to the perspectives of using such materials as ultra-fast optical switching systems and optical limiters. These composite materials with metal nanoparticles possess high values of third-order nonlinear susceptibilities, especially in the spectral range of the SPRs [64, 66]. The SPRs of metal nanoparticles are located in a broad spectral range depending on the metal type, particle shape, structure and size, as well as on the properties of the surrounding insulator.

The nonlinear optical parameters of ellipsoidal silver nanoparticles in the longitudinal surface plasmon absorption region (800 nm) were analyzed in [67]. It has been observed that the nonlinear absorption and nonlinear refractive indices of silver nanorods embedded in borosilicate glass differ remarkably from those of spheroids. Third-order nonlinearities of glasses doped with a large number of copper or silver particles were studied by Uchida et al [51] using degenerate four-wave mixing with an emphasis on particle-size dependence. They observed the growth of $\chi^{(3)}$ at a wavelength of the absorption peak (up to 10^{-7} esu). The increase of the $\chi^{(3)}/\alpha$ value (α is the linear absorption coefficient) with the growth of particle radius was explained by the size dependence of the imaginary part of the dielectric constant of metal particles and the local-field factor. Analogous size dependence of the third-order susceptibility was reported by Yang et al [68] in their experiments with copper nanoparticles embedded in fused silica.

All these measurements were carried out using UV [69], visible (532 nm [68], 570 nm [70], 420 nm [51]) or 800 nm radiation [67]. At the same time it is interesting to analyze the behavior of such nanoparticle-doped glasses in the infrared range using photons with the energy far below the SPRs of these materials. The attention to this spectral range can be explained by potential applications of such media in optical limiting and optical switching devices. In this subsection, the studies of nonlinear optical characteristics of the silver nanoparticles implanted in glass matrices are presented using the z-scan technique and infrared laser radiation ($\lambda = 1064$ nm). The optical limiting properties of these materials are discussed in the framework of two-photon absorption [71].

The silica glasses containing 100 % SiO_2 and soda-lime silicate glasses containing 70 % SiO_2, 20 % Na_2O and 10 % CaO were used for the preparation of composite materials. The linear refractive indices of silica glasses (SGs) and soda-lime silicate glasses (SLSGs) were 1.5 and 1.54, respectively. The glasses were prepared as $2 \times 2\,\text{cm}^2$ plates with a thickness of 1 mm (SG) and 3 mm (SLSG). The energy used for ion implantation was 60 keV at dose of 4×10^{16} ions cm^{-2} in the case of Ag^+. The depth of the location of metal nanoparticles in the surface glass layer was 60 nm [72]. X-ray reflectometry was used in order to estimate the density and thickness of thin layers [73]. Composite materials were also analyzed by the grazing incidence small-angle X-ray scattering (GISAXS) technique. Optical transmission measurements were carried out using a fiber spectrophotometer. The z-scan technique was used for the determination of the nonlinear optical characteristics of investigated samples. A Nd:YAG laser used in these studies was operated in the mode-locked regime. Output characteristics of Nd:YAG laser radiation ($\lambda = 1064$ nm) were as follows: pulse duration 35 ps, pulse energy 1 mJ, and 2 Hz pulse repetition rate. The maximum laser intensity at the focal point was $3 \times 10^{10}\,\text{W cm}^{-2}$, whereas the intensity of optical breakdown was $8 \times 10^{10}\,\text{W cm}^{-2}$ for the glasses with silver nanoparticles.

The ion-implantation technique applied for preparation of metal-doped glasses [74], being capable of sensitizing nanoparticles with a high filling factor, forms in some cases layers with a broad distribution of particle sizes [75]. The growth of metal nanoparticles in glasses during ion implantation occurs when the concentration of metal atoms becomes larger than the solubility level, which is determined by an ionic dose of 1×10^{16} ion cm^{-2} at the ion energy of 60 keV [76]. X-ray and optical spectroscopic measurements proved the formation of metal nanoparticles. These studies (GISAXS, X-ray reflectometry) have shown a broader size distribution of nanoparticles (from 2 to 18 nm) that was also observed in the transmission electron microscopy measurements. The depth of the location of nanoparticles in glasses was 60 nm. The formation of metal nanoparticles was also proven by the appearance of selective absorption lines in optical transmission spectra with the transmission minima in the ranges of 400–450 nm corresponding to the SPR of silver nanoparticles (Fig. 5.7). Radiation defects stimulate additional light absorption in glasses in the

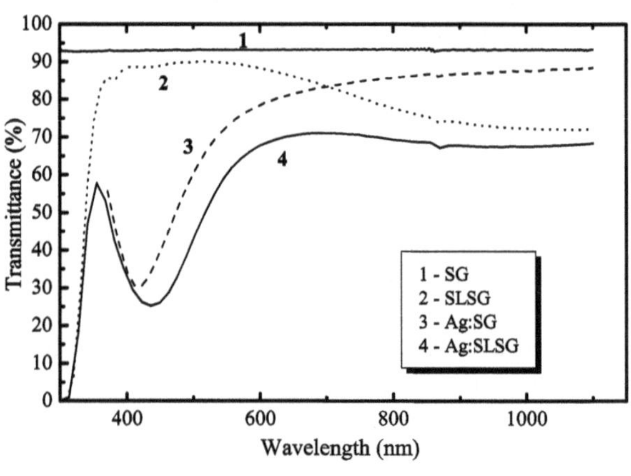

Fig. 5.7 The transmission spectra of (*1*) SG, (*2*) SLSG, (*3*) Ag:SG, and (*4*) Ag:SLSG. Adopted from [71] with permission from John Wiley and Sons

short-wavelength spectral range. However, it should be noted that those nonlinear optical studies were performed using the 1064 nm radiation, which lies far from both SPRs and intra-band transitions.

No considerable differences in normalized transmittance dependences measured both from implanted and non-implanted sides were observed. In fact, as the quantum dots are not uniformly implanted but occupy only a thin layer of the glasses, it can be guessed that the linear refractive indices of the implanted and non-implanted surfaces are different. This variation could induce a small difference in the reflection losses and hence in the net intensity of the laser beam entering the sample [77].

The value of the nonlinear refraction index can be varied in a large range depending on interaction type (resonant or non-resonant). Non-resonant contributions to the refractive index are usually positive [78]. In the case of quasi-resonant (one- or two-photon) interaction, the sign of the nonlinear refraction index is determined by a correlation between fundamental frequency, doubled frequency (in the case of two-photon quasi-resonance) and resonant frequency of the investigated medium. A considerable nonlinear contribution to the refractive index can also be due to the thermal effect caused by heat transfer from metal nanoparticles to the lattice. The nonlinear thermal effect is due to matter density changing after local heating. The rise time (τ) of refractive index variations is determined by $\tau = R/V_s$ (R is the beam-waist radius and Vs is the sound velocity in the lattice). At the used experimental conditions ($R = 75\ \mu$m, $V_s \sim 5500\,\text{ms}^{-1}$) $\tau \approx 14$ ns, i.e. three orders of magnitude larger than the laser pulse duration (35 ps), so that the thermal effect can be excluded from the consideration.

The optical properties of nanoparticles can be considered in the framework of an effective medium [79] because the wavelength of the laser radiation used in these studies was larger than the sizes of metal nanoparticles. Such method allows considering the composite materials as quasi-homogeneous media. One can consider the simple two-level model for homogeneous condensed matter [63], from which the nonlinear refractive index can be presented as follows:

$$n_2 = -2\pi N \mid \mu_{i0} \mid 4\{n_0[\hbar(\omega_{i0} - \omega_{\text{spr}})]^3\}^{-1}. \tag{5.1}$$

Here ω_{spr} and ω_{i0} are the frequencies of SPR and laser radiation. Index i determines one- or two-photon processes, N is the concentration of nanoparticles and μ_{i0} are the transition dipole moments at the frequencies ω_{i0}. The sign of the nonlinear refractive index is determined by the sign of detuning of the laser radiation frequency (or their harmonics) from the SPRs of metal nanoparticles. The real part of the third-order nonlinear susceptibility has a linear dependence on n_2 [63]. The spectral dependence of $\chi^{(3)}$ can be written as follows:

$$\text{Re}\chi^{(3)} \propto -(\omega_{\text{i0}} - \omega_{\text{spr}})^{-3}. \tag{5.2}$$

The self-defocusing from a two-photon resonance has been reported by Lehmberg et al. [80]. The two-photon energy of their laser radiation (1064 nm, 30 ps) was

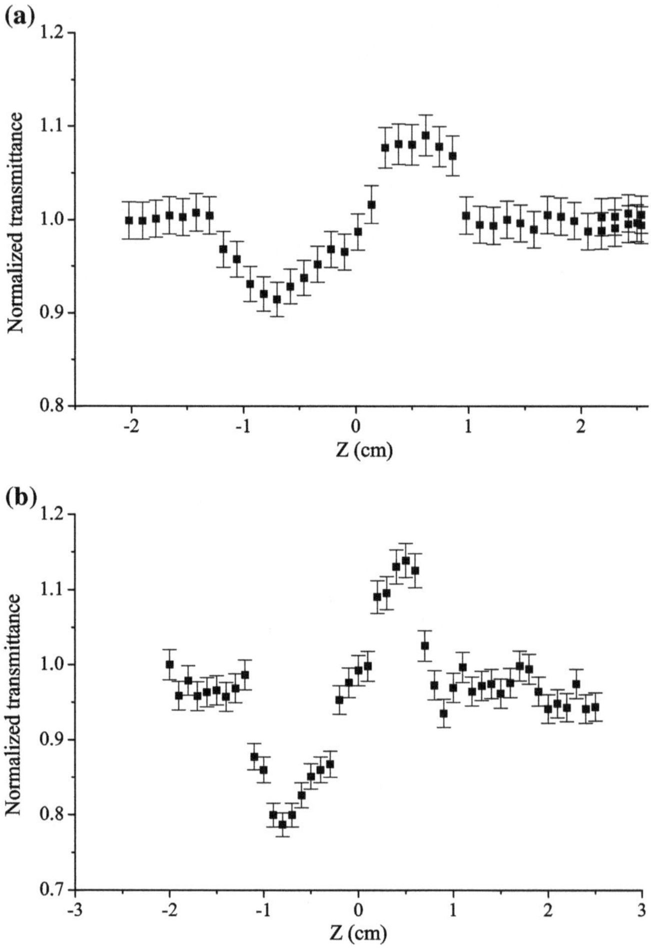

Fig. 5.8 The normalized transmittance dependences for **a** Ag:SG and **b** Ag:SLSG in closed aperture scheme. $\lambda = 1064$ nm. Reproduced from [71] with permission from John Wiley and Sons

250 cm^{-1} above the 7*s* level of Cs vapors, giving a two-photon enhancement and a negative contribution to the refractive index.

Figure 5.8 presents the dependences of the normalized transmittance in the closed aperture scheme for Ag:SG and Ag:SLSG samples. As one can see from the figure, the positive nonlinearity (self-focusing effect) was observed for both samples. The composites containing silver nanoparticles are characterized by different locations of SPR maxima depending on substrate type. SPRs of Ag:SG and Ag:SLSG were located at 415 nm ($\omega_{SPR} = 24096.4\,\text{cm}^{-1}$) and 440 nm ($\omega_{SPR} = 22727.3\,\text{cm}^{-1}$), respectively. The doubled frequency of the laser radiation is smaller than the SPRs of Ag nanoparticles in both matrices, which corresponds to the positive sign of the refractive nonlinearity. The corresponding n_2 and Re $\chi^{(3)}$ for Ag:SG were measured to be 1.5×10^{-8} esu and 3×10^{-9} esu, while in the case of Ag:SLSG these parameters were measured to be 3.5×10^{-8} esu and 6×10^{-9} esu. The studies of silver-doped samples using the open aperture scheme have shown no nonlinear absorption in those media at $\lambda = 1064$ nm. Analogous studies were carried out using the rz-scan technique [81]. The normalized transmittance curves presented in Fig. 5.8 show some instability. This phenomenon can be attributed to the broad size distribution of Ag nanoparticles. Inhomogeneous distribution of nanoparticles in the area of the focal volume can also play an important role in the observed instabilities of the $T(z)$ dependences.

5.2.1.2 Saturable Absorption and Nonlinear Refraction of Silicate Glasses Doped with Silver Nanoparticles at 532 nm

The metal nanoparticles demonstrate high values of nonlinear susceptibility in the vicinity of SPR. The resonance occurs at the frequency where the local electric field inside of metal nanoparticle becomes maximal. The SPR maximum is determined by the following equation [79]:

$$\varepsilon_1(\omega_{\text{spr}}) + 2\varepsilon_{\text{m}}(\omega_{\text{spr}}) = 0 \tag{5.3}$$

where ε_1 and ε_{m} are real dielectric constants of metal nanoparticles and matrix glass. The SPR of spherical silver nanoparticles is located in the visible range (400–425 nm). The $\chi^{(3)}$ of silver nanoparticles at the wavelength of second harmonic radiation of Ti:sapphire laser ($\lambda = 400$ nm) was measured to be 1.5×10^{-7} esu [59], which is one of the highest values among the reported results of nonlinear susceptibility measurements. It was shown previously that silver-doped glasses possess by saturable absorption. Hamanaka et al have analyzed the spectral dispersion of the imaginary part of the third-order susceptibility of silver-doped glass matrices [82]. It was shown that Im $\chi^{(3)}$ was negative in the range of 385–436 nm. It should be noted that previous studies of the nonlinear optical parameters of silver nanoparticles-doped glasses were mostly focused on determination of $\chi^{(3)}$. In this subsection, we discuss the results of study of the saturable absorption and nonlinear refraction in silicate glasses doped with Ag nanoparticles at $\lambda = 532$ nm and their dependence on the laser radiation intensity [83]. Silver nanoparticles were implanted in 1 mm thick silicate glass and soda lime silicate glass plates using the ion implantation technique. The output characteristics of the second harmonic radiation of Nd:YAG laser were as follows: pulse duration $t = 55$ ps, pulse energy $E = 0.26$ mJ, and wavelength $\lambda = 532$ nm.

Figure 5.9 shows the normalized transmittance dependences of Ag:SLSG and Ag:SG measured using open aperture z-scan scheme at $I = 2.5\times10^9$ W cm^{-2}. The transmission of samples was increased due to saturable absorption as they approached close to the focal plane. The normalized transmittance in the case of open aperture scheme can be determined by Eq. (3.10). After fitting of Eq. (3.10) with experimental data the nonlinear absorption coefficients of Ag:SLSG and Ag:SG were calculated

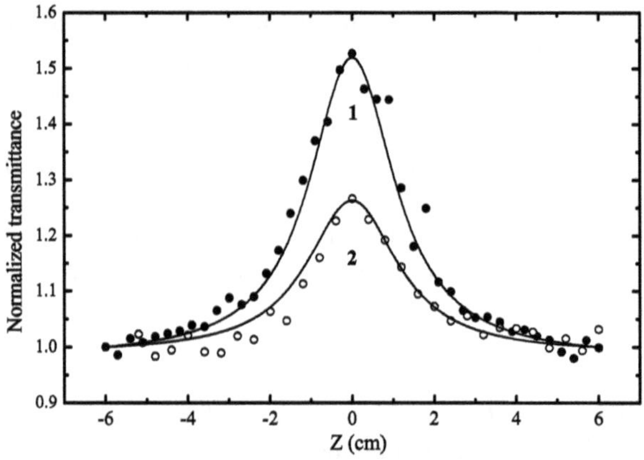

Fig. 5.9 The normalized transmittance dependences of (*1*) Ag:SLSG, and (*2*) Ag:SG in the case of open aperture z-scan scheme at laser radiation intensity of 2.5×10^9 W cm^{-2}. *Solid lines* show the theoretical fits. Reproduced from [83] with permission from Springer Science + Business Media

to be -6.7×10^{-5} and -3.6×10^{-5} cm W^{-1}, respectively. Parameter β can be presented as $\beta = -\alpha_0/I_s$ that is a ratio between linear absorption coefficient and saturable intensity. The I_s were evaluated for two samples to be 1.1×10^9 W cm^{-2} (Ag:SLSG) and 1.4×10^9 W cm^{-2} (Ag:SG). The Im $\chi^{(3)}$ of Ag:SLSG and Ag:SG at $I_0 = 2.5 \times 10^9$ W cm^{-2} were calculated to be -2.4×10^{-8} and -1.3×10^{-8} esu, respectively.

The saturable absorption has been observed from various materials such as dyes [84], polythiophene thin films [68], semiconductor-doped glasses [85], etc. The saturable absorption of Ag nanoparticles embedded in silica glass matrices has been observed as far as 355 nm [86]. Such process was also observed in silver-doped glasses in the vicinity of SPR of Ag nanoparticles [82, 87]. Analogous features of silver-contained glasses were reported in [42, 85, 88, 89]. The saturable absorption was also reported previously in the modified $LiNbO_3$ crystals containing Ag nanoparticles close to plasmon band shifted to 550 nm after heat treatment [90].

The intensity-dependent variations of β of the samples were observed in these studies. The measurements were carried out at laser intensity varied from 10^9 to 2×10^{10} W cm^{-2}. The $\beta(I_0)$ dependences of Ag:SLSG and Ag:SG are presented in Fig. 5.10. A decrease of β was observed for higher intensities. In particular, a 21- and 12-fold decrease of β was measured at $I = 1.15 \times 10^{10}$ W cm^{-2} for Ag:SLSG and Ag:SG, respectively, compared to β measured at $I = 1 \times 10^9$ W cm^{-2}. The variations of transmission and nonlinear refraction in such structures were attributed in some cases to the fragmentation [91–93], or fusion [94] of nanoparticles following the photothermal melting. Osborne et al [70] reported the alteration of the sign of the nonlinear refractive index of the small silver clusters embedded in soda-lime glass. They noted that thermal effects could change the properties of nanoclusters. The transparency in these samples was associated with oxidation of Ag nanoparticles. However, no irreversible changes of transmittance were observed in the dis-

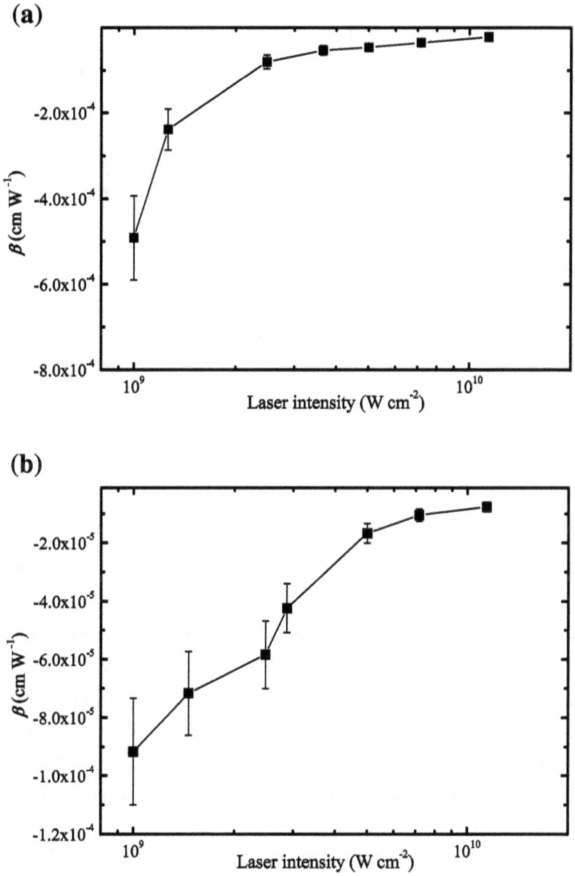

Fig. 5.10 The $\beta(I)$ dependences in the cases of **a** Ag:SLSG, and **b** Ag:SG. Reproduced from [83] with permission from Springer Science + Business Media

cussed experiments. These measurements were repeated at several times to ensure reproducibility, and were also performed in pure silica glasses, which showed no measurable nonlinear response under the same conditions.

The reverse saturable absorption can be responsible for the decrease of the negative nonlinear absorption of Ag nanoparticles. The common feature of the dispersion curve of the Im $\chi^{(3)}$ of different metal nanoparticle-contained compounds near their SPRs is a negative sign of nonlinear absorption [82]. One can assume that in the case of picosecond pulses the reverse saturable absorption starting to play an important role in the overall dynamics of nonlinear optical transmittance of metal nanoparticles-contained compounds, taking into account the saturation of intermediate transitions responsible for saturable absorption [83].

The theoretical dependences (3.5) were fitted with the experimental data obtained in the case of closed aperture z-scans. The best correlation between theory and experiments was observed at $\Delta\Phi_0 = 0.8$ and 1.08 for Ag:SLSG and Ag:SG, respectively. From these data, the values of γ and Re$\chi^{(3)}$ were calculated to be

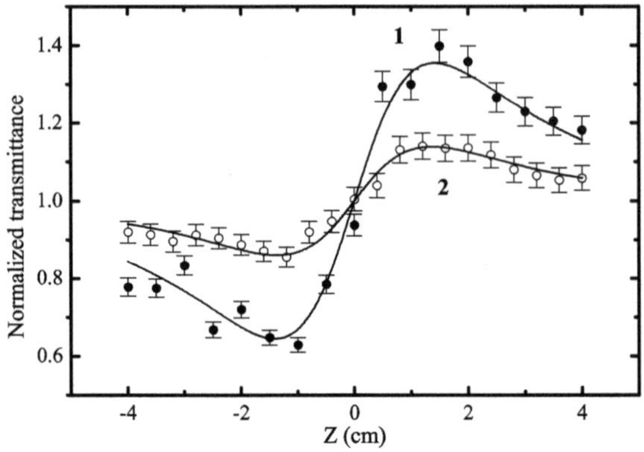

Fig. 5.11 The normalized transmittance dependences of (*1*) Ag:SLSG, and (*2*) Ag:SG in closed aperture z-scan scheme at $I = 1.4 \times 10^{10}\,\mathrm{W\,cm^{-2}}$. $\lambda = 532\,\mathrm{nm}$. Reproduced from [83] with permission from Springer Science + Business Media

$-4.1 \times 10^{-10}\,\mathrm{cm^2W^{-1}}$ and -2.4×10^{-8} esu for Ag:SLSG, and $-6.2 \times 10^{-10}\,\mathrm{cm^2W^{-1}}$ and -3.5×10^{-8} esu for Ag:SG. During these studies, the variation of nonlinear optical parameters with growth of laser intensity was observed leading to the change of the sign of nonlinear refractive index from negative to positive (Fig. 5.11). The maximum positive values of γ and Re $\chi^{(3)}$ of these samples were observed at the laser intensity of $1.4 \times 10^{10}\,\mathrm{W\,cm^{-2}}$ and were calculated to be $1.7 \times 10^{-10}\,\mathrm{cm^2\,W^{-1}}$ and 9.6×10^{-9} esu for Ag:SLSG, and $7 \times 10^{-11}\,\mathrm{cm^2\,W^{-1}}$ and 4×10^{-9} esu for Ag:SG. The measurements of the nonlinear optical parameters of samples are summarized in Table. 5.1.

The possible mechanism of the decrease of Im$\chi^{(3)}$ is the influence of nonlinear optical processes with opposite dependences on laser intensity, such as reverse saturable absorption and two-photon absorption [22, 95]. The wavelength range corresponding to the interband transitions in silver is located below 320 nm, so the two-photon absorption connected with interband transitions can be involved in the case of 532 nm radiation. The possibility of two-photon absorption due to interband

Table 5.1 Nonlinear optical parameters of silicate glasses doped with Ag nanoparticles at different intensities of 532 nm radiation

Sample	I, 10^9 ($\mathrm{W\,cm^{-2}}$)	β, 10^{-5} ($\mathrm{cm\,W^{-1}}$)	Im$\chi^{(3)}$ 10^{-8} (esu)	γ 10^{-10} ($\mathrm{cm^2\,W^{-1}}$)	Re$\chi^{(3)}$, 10^{-8} (esu)
	2.5	−3.6	−1.3	−6.2	−3.5
Ag:SG	11.5	0.48	−0.17		
	14			0.7	0.4
	2.5	−6.7	−2.4	−4.1	−2.4
Ag:SLSG	11.5	−1.7	−0.6		
	14			1.7	0.96

Reproduced from [83] with permission from Springer Science + Business Media transition of photoexcited electrons has been demonstrated for silver [96] and copper [97] nanoparticles. The three-photon absorption connected with interband transition for silver nanoparticles was analyzed in [67]. At the same time the reverse saturable absorption can also be involved into the consideration. The only contribution to the nonlinear refraction and absorption processes was caused by metal nanoparticles since no variations in the normalized transmittance dependences of pure substrates were observed in experiments. However, the defects in matrix can also play an important role in overall change of the optical and nonlinear optical properties of metal-doped glasses. The doping of accelerated ions into the glass substrates leads to the appearance of defects, which can stimulate the reversible and irreversible changes in material structure [74]. Different types of defects were registered in previous studies of metal-implanted glasses. Among them are the prolonged and dotted defects, local crystalization and amorphization, etc. In particular, the growth of metal nanoparticles leads to the increase of the volume of implanted layer and appearance of the internal tension.

Note that the negative value of $\mathrm{Re}\chi^{(3)}$ was previously observed in Ag-doped materials at the wavelength range of $\lambda > \lambda_{spr}$ [82]. The observed sign change of $\mathrm{Re}\chi^{(3)}$ can be caused by several processes:

(1) The influence of interband transitions taking into account the possibility of two-photon processes. The decrease of the imaginary part of $\chi^{(3)}$ confirms such assumption.
(2) The influence of thermal effect ($dn/dT > 0$ for glasses). Positive contribution to the nonlinear refractive index can be caused by thermal effect, which can be considered as a result of energy transfer from heated metal nanoparticles to surrounding dielectric matrix [54]. However, the time, which is necessary for this process to be important equal to a few nanoseconds, whereas the pulse duration was three orders of magnitude shorter (55 ps), that diminishes the influence of the thermal effect causing the acoustic-induced variation of the density and refractive index of the matrix.
(3) The irreversible change of γ caused by laser radiation. This opportunity was analyzed in several studies of composite materials doped with copper [98] and silver [70] nanoparticles. The sign change of both refractive and absorptive nonlinearities was observed in the Cu-doped glass with the growth of laser intensity, and the mechanism of modification was found to be a thermally activated enlargement of Cu nanoparticles [98]. The sigh change of nonlinear refractive index from positive to negative was observed also in Ag-doped glasses [70]. The mechanism responsible for sign change was a photochemical reaction, which produced a silver-oxide layer on the surface of nanoclusters. The irreversible changes in both these studies were caused by a thermal influence produced by a high pulse repetition rate radiation (3.8 MHz). In the reported case, the z-scans had a good reproducibility, so the influence of irreversible changes can be easily excluded.

Based on above-mentioned analysis one can attribute both nonlinear absorptive and refractive changes observed at the growth of laser intensity to the influence of interband transitions.

In order to be useful in optical switching processing devices a third-order nonlinear material must simultaneously satisfy the following conditions [99]:

(1) The excitation time of the nonlinear effect must be less than the pulse width. In the reported case the nonlinear refraction is caused by a Kerr-related phenomenon characterized by femtosecond range excitation time.
(2) The effect of linear absorption must be weak compared to the effect of nonlinearity. This condition quantified in terms of Stegeman figure of merit W:

$$W = n_2 I / \alpha_0 \lambda > 1. \tag{5.4}$$

In present case $W = 0.7$ (for Ag:SLSG sample), i.e., close to chosen limit.
(3) The effect of two-photon absorption must be weak compared to the nonlinear effect. This condition quantifies in terms of Stegeman figure of merit T:

$$T = 2\beta\lambda / n_2 < 1 \tag{5.5}$$

In present case $T = 40$ (for Ag:SLSG sample), that is quite far from chosen limit.

Based on above-presented estimations, these samples fell short to the conditions of effective application as optical switching devices at 532 nm. The comparison of observed processes in Ag nanoparticles-contained glasses with those in other nonlinear optical materials seems important in spite of difference in physical mechanisms playing an appropriate role in each of these particular cases. Such a comparison is valuable when aimed estimating the practical applicability of investigated material in visible range compared to other materials.

5.2.1.3 Nonlinear Optical Susceptibilities of Silver-Doped Silicate Glasses in the Ultraviolet Range

As it was mentioned earlier, the interest in metal nanoparticles (MN) embedded in dielectric materials is connected with their applications as optical switchers with ultrashort time response [59], and optical limiters of intense laser radiation [100, 101]. The saturable absorption observed in such structures makes them perspective mode-lockers [102]. Analogous studies were generally carried out in the visible and near IR ranges. At the same time such structures can possess interesting nonlinear optical properties in the UV range. There are only few studies of the nonlinear optical properties of nanoparticle composites in that spectral range. In particular, Si nanoparticles were investigated at the wavelength of the Nd:YAG laser third harmonic radiation ($\lambda = 355$nm, $\tau = 8$ns) and the giant value of nonlinear susceptibility was measured to be 2.3×10^{-5} esu [103].

Below, we discuss the nonlinear optical properties of Ag nanoparticles embedded in silicate glasses using the near UV radiation [69]. The parameters of third harmonic radiation of the Nd :YAG laser were as follows: pulse duration 55 ps, pulse energy

0.1 mJ, wavelength $\lambda = 354.7$ nm. The laser beam waist radius at the focal plane was 42 μm. The intensity of laser radiation was varied in the range of 10^9–5×10^9 W cm^{-2}, whereas the optical damage threshold of samples was measured to be $\sim 10^{10}$ W cm^{-2}. The closed aperture z-scan technique was used to determine the value of the nonlinear refractive index n_2 of the samples. The glasses without nanoparticles did not show any changes in transmission experiments at the intensities used. In Fig. 5.12, the normalized transmittance dependence $T(z)$ in the closed aperture z-scan scheme is shown for Ag:SG indicating the negative sign of nonlinear refraction. An observed asymmetry of this dependence, when peak exceeds valley, characterizes the saturable absorption in the sample. When nonlinear refraction and nonlinear absorption are simultaneously present in the sample, the $T(z)$ dependence in the closed aperture scheme can be described by Eq. (3.5). Using Eq. (3.5), the numerical calculations of $T(z)$ fits were carried out for the experimental curve presented in Fig. 5.12. The best fit was obtained at $\rho = -0.07$ and $\Delta\Phi_0 = -0.75$. From this fit the nonlinear optical parameters were found to be $n_2 = -2.7 \times 10^{-7}$ esu, and $\beta = -1.4 \times 10^{-5}$ cm W^{-1} for Ag:SG. Re$\chi^{(3)}$ and Im$\chi^{(3)}$ of Ag:SG were calculated to be -6×10^{-8} and -6.1×10^{-9} esu, respectively.

5.2.2 *Characterization of Optical and Nonlinear Optical Properties of Silver Nanoparticles Prepared by Laser Ablation in Various Liquids*

Here we present the studies of the optical, structural, and nonlinear optical properties of the Ag nanoparticles prepared by laser ablation in various liquids. We analyze the

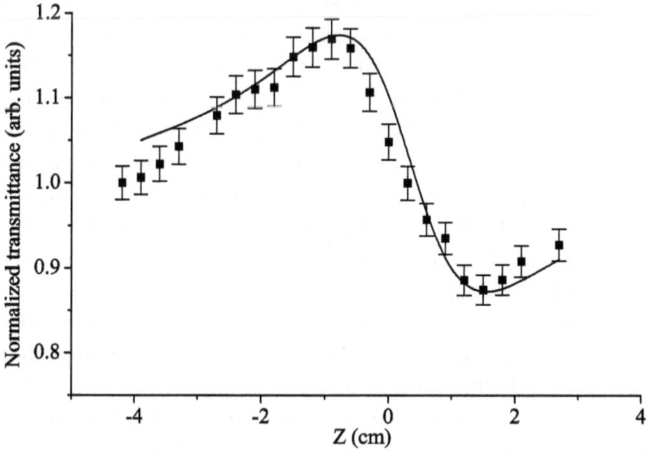

Fig. 5.12 Normalized transmittance dependence for Ag:SG in closed aperture z-scan scheme. Solid line is the theoretical fit. $\lambda = 355$ nm. Reproduced from [69] with permission from John Wiley and Sons

nonlinear refraction and nonlinear absorption of these media using the radiation of different wavelength, pulse repetition rate, and pulse duration [41]. Laser ablation of silver in various liquids was performed using second harmonic radiation of Nd:YAG laser. Laser radiation (wavelength: $\lambda = 532$ nm, pulse duration: $t = 9$ ns, pulse energy: $E = 30$ mJ, 10 Hz pulse repetition rate) was focused by a 50 mm focal length lens at normal incidence to the surface of silver block placed inside the 10 mm thick cell filled with the liquids of different viscosity (ethylene glycol, water, or ethanol). The fluence of 532 nm radiation at target surface was measured to be 20 J cm^{-2}. The ablation was carried out during 15–30 min. The ablated solution was constantly stirred during interaction of laser radiation with Ag target. The prepared suspensions were analyzed by TEM and absorption spectroscopy to confirm the appearance of Ag nanoparticles.

The ablated silver suspensions were then studied using z-scan technique to measure their nonlinear optical characteristics. The 2, 5, and 10 mm thick silica glass cells filled with investigated suspensions were used in these studies. The Ti:sapphire laser ($t = 110$ fs, $\lambda = 795$ nm, $E = 10$ mJ, 10 Hz pulse repetition rate) was used in these z-scan studies. The output radiation with variable pulse duration (from 110 fs to 1.6 ps) was also available from this laser. The laser also operated at Q-switched regime delivering 8 ns pulses. The 80 MHz pulses from the seeding oscillator (Spectra-Physics, Tsunami) were used for the investigation of thermal-induced self-defocusing in the suspensions. The 1054 nm, 100 MHz, 280 fs, 100 mW radiation (Time Bandwidth, GLX-200) was also used for the investigation of the thermal-induced nonlinear optical processes caused by IR radiation. Second harmonic radiation of Nd:YAG laser ($\lambda = 532$ nm, $t = 8$ ns, 1 Hz pulse repetition rate) was applied in the z-scan measurements using long pulses.

Previously observed influence of the thermal accumulative effects on the behavior of nonlinear optical refraction in various media at high repetition rates of femtosecond pulses (of order of 10 MHz [104, 105]) has shown the importance of heat accumulation between pulses due to various (linear and/or nonlinear) mechanisms of optical losses. There are different mechanisms of absorption of 795 nm radiation in Ag nanoparticle-contained medium. The first one is associated with Mie scattering and absorption and the second one can be caused by two-photon absorption due to the closeness of the energy of two photons and the SPR of silver nanoparticles. The radiation of Ti:sapphire oscillator (Spectra-Physics, Tsunami; $\lambda = 795$ nm, $P = 300$ mW, $t = 100$ fs, 80 MHz pulse repetition rate) was used to analyze the influence of the thermal-induced nonlinearities on the propagation of laser radiation through the Ag-contained suspensions.

Figure 5.13 shows the beam shape variations in the far field in the case of Ag:ethanol solution using z-scan-like configuration. The arrows show the positions of 5 mm thick cell filled with this solution and CCD images in the far field, respectively. A strong thermal-induced self-defocusing caused by absorbed radiation clearly seen in these images. The self-defocusing caused by thermal lens concentrates radiation in the far field (position $z = z_1$, CCD image No. 2) relative to the reference beam when the cell was placed far from the focal plane (Nos. 1 and 4). At $z = z_2$, the self-defocusing leads to the appearance of ring structure (No. 3). The positions z_1

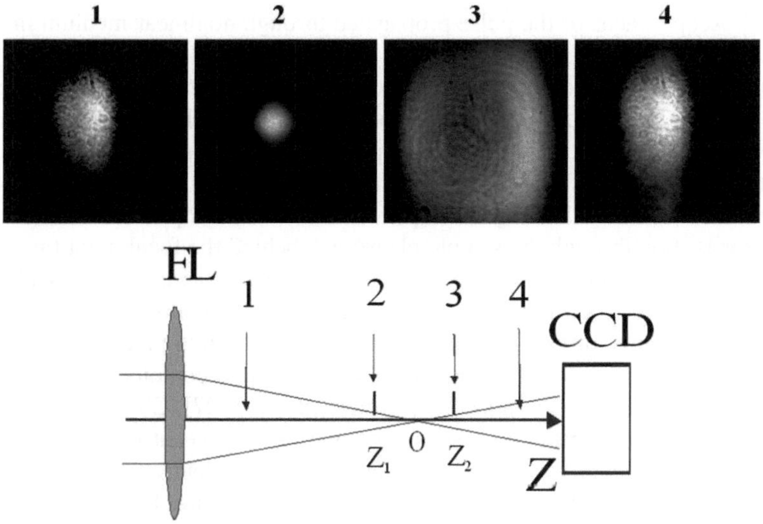

Fig. 5.13 The CCD images of beam shapes of 795 nm radiation operated at high pulse repetition rate and propagated through Ag:EG solution at different positions of cell in z-scan-like configuration. Reproduced from [41] with permission from Elsevier

and z_2 correspond to the maximum and minimum of normalized transmittance in the case of closed aperture z-scan measurements.

The thermal-induced nonlinear refractive index of Ag:EG solution in the case of high pulse repetition rate was measured to be $\gamma = -8 \times 10^{-12}\,\mathrm{cm^2\,W^{-1}}$. The pure ethylene glycol has also shown self-defocusing properties caused by thermal lens appearance due to the nonlinear, probably three-photon, absorption of radiation. In that case the thermal-induced self-defocusing was considerably weaker in comparison with silver nanoparticles-contained solution. Those observations have also shown the insignificant influence of accumulative thermal effects on the variations of refractive index in the cases of pure water and ethanol. The observation of similar variations of beam shape in the case of 1054 nm radiation ($t = 280$ fs, $P = 100$ mW, 100 MHz) propagated through silver-contained suspensions confirmed a decisive role of the first mechanism of absorption. No absorption and beam shape variation was observed in the case of pure ethylene glycol, ethanol, and water at $\lambda = 1054$ nm.

In the next set of nonlinear optical studies the radiation operated at 1 and 10 Hz pulse repetition rates was used in order to exclude the influence of thermal accumulative processes. In the case of nanosecond pulses of Nd:YAG laser second harmonic radiation ($\lambda = 532$ nm, $t = 8$ ns, $E = 2$ mJ, 1 Hz pulse repetition rate) the fast Kerr-induced nonlinear optical processes in Ag nanoparticle-contained solutions were not observed up to maximum intensities of $I = 8 \times 10^9\,\mathrm{W\,cm^{-2}}$. These studies were carried out using closed aperture z-scan and time-integrated photodiodes. However, the self-defocusing was observed when the time waveform of propagated pulses was analyzed using the fast p–i–n diodes. The idea of this method is based on the analysis of oscilloscope traces of the pulse propagated through nonlinear medium in closed aperture z-scan scheme.

The avalanche photodiode with a rise-time of 1 ns was used as a detector in these studies. The detected signal was displayed on a 10 GHz bandwidth sampling oscilloscope. The variations of the trailed part of laser pulse registered in the far field after the propagation of limiting aperture depending on cell position were observed in those studies. Curve 1 of Fig. 5.14a presents the oscilloscope trace of such a pulse after propagation through the sample placed just behind the focal point (at $z = z_2$). One can see a suppression of the trailing part of pulse as compared with initial pulse shape (Fig. 5.14a, curve 2). Such a variation of temporal waveform points out on the influence of time-dependent self-defocusing in the medium. Analogous observations were previously reported in various liquid-contained media (methyl nitroaniline solutions [106], aqueous colloidal metals [107], CS_2 [38]). The origin of observed self-interaction process is a thermal-induced negative lens appearance. However, this process was caused by the physical mechanisms other than those that were observed in the case of high pulse repetition rate. In the latter case the appearance of negative lens was caused by slow heat accumulation, whereas in the former case the self-defocusing was caused by acoustic wave propagation through the beam waist area.

The acoustic wave was generated as a result of optical absorption of laser radiation. The influence of absorption was clearly seen in the case of 10 mm thick cell filled with Ag:EG. In that case the optical breakdown of solution at high intensity of laser radiation was observed (Fig. 5.14b, curve 1). Analogous studies of waveform variations were carried out using the 795 and 397.5 nm, 8 ns radiation. In that case the narrowing of pulse duration of the radiation propagated through limiting aperture was also observed. However, in the case of 397.5 nm radiation this process was seen only when the single pulses were interacted with the Ag nanoparticles-containing suspensions (Fig. 5.14c, curve 1). In the case of 10 Hz repetition rate, the pulse shape after first 3–5 shots became the same as initial one (curve 2). No changes of pulse shape were observed in these studies in the case of Ag nanoparticle-free liquids. The observed peculiarity could be attributed to the fragmentation, or fusion of nanoparticles following the photo-thermal melting. Such processes lead to the change of particle size and decrease of thermal-induced self-interaction processes. Another process that accompanied the laser irradiation of colloidal solutions is a formation of nanowires, as was reported by Tsuji et al [108] in their studies of the spherical silver colloids dissolved in water.

The open aperture z-scans of Ag nanoparticles-containing suspensions were carried out using 397.5 nm, 8 ns radiation (Fig. 5.15a). In that case the reverse saturable absorption can be responsible for the nonlinear absorption of Ag nanoparticles. The common feature of the dispersion curve of Im $\chi^{(3)}$ of different nanoparticle-containing compounds near their SPRs is a negative sign of nonlinear absorption [82], which was also confirmed in [41] using short laser pulses. However, in the case of long laser pulses the positive nonlinear absorption in Ag-contained samples was observed. One can assume that in the case of long pulses the reverse saturable absorption plays an important role in the overall dynamics of the nonlinear opti-

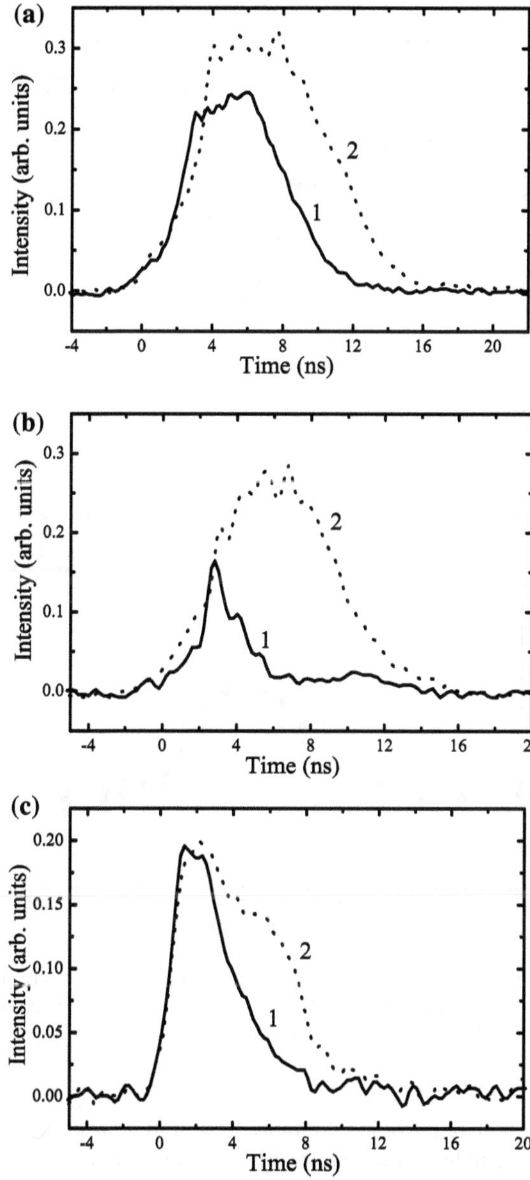

Fig. 5.14 The oscilloscope traces of nanosecond pulses propagated through the aperture in z-scan-like configuration in the cases of **a** 532 nm radiation, 2 mm thick Ag:EG-contained cell, **b** 532 nm radiation, 10 mm thick Ag:EG-contained cell, and **c** 397.5 nm radiation, 2 mm thick Ag:water-contained cell. *1* the oscilloscope traces in the case when the cells were placed behind focal point, close to z_2 position; *2* the oscilloscope traces in the case when the cells were placed far from focal point (the initial waveforms of laser pulses). z_2 corresponds to the minimum of normalized transmittance in closed aperture scheme. Reproduced from [41] with permission from Elsevier

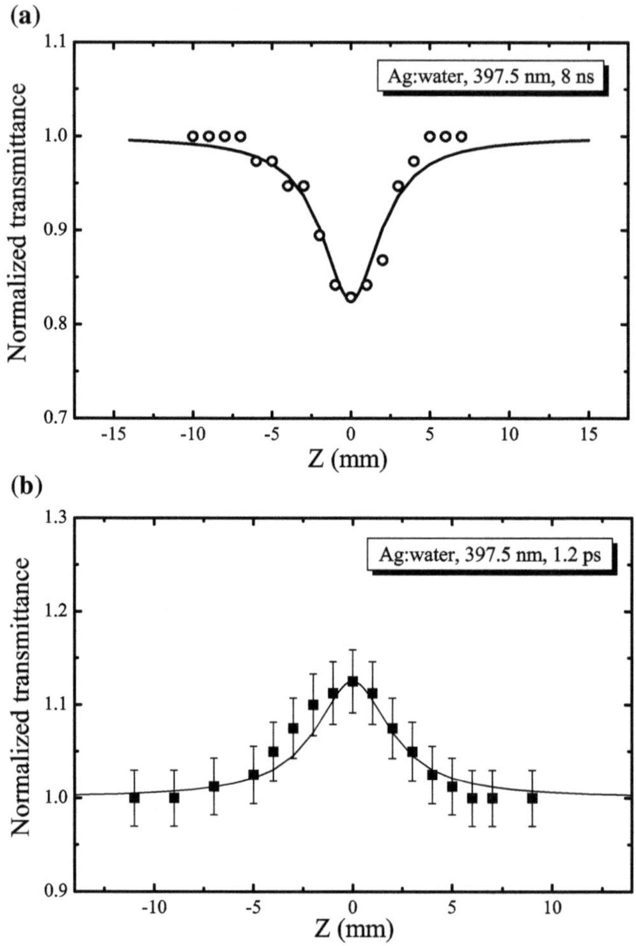

Fig. 5.15 The normalized transmittances of Ag:water solution in the cases of **a** long pulses (8 ns), and **b** short pulses (1.2 ps). Open aperture scheme, $\lambda = 397.5$ nm. *Solid lines* show the theoretical fits. Reproduced from [41] with permission from Elsevier

cal transmittance of nanoparticles-contained compounds, taking into account the saturation of intermediate states responsible for saturable absorption in the case of femtosecond and picosecond pulses. After fitting of Eq. (3.10) with experimental data (Fig. 5.15a, solid curve) the nonlinear absorption coefficient of Ag:water solution was calculated to be 3×10^{-9} cm W^{-1}. The imagine part of the nonlinear susceptibility of Ag nanoparticles was estimated to be $\mathrm{Im}\chi^{(3)} = 3 \times 10^{-8}$ esu taking into account the small volume ratio of silver nanoparticles in water solution ($p \approx 4 \times 10^{-5}$).

The application of short laser pulses in z-scan studies allows excluding the influence of slow thermal-related nonlinear optical processes and analyzing the self-interaction mechanisms caused by electronic response of Ag nanoparticles.

The z-scan scheme was calibrated using carbon disulfide. The nonlinear refractive index of carbon disulfide was measured to be $(3.5 \pm 0.7) \times 10^{-15}\,\text{cm}^2\,\text{W}^{-1}$ [40] using 300 fs pulses. The studies using 795 nm radiation were performed at laser intensities up to $1 \times 10^{11}\,\text{W cm}^{-2}$. However, the influence of Ag nanoparticles on the nonlinear refractive properties of Ag:liquid solutions was negligible, probably due to small volume ratio of those clusters. The only nonlinear optical process observed at 795 nm in silver nanoparticles-containing liquids was a nonlinear absorption caused by the two-photon processes in these compounds. The nonlinear absorption coefficient of nanoparticles in Ag:water suspension using 795 nm, 800 fs radiation was measured to be $8 \times 10^{-9}\,\text{cm W}^{-1}$.

In the case of 397.5 nm, 1.2 ps radiation, a positive nonlinear refraction induced by the influence of silver nanoparticles was registered in closed aperture z-scans. No nonlinear optical processes were observed in this case in silica glass cells filled with pure water at the intensities up to $8 \times 10^{9}\,\text{W cm}^{-2}$. The γ and $\text{Re}\chi^{(3)}$ of Ag:water suspension were calculated to be $3 \times 10^{-13}\,\text{cm}^2\text{W}^{-1}$ and 2×10^{-12} esu, respectively. In the case of quasi-resonant interaction the sign of nonlinear refractive index determines by a relation between the fundamental and doubled frequencies of laser radiation and the SPR frequency of investigated medium [87]. The analysis of the sign of γ was based on this consideration and confirmed the positive value of this parameter, in agreement with experimental data. The normalized transmittance in that case is given by Eq. (3.5) with negative value of β. The nonlinear absorption coefficient in that case can be presented as $\beta = \alpha / I_s$ that is a ratio between linear absorption coefficient and saturable intensity. The solid line fitted to the data in Fig. 5.15b was calculated by setting β equal to $-1.5 \times 10^{-9}\,\text{cm W}^{-1}$. I_s was consequently evaluated to be $6 \times 10^{8}\,\text{W cm}^{-2}$. The studies of Ag:water suspension have shown the dominance of $\text{Re}\chi^{(3)}$ over $\text{Im}\chi^{(3)}$ at $\lambda = 397.5$ nm $\left(\left|\text{Re}\chi^{(3)}\right| = 2 \times 10^{-12}\,\text{esu}, \left|\text{Im}\chi^{(3)}\right| = 6 \times 10^{-13}\,\text{esu}\right)$.

The third-order susceptibility of nanoparticle-containing dielectrics can be written as [109]

$$\chi^{(3)} = p \mid f_e \mid 2 f_e 2 \chi^{(3)}{}_{\text{np}}, \tag{5.6}$$

where $\chi^{(3)}_{\text{np}}$ is the intrinsic third-order nonlinear optical susceptibility of nanoparticles, and p is the filling factor (i.e., volume ratio). Factor f_e is the local field enhancement of polarization. The electric field that actually polarizes nanoparticle can be much larger than the external applied electric field. This effect is known as dielectric confinement. The most conspicuous manifestation of confinement in the optical properties of nanoparticles is the appearance of the SPR that strongly enhances their linear and nonlinear optical characteristics in the vicinity of SPR wavelength. This resonance is a direct consequence of dielectric confinement and can be interpreted in term of a collective motion of the electrons in the clusters. Therefore the impact of dielectric confinement on the nonlinear optical response of nanoparticles has been extensively studied using different technique in the vicinity of SPR.

A variation of normalized transmittance in the case of 397.5 nm, 800 fs pulses operated at 10 Hz repetition rate was observed in those studies. This regime of laser-colloids interaction at high irradiation leads to the fragmentation of Ag nanoparticles. As a result, the saturable absorption presented by a normalized *T*(z) dependence in the case of open aperture configuration (analogous to z-scan trace presented for single pulses in Fig. 5.15b) was less pronounced at 10 Hz pulse repetition rate compared with single-shot interaction.

5.2.3 Role of Aggregation in Variations of Nonlinear Optical Parameters of Silver Nanoparticle Suspensions

Now we come to the discussion of the role of aggregation rate on the nonlinear optical response of Ag clusters. Colloidal silver is of interest due to the perspectives of its application in OL, optoelectronics, etc [64, 110–112]. Aggregation of such samples leads to the increase of nonlinear optical response, in particular, in processes of degenerate four-wave mixing due to the growth of local field amplitude in fractal clusters with high polarizability [113]. The optical characteristics of aggregated silver are determined by the variations of surface plasmon frequency located near the region of 415 nm (in the case of non-aggregated silver) and long-wave absorption wing with the growth of aggregation rate [114]. This parameter was proposed in [115] and was based on a variety of theoretical and experimental studies [116–118]. It connects the dynamics of colloidal silver absorption spectrum with the spatial parameters of clusters. Thus, the aggregation rate allows determining the sizes of clusters without electron microscopic methods.

Here we compare the mechanisms of self-action in colloidal silver for various aggregation rates using laser pulses ($\lambda = 1064$ nm) of different durations (picosecond and nanosecond) [62]. Figure 5.16 shows the set of measurements of *T*(z) dependences under the action of picosecond pulses for different aggregation rates of colloidal silver. It is worth noting the sign change of n_2 (from the positive (Fig. 5.16, curve 1) to the negative (Fig. 5.16, curves 2–4)). The aggregation rate was an alternating parameter of the investigated medium. This parameter (A) was determined using the absorption spectra of the colloidal solution [115]. For this purpose, one can consider the difference of integral absorption in the long-wave wing region between the aggregated and non-aggregated suspensions and normalized this parameter with respect to the spectrum of maximally aggregated colloidal silver. The variations of absorption spectra for each of the aggregated states are shown in Fig. 5.17. One can see that the growth of the aggregation rate leads to the broadening of absorption spectra and the appearance of the long-wave wing near the surface plasmon resonance of silver. The linear absorption coefficients were increased (from 0.8 cm^{-1} to 2 cm^{-1}) with the growth of the aggregation rate, so it was necessary to decrease the power density of picosecond radiation to prevent an optical damage of the sample. These changes have been taken into account for $\chi^{(3)}(-\omega; \omega, -\omega, \omega)$, n_2, and β calculations. The nonlinear absorption has reached its maximum ($\beta = 3.8 \times 10^{-11}$ cm W^{-1}) for maximal aggregation rate (A = 0.8).

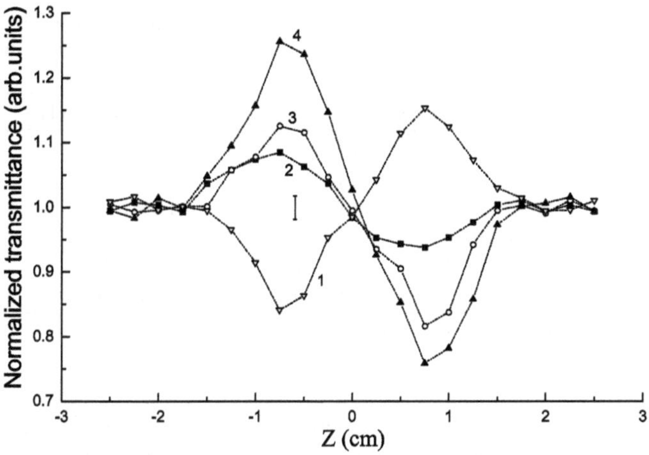

Fig. 5.16 Normalized transmittances as the functions of cell position in closed aperture scheme for different aggregation rates (A) of colloidal silver. (*1*) A = 0, (*2*) A = 0.16, (*3*) A = 0.54, (*4*) A = 0.80. Reproduced from [62] with permission from IOP Publishing

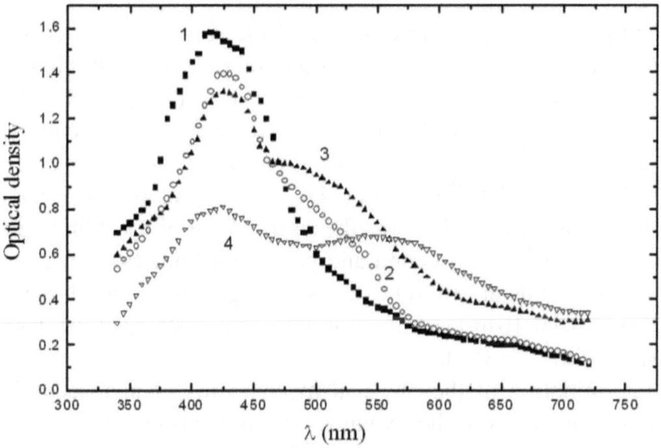

Fig. 5.17 Absorption spectra of colloidal silver for different aggregation rates. (*1*) A = 0, (*2*) A = 0.16, (*3*) A = 0.54, (*4*) A = 0.80. Reproduced from [62] with permission from IOP Publishing

It should be noted that the nonlinear absorption for non-aggregated colloidal silver was not detected. The dependences between the aggregation rates of colloidal silver and the nonlinear susceptibilities and nonlinear refractive indices for picosecond pulses are presented in Table 5.2. Maximum values of $\chi^{(3)}(-\omega; \omega, -\omega, \omega)$ and n_2 were found to be -1.5×10^{-14} esu and -1.1×10^{-13} esu, respectively. Here, the z-scan measurements of the self-defocusing of nanosecond pulses, which is due to the thermal energy transition from the nanoparticles to the surrounding solvent (water), are presented. It should be noted that this nonlinearity has a 'slow' character, which

Table 5.2 The dependences between the aggregation rates of colloidal silver and its nonlinear susceptibilities and nonlinear refractive indices for picosecond and nanosecond radiation (λ = 1064 nm)

Aggregation rate	Picosecond pulses		Nanosecond pulses	
	n_2, esu	χ^3 esu	n_2, esu	χ^3 esu
0	1.43×10^{-14}	1.97×10^{-15}	1.1×10^{-11}	1.5×10^{-12}
0.16	-3.81×10^{-14}	-5.2×10^{-15}	-4.65×10^{-11}	-6.4×10^{-12}
0.54	-5.69×10^{-14}	-7.85×10^{-15}	-15.9×10^{-11}	-2.19×10^{-11}
0.80	-11×10^{-14}	-1.5×10^{-14}	-31.9×10^{-11}	-4.4×10^{-11}

Reproduced from [62] with permission from IOP Publishing

plays a negligible role in experiments with picosecond pulses. Those experiments have shown that the nonlinear optical response of a colloidal structure has a 'slow' component, due to the thermal nonlinearity of surrounded dielectric, and a 'fast' component, due to the Kerr-induced nonlinearity of aggregated particles [44].

The same measurements as for picosecond pulses were performed using the nanosecond pulses (see Table 5.2). The dynamics of nonlinear optical parameters with the growth of aggregation rate remained the same as for picosecond pulses. It should be noted that, in the case of nanosecond pulses, the nonlinear absorption was not observed up to the maximum intensity used ($I_{max} = 8 \times 10^9\,W\,cm^{-2}$), which was much smaller than the maximal intensity of picosecond pulses ($I = 4 \times 10^{11}\,W\,cm^{-2}$). Maximum values of n_2 and $\chi^{(3)}(-\omega; \omega, -\omega, \omega)$ in that case were found to be -3.2×10^{-10} esu and -4.4×10^{-11} esu, respectively. The characteristic peculiarity of these studies was the sign change of n_2 with the growth of the aggregation rate for both nanosecond and picosecond pulses. Electron microscopy analysis of silver fractals at different stages of aggregation has shown that particle sizes were varied from 10 nm (for the non-aggregated state) to 40–100 nm (for the aggregated state with A = 0.8).

For the determination of the physical mechanisms of nanosecond pulse' self-defocusing in colloidal silver, the dynamics of laser pulses passed through the cell in closed aperture experiments was analyzed for different cell positions with respect to the focal point (z = 0). Part of the radiation was defocused inside the cell when the energy density reached its definite value. As a result of this, a smaller part of the whole radiation passed through the aperture and was registered by the photodiode. The temporal trace of input radiation passed through the cell and aperture had reached its maximal value before the pulse maximum, which was detected in the case when a cell was placed far from the focal point. Such behavior characterizes the thermal nature of self-defocusing [38].

Now we discuss the thermal mechanism of self-defocusing in the nanosecond regime (heat transfer to the water, characteristic times, comparison with others results). As was mentioned, the observations of the temporal behavior in the nanosecond regime are based on the assumption of the thermal nature of lensing. The thermal lens is built-up with the energy absorption growth. Because of the delay of this process due to the transfer of energy to the solvent, this lensing is more effective for the trailing part of the pulse. It was seen in temporal traces of propagated pulses that its trailing part changes depending on sample position with respect to the focal point. The defocusing effect is more distinctly seen in the cases when the cells were placed close to the T_{min} position (for the closed aperture z-scan scheme). An important parameter of self-defocusing in that case is $M = 1/C\rho(dn/dT)$, which is equal to $-1.04 \times 10^{-4}\,cm^3\,cal^{-1}$ for water [119]. Here, C is the specific heat and ρ is the density of the solvent. This parameter is crucial for determining the variations of refractive index as a function of absorbed energy per unit volume ($\Delta n = M\Delta E$).

Usually, the z-scan experiments are performed in homogeneous and long-term stable samples, which can be characterized by some refractive index. Colloidal systems, in particular when they are aggregated, are inhomogeneous on the λ-scale and, hence, require 'effective medium' models to introduce some effective refractive index n_{eff}. This holds true, both for linear and nonlinear contributions. As we mentioned earlier, the sizes of the investigated nanoparticles have changed from four to ten times (from 10 to 40–100 nm). Further growth of particle size led to the increase of absorption and the sedimentation of nanoparticles. These changes do not influence the propagation of laser beam near the focal point due to variations of n_{eff}. It follows from those studies that the nonlinear refractive index in the case of thermal contribution (nanosecond pulses) depends on the aggregation rate (or in other words, on particle size) but not on the beam size. The studies of colloidal solutions in the cases of different beam waists have shown the same values of nonlinear optical parameters within the accuracy of measurements.

5.3 Influence of Laser Ablation Parameters on the Optical and Nonlinear Optical Characteristics of Colloidal Suspensions of Semiconductor Nanoparticles

An interest in the synthesis, characterization, and application of colloidal semiconductor "quantum dot" materials has grown markedly since first reports on these topics [120, 121] due to strong size-related dependence of their optical and electronic properties. However, the dependence of the optical nonlinearities on the sizes of the semiconductor nanoparticles has yet to be determined. In this section, the laser ablation method used for the preparation and investigation of aqueous colloidal solutions of As_2S_3 and CdS nanoparticles is discussed [122]. The measurements of nonlinear refractive indices, nonlinear absorption coefficients, and third-order nonlinear susceptibilities of these suspensions by the z-scan technique are presented using picosecond radiation at the wavelength of $\lambda = 532$ nm. We compare these data with those obtained for As_2S_3 and CdS thin films [123].

The Q-switched Nd:YAG laser ($\lambda = 1064$ nm, $t = 20$ ns, $E = 15$ mJ) at 10 Hz pulse repetition rate was used for the laser ablation of bulk As_2S_3 and CdS semiconductors. The samples (bulk As_2S_3 or CdS) were immersed in a quartz cell containing distilled water (cell thickness of 5 cm). Laser radiation was focused by an 8 cm focal length lens onto the surface of the sample that was kept close to the cell's back window to prevent the optical breakdown of the front window. This method allowed synthesizing semiconductor nanoparticles, such as cadmium sulfide, in colors ranging from water white to orange depending on the particle's diameter in aqueous solutions (from 2 to 8 nm) with no change in chemical composition. Other optical properties of the particles, such as refractive index and nonlinear optical response, exhibited similar changes with particle size. Weight volume of nanoparticles in suspension was estimated to be (3–5)$\times10^{-5}$. Chalcogenide CdS and As_2S_3 thin films were used for comparison with CdS and As_2S_3 nanoparticle suspensions. The films were prepared by evaporation in vacuum of the chalcogenide components (CdS, As_2S_3) onto the surface of BK7 glass substrates. The thickness of films was 10 μm. The z-scan technique was applied for the investigation of the nonlinear optical characteristics of the samples. The detailed description of the experimental setup was published elsewhere [124] and is discussed in Chap. 3. The output laser characteristics were as follows: pulse duration 55 ps, $\lambda = 532$ nm, $E = 0.2$ mJ, 2 Hz pulse repetition rate. The experiments were carried out at the intensities of up to 5×10^{10} W cm^{-2}.

In Figure 5.18, the normalized transmission dependences of As_2S_3 and CdS colloidal solutions are presented for a closed aperture scheme. One can conclude from these dependences about the positive value of nonlinearity that is a self-focusing of laser radiation. Relatively symmetric curves (respectively to the focal point, z = 0) indicate an insignificant influence of nonlinear absorption at the intensity level of 4×10^9 W cm^{-2}, thus causing to make the nonlinear absorption measurements using an open-aperture scheme at higher intensities when this nonlinear optical process appeared to be more strong.

The values of n_2 and Re$\chi^{(3)}$ of the CdS solution were calculated to be 1.72×10^{-12} and 2.74×10^{-13} esu, respectively, at the intensity of 4×10^9 W cm^{-2}. The solid lines in the figure are the theoretical fits calculated for the medium with third-order nonlinearity. Intensity-dependent measurements of n_2 and Re$\chi^{(3)}$ have shown the decrease in their values with intensity growth. Thus, at intensity increasing up to 1.17×10^{10} W cm^{-2}, the values of n_2 and Re$\chi^{(3)}$ were measured to be 6.58×10^{-13} and 1.05×10^{-13} esu that are 2.6 times smaller as compared with the previous case [125].

The unfixed value of the nonlinear refractive index in those studies indicates the influence of higher orders of nonlinearity induced by the appearance of free carriers. The same feature was reported by Du et al. [126]. The importance of free carrier influence on the variations of nonlinear refraction was also discussed by Li et al. [127] in their studies of bulk CdS. The intensity-dependent variation of the nonlinear refractive index is the proof of the influence of the fifth-order nonlinear optical properties.

The same measurements were carried using As_2S_3 suspension. In that case, the dependence of n_2 on laser intensity was also observed. The values of n_2 and Re$\chi^{(3)}$ at $I = 2.9\times10^9$ W cm^{-2} were calculated to be 2.2×10^{-12} and 3.5×10^{-13} esu, while at the intensity of 3.3×10^{10} W cm^{-2} the 1.5 times decrease of n_2 and Re$\chi^{(3)}$ ($n_2 = 1.5\times10^{-12}$ esu, Re$\chi^{(3)} = 2.3\times10^{-13}$ esu) was observed.

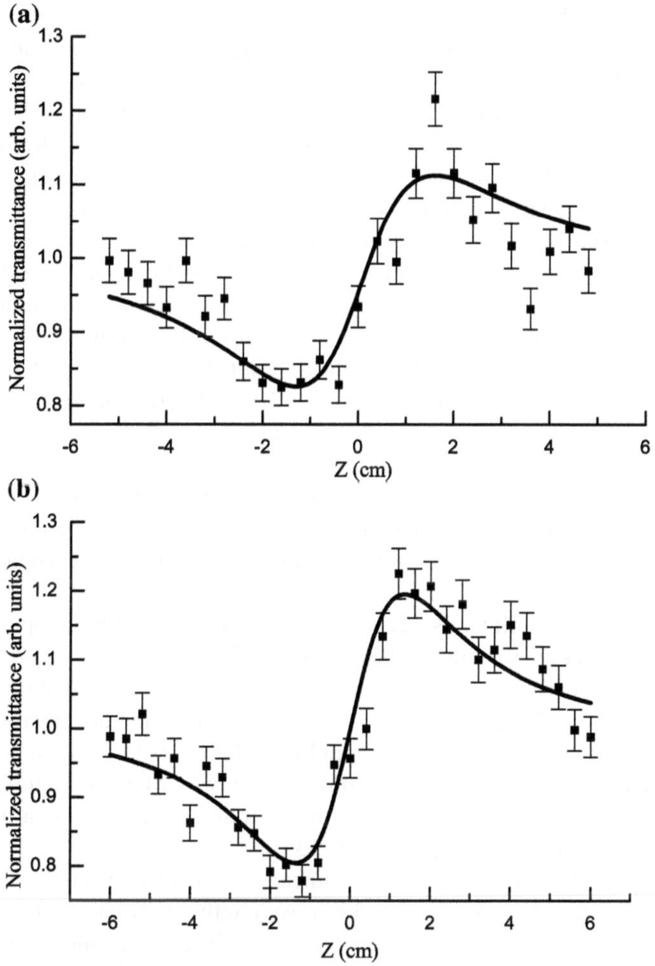

Fig. 5.18 Normalized transmission dependences of **a** As_2S_3 and **b** CdS nanoparticles solutions in the case of closed aperture scheme at $\lambda = 532\,\text{nm}$. *Solid lines* are the theoretical fits. Reproduced from [125] with permission from IOP Publishing

Small values of nonlinear optical parameters of these suspensions were predominantly due to low concentration of nanoparticles. Taking into account the volume part of semiconductor nanoparticles (4×10^{-5}), one can estimate that their nonlinearities exceed the ones of bulk materials. In particular, n_2 of bulk CdS has previously been measured to be -2×10^{-10} esu ($\gamma = -5 \times 10^{-13}\,\text{cm}^2\,\text{W}^{-1}$ [127]). The reviewed data show that this parameter for semiconductor nanoparticles is of the order of 10^{-7} esu, taking into account the volume part of the active nonlinear medium.

One of the possible explanations of observed high nonlinearities of semiconductor nanoparticles is the influence of field localization in such structures. Surface-enhanced nonlinear spectroscopy of various nanosized structures gives valuable information about the nature of considerable growth of both optical and nonlinear optical characteristics of such media. Surface-enhanced hyper-Raman scattering and anti-Stokes Raman scattering, second- and third-harmonic generation, as well as other nonlinear optical responses of fractals and nanofilms, were studied in [128–130]. The fractal geometry results in the localization of Plasmon excitation in the "hot" spots, where the local field can exceed the applied field by several orders of magnitude. The high local fields of the localized fractal modes result in dramatic enhancement of nonlinear optical responses. The estimations of nonlinear indices of CdS nanoparticles (taking into account their volume part in solutions) have shown that their n_2 exceeds the one of bulk CdS by more than 3 orders of magnitude ($1.72\times10^{-12}\times 4\times10^{-5} \approx 7\times10^{-7}$, and 2×10^{-10}esu, respectively). This value (3.5×10^{-3}) is comparable with calculations by various authors predicting the 10^3enhancement of nanoparticle nonlinearities with respect to the ones of bulk material. These estimations of n_2 of dissolved CdS nanoparticles can be considered as a quantitative confirmation of an assumption on local field induced enhancement of observed nonlinearities.

Large third-order nonlinear susceptibilities of semiconductor nanoparticles were reported previously in various studies. Wang and Mahler [131] reported on large third-order nonlinearities of CdS nanocrystallites embedded in Nafion films. The self-focusing ($\lambda = 514.5$ nm) in colloidal CdS semiconductor quantum dots was observed by Shen et al. [132]. n_2 was found to be 10^5times larger than the nonlinear refractive index of CS_2 (that has the nonlinearity of two orders of magnitude smaller than bulk CdS). Analogous results indicating extremely high nonlinearities of such nanoparticles were reported in [126, 133, 134].

Previously, measurements of the nonlinear optical characteristics of semiconductor As_2S_3 colloidal solution at the wavelength of Nd:YAG laser radiation ($\lambda = 1064$ nm, $t = 25$ ns) were reported in [122]. In the case of nanosecond radiation, the nonlinear refractive index was measured to be -3×10^{-11} esu. The appearance of nonlinear refraction in those studies was attributed to the thermal effect. The prevailing mechanism contributing to the refractive index variation was the thermal lens due to nanosecond laser radiation absorption in the investigated solution. Such assumptions of the self-defocusing process were confirmed in the experiments with colloidal metals, when the temporal shape of nanosecond pulses propagating through the aperture was analyzed [32].

In the case of picosecond or shorter pulses, the main origins of the nonlinearities in such semiconductor compounds are known to be two-photon and Raman transitions, ac-Stark shift, optical Kerr effect, and photoexcited free carriers. Picosecond z-scan and degenerate four-wave mixing measurements showed that the first four effects are responsible for the nonlinearities at the intensities up to 5×10^8 W cm^{-2}, while at higher intensities, the nonlinear effects originating from the two-photon excited free carriers proved not to be neglected [127]. Those observations also confirm the assumptions on the influence of free carrier generation on the decrease of n_2 at the intensities used (i.e., well above the mentioned intensity level).

Let us now discuss the sign of the nonlinear refraction of investigated semiconductor nanoparticle suspensions. Such nanostructures, for instance CdS, can exhibit another sign of n_2 in comparison with that of bulk CdS at $\lambda = 532$ nm. A scaling rule between the nonlinear refractive index and the ratio of the photon energy to the band gap energy ($h\omega/E_g$) was analyzed previously in different materials. The nonlinear refractive index was found to change its sign at 0.69 value of this ratio under different E_g or laser frequency ω [135]. Those studies of bulk CdS showed the spectral dispersion of the Kerr nonlinearity sign. Thus, n_2 had a negative sign at $\lambda = 610$ nm, and changed to positive at $\lambda = 780$ nm, that was in agreement with the calculations using the Kramers–Kronig relation. In the described case the ratio between laser frequency and surface plasmon resonance of CdS nanoparticles is a crucial factor in the definition of the sign of nonlinear refraction.

Some previous studies of such nanoparticles show that they possess the analogous sign of n_2 as bulk media. Du et al. [126] demonstrated the negative sign of nonlinear refractive index for CdS nanoparticles at $\lambda = 532$ nm using 7 ns pulses and similar results were also reported by Rakovich et al. [134]. Large crystallites (>10 nm) show an optical absorption close to that of bulk crystalline material. However, smaller crystallites show a large blueshift in absorption edge that lead to the variations of effective band gap. These observations can be understood as quantum size effects resulting from confinement of the electron and hole in a small volume.

As described above, the change in the refractive index can be determined via the Kramers–Kronig relationship. The relationship between particle sizes and the peak energy of excitons was demonstrated previously in [136]. As the radii of semiconductor particles are decreased below or close to the Bohr exciton radius, their optical and electronic properties can be tuned by changing the size of the particles. The linear properties, such as blueshift in absorption and fluorescence, have been well characterized for II–VI semiconductor nanoparticles for several years. However, the nonlinear optical responses in relation to particle size variations and charge carrier dynamics are less understood. The intrinsic measurement of size, at which variations in nonlinear optical processes begin to become important, is given by the diameter of the 1S exciton in bulk crystalline material. For CdS, this diameter is 6 nm. Rossetti et al. [137] calculated the shift in band gap energy at different sizes of CdS nanoparticles. Their calculations predict a wide range of absorption thresholds (effective band gaps), ranging from 490 nm ($E_g = 2.53$ eV for bulk CdS), then to 471 nm ($E_g = 2.63$ eV, at 6.6 nm diameter of CdS nanoparticles), 367 nm ($E_g = 3.36$ eV, at 3 nm diameter), and further to 286 nm ($E_g = 4.33$ eV at 2.1 nm diameter). The absorption spectrum of CdS nanoparticle suspension ($E_g = 3.86$ eV) was a sum of various spectra corresponding to the different sizes of CdS nanoparticles. The influence of small nanoparticles (with size diameters between 3.5 and 4.5 nm) can change the sign of the self-interacting process (from self-defocusing to self-focusing). In the discussed research, the $h\omega/E_g$ value becomes equal to 0.65. If applicable, these assumptions and observations show the change of the sign of nonlinear optical process (from self-defocusing for bulk material to self-focusing for semiconductor nanoparticle solutions).

Analogous consideration was carried out for As_2S_3 solution. Its energy band gap shifted in accordance with the absorption spectrum from 2.37 (bulk As_2S_3) to 3.19 eV (semiconductor solution). The absorption cutoff in that case was not so expressed as in the case of the CdS solution, indicating the broader size distribution of As_2S_3 nanoparticles.

The CdS nanoparticle sizes were estimated to be between 4 and 6 nm taking into account the shift of the absorption edge. The sizes of As_2S_3 nanoparticles were found to be between 4 and 9 nm. The TEM measurements of investigated suspensions were carried out for comparison with the estimations of nanoparticle sizes depicted from absorption spectra. The size distribution of nanoparticles measured from the TEM images was found to be between 2.5 and 6 nm and mostly centered at 4 nm. The same measurements were carried out for As_2S_3 nanoparticles. It was found to be of a broader size distribution (4–12 nm) in comparison with CdS nanoparticles and with the one depicted from spectral measurements.

Small nanoparticles (with sizes of 4 nm and smaller) can play a predominant role in the overall nonlinear refractive index of such solutions because of quantum confinement effects. Their positive sign of n_2 and the contribution in overall nonlinear refraction can prevail with respect to the negative sign of n_2 of higher sized nanoparticles. This peculiarity of small-sized nanoparticles can be considered as an explanation of the experimentally observed positive sign of chalcogenide solutions.

Local elemental characterization of the nanoparticles was carried out using EDX spectroscopy. In particular, the quantitative result of EDX had shown that the ratio of Cd and S was 55 : 45 (at.%), whereas the same for As and S was 35 : 65. The expected result for CdS was 50 : 50 because the EDX technique allows measuring the atomic ratio of the investigated compound. The same parameter for As_2S_3 was expected to be 2:3 or 40:60. The deviation from the atomic ratio of such semiconductor nanoparticles was also reported in [138].

The open aperture z-scan scheme was used for the investigation of nonlinear absorption as well. The reason for these studies was the asymmetric dependence observed at high intensities in the closed aperture z-scan scheme, indicating the presence of nonlinear absorption. Previously, the open aperture normalized transmittance dependence has shown the characteristic appearance of nonlinear absorption in the case of nanosecond 1064 nm pulses [122]. The same measurements were performed in the case of picosecond pulses at $\lambda = 532$ nm. The results of these investigations are presented in Fig. 5.19. The normalized transmittance for the open aperture conditions is given by Eq. (3.10). The fits of Eq. (3.10) to the experimental data are depicted in Fig. 5.19 by the solid lines. From these fits the β values were calculated to be 6×10^{-10} and 2.9×10^{-10} cm W^{-1} for As_2S_3 and CdS solutions, respectively. The corresponding values of Im$\chi^{(3)}$ for As_2S_3 and CdS solutions were calculated to be 2.2×10^{-13} and 1×10^{-13} esu, respectively. Analyzing the obtained results for real and imaginary parts of $\chi^{(3)}$, one can conclude about their equal contribution in absolute value of $\chi^{(3)}$.

The closed aperture data in some cases appeared to be sufficient for analysis of both the nonlinear refraction and nonlinear absorption. The open aperture *T*(z) curves (Fig. 5.19) allowed calculating the β within an accuracy of 30 %, which was better than the data depicted from the closed aperture scheme (50 %). The previously measured value of β for bulk cadmium sulfide ($\lambda = 532$ nm) was reported to be 5.4×10^{-9} cm W^{-1}[128]. Taking into account a small volume of active particles one can again conclude about strong confinement effects in such semiconductor solutions, leading to the considerable enhancement of nonlinear absorption characteristics in comparison with bulk media. In discussed studies, the CdS and As_2S_3 concentrations in solutions were approximately 0.06 g l^{-1}. Note that the nonlinear parameters of solutions can be increased by increasing the volume fraction of nanoparticles. The influence of the volume fraction of nanoparticles on their nonlinear optical properties was analyzed theoretically in [139] for semiconductor nanoparticles in various films, but in the case of semiconductor solutions an additional investigation should be done.

It is worth noting the opportunity of using the organic solvents instead of water due to various reasons. One of the advantages of preparing nanoparticles in organic solvents is that the distribution of particle sizes is narrower than that in water [137], as it was shown using PVP solutions of CdS nanoparticles [133]. Stabilization of metal nanoparticle nonlinearities was shown previously also using PVP stabilizers [101]. The nonlinear parameters of semiconductor nanoparticle solutions in the case of organic solvents can be considerably increased through the formation of clusters with controlled sizes.

5.4 Studies of Low-Order Nonlinear Optical Properties of $BaTiO_3$ and $SrTiO_3$ Nanoparticles

There is a new emerging class of nanoparticles–nanocrystallites–that has attracted much interest. Among them, the nanoclusters of photorefractive crystals possess the advanced properties both from the point of view of low- and high-order nonlinearities. The photorefractive effect is a phenomenon in which the local index of refraction of a medium is changed by illumination, using a laser beam with spatial variation of the intensity. The crystals possessing photorefractive properties have attracted interest because of their various potential applications, such as optical limiting, spatial light modulation, online holographic recording, and so on. However, nonlinear optical properties of such nanoparticles are less explored.

Barium titanate is one of the most prominent ferroelectric photorefractive materials. The $BaTiO_3$ and $SrTiO_3$ crystals have often been used as the matrices for the nanoparticles [140, 141]. These materials possess strong dielectric constants, a large nonlinear optical effect, and ferroelectricity. At the same time, these materials could be prepared as the nanoparticles for various applications. In the past, the third-order nonlinear optical properties of bulk $BaTiO_3$ and $SrTiO_3$ crystals were analyzed [142–144]. In particular, the two-photon absorption coefficient of strontium titanate at the

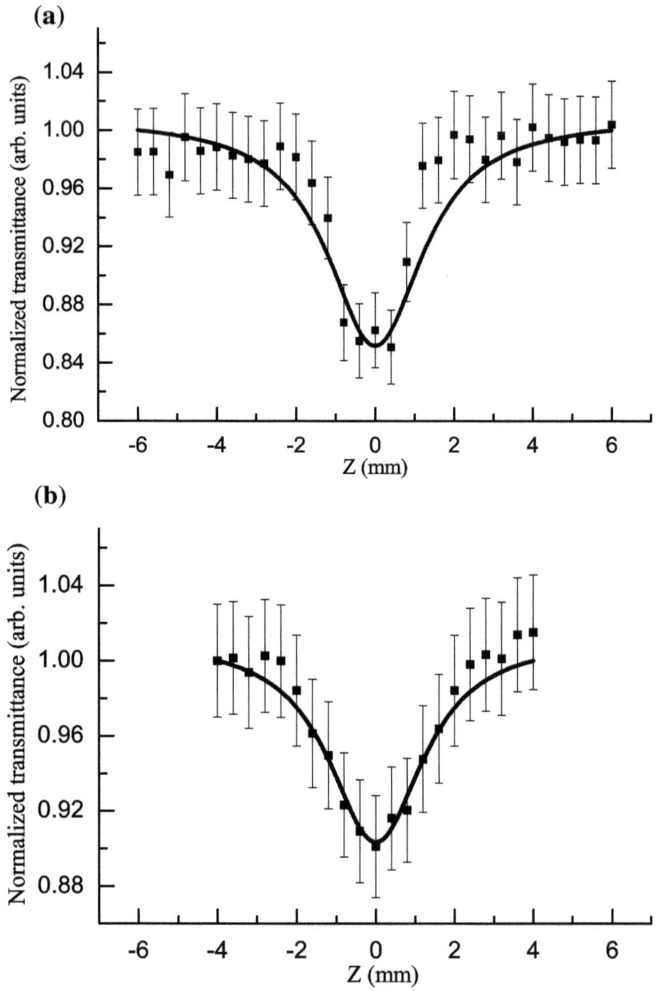

Fig. 5.19 Normalized transmittance dependences for **a** As_2S_3 and **b** CdS in the case of open aperture scheme at $\lambda = 532\,\text{nm}$. *Solid lines* are the theoretical fits. Reproduced from [125] with permission from IOP Publishing

wavelength $\lambda = 0.69\,\mu\text{m}$ was found to be $3\times10^{-9}\,\text{cm W}^{-1}$[142], while the nonlinear refractive index at $1.064\,\mu\text{m}$ was measured to be $48.5\times10^{-16}\,\text{cm}^2\text{W}^{-1}$[143]. In the case of bulk barium titanate, the two-photon absorption coefficient at $\lambda = 0.596\,\mu\text{m}$ was reported to be $0.1\times10^{-9}\text{cm W}^{-1}$[144]. However, no studies of the low-order nonlinear optical characteristics of $BaTiO_3$ and $SrTiO_3$ nanoparticles were reported so far. The same can be said about the investigation of high-order nonlinearities of these nanostructures, particularly with regard to their application for the high-order harmonic generation, which will be discussed in the following chapter.

In this section, we discuss the systematic studies of the nonlinear optical properties of $BaTiO_3$ and $SrTiO_3$ nanoparticles, as well as their structural characteristics [145]. The mean size of $BaTiO_3$ nanoparticles was defined to be 92 nm, with the particle size distribution varying between 50 and 160 nm. The TEM images of $SrTiO_3$ nanoparticles showed smaller sizes, with the mean size of 38 nm, and narrower size distribution compared to the $BaTiO_3$ nanoparticles. In the case of the studies of third-order nonlinearities, these nanoparticles were dissolved in ethylene glycol at different concentrations varying from 5×10^{-4} to 5×10^{-3} weight parts (i.e., at the molar concentrations between 2 and 20 mM). Low-order nonlinear optical characteristics of $BaTiO_3$ and $SrTiO_3$ nanoparticles suspensions were studied using the conventional z-scan technique. The 1 mm thick silica glass cells filled with these suspensions were used in these studies. The 790 nm laser radiation of different pulse durations (120 fs or 210 ps) was focused by a 300 mm focal length lens. The samples were moved along the z-axis through the focal area to observe the variations of the phase and amplitude of propagated radiation in the far field.

The pulse duration of laser radiation has proved to be an important parameter for the determination of the nonlinear optical properties of different materials [40]. This peculiarity seems to be of special importance in the case of the photorefractive media where both the electronic Kerr effect and two- and three-photon absorption can be concurrent with the photorefractive effect, dynamic grating, and free carrier generation. The diffusion time, recombination time, diffusion length of photoexcited charge carriers, and pulse duration are the main parameters that can influence the nonlinear optical characteristics of photorefractive bulk materials. Although the influence of above processes remains unclear in the case of the nanocrystallites of photorefractive materials, the temporal characteristics of laser pulse should play a role in the definition of the nonlinear absorption and refraction.

The parameters of these two nonlinear optical processes in the case of $BaTiO_3$ and $SrTiO_3$ nanoparticle-containing suspensions were measured using the 120 fs and 210 ps pulses. The fresh suspensions possessed the higher values of nonlinear optical parameters; however, these measurements have shown poor reproducibility. The sedimentation of higher sized nanoparticles caused variations of the nonlinear optical properties of suspensions. The nonlinear optical characteristics of $BaTiO_3$ and $SrTiO_3$ nanoparticle-containing suspensions were analyzed after the stabilization of their structural characteristics. These studies revealed that a liquid with higher viscosity (ethylene glycol) is preferable from the point of view of the structural stabilization of nanoparticle suspensions. The z-scan technique suffers from being highly sensitive to sample scatter. The background scans were done to correct for this possibility. They did not show the influence of scattering on the definition of the nonlinearities of the particles under investigation, when one used the moderate concentrations of the nanoparticles in the suspensions (less than 10 mM).

Figure 5.20a shows the closed aperture z-scan of $BaTiO_3$ nanoparticle-containing suspension measured using the 210 ps pulses. It demonstrates the positive nonlinear refraction, with some indication of the appearance of nonlinear absorption. The same pattern was observed in the case of the $SrTiO_3$ nanoparticle-containing suspension (Fig. 5.20b). However, the asymmetry of the normalized transmittance curve in the

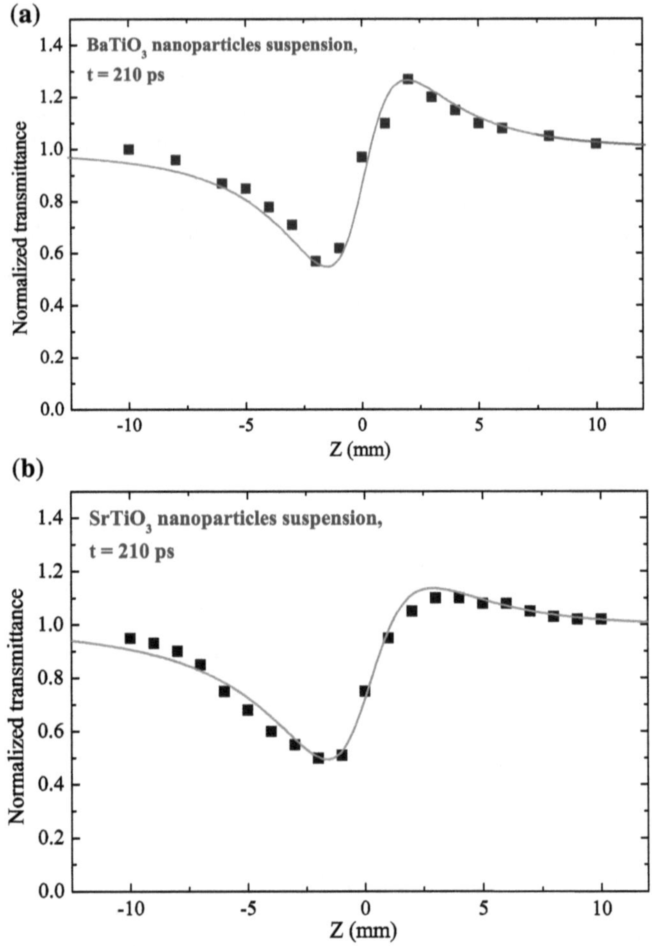

Fig. 5.20 Normalized transmittance dependences in the cases of **a** $BaTiO_3$ and **b** $SrTiO_3$ nanoparticles-containing suspensions measured using the 210 ps pulses. *Solid curves* are the theoretical fits. Reproduced from [145] with permission from Optical Society of America

latter case was more pronounced compared with the $BaTiO_3$ nanoparticle-containing suspension, pointing out the significant influence of the nonlinear absorption.

Equation (3.5) was used for fitting the theoretical dependencies and experimental data. The γ and β of samples were measured to be $3.4\times10^{-15}\,cm^2\,W^{-1}$ and $7\times10^{-13}\,cm\,W^{-1}$, respectively, in the case of the $BaTiO_3$ nanoparticles suspended in ethylene glycol, and $9\times10^{-15}\,cm^2\,W^{-1}$ and $2\times10^{-12}\,cm\,W^{-1}$, respectively, for $SrTiO_3$nanoparticle contained suspension (at 5 mM concentration of nanoparticles), when the 210 ps pulses were used as a probe radiation. The nonlinear optical response of ethylene glycol ($\gamma < 5\times10^{-16}\,cm^2\,W^{-1}$) was insignificant compared with the nonlinearities caused by the nanoparticles. No difference in the nonlinear optical

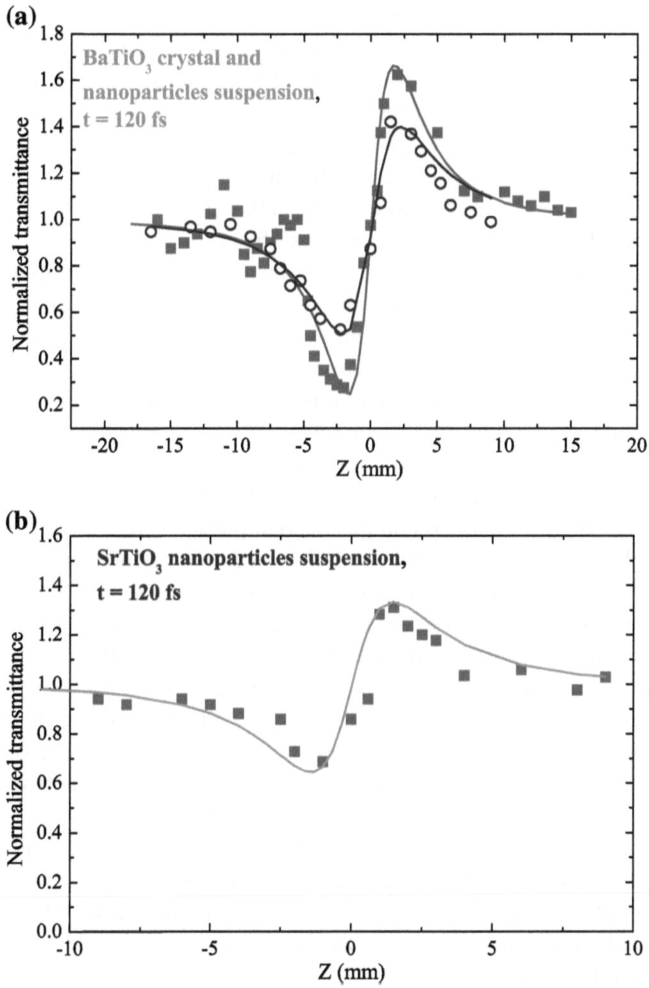

Fig. 5.21 Normalized transmittance dependences of the **a** $BaTiO_3$ nanoparticles-containing suspension (*open circles*) and $BaTiO_3$ crystal (*filled squares*) and **b** $SrTiO_3$ nanoparticles-containing suspension measured using the 120 fs pulses. *Solid curves* are the theoretical fits. Reproduced from [145] with permission from Optical Society of America

properties of suspensions was found in the case of the 10 Hz pulse repetition rate and single pulses, thus indicating the insignificant role of accumulative processes. The application of short laser pulses (in picosecond and femtosecond timescales) at low pulse repetition rate allows excluding the influence of slow thermal-related nonlinear optical processes and analyzing the self-interaction processes caused by the response from the nanoparticles.

Analogous measurements of the nonlinear optical characteristics of photorefractive nanoparticle suspensions were performed using the shorter pulses (Fig. 5.21).

The calculations of the γ of these suspensions in the case of 120 fs pulses gave approximately the same results as in the case of longer pulses, while the nonlinear absorption became almost insignificant. These studies revealed that in the case of nanoparticle medium, the difference between the nonlinear refractive indices at different time scales (210 ps and 120 fs) is insignificant, thus pointing out the involvement of only the fast processes in this nonlinear optical process, contrary to the case of bulk materials (see [40, 146] and references therein).

To compare the nonlinear optical response from nanoparticles and bulk sample of the same material, the nonlinear refractive characteristics of a 2 mm thick $BaTiO_3$ crystal were measured using the 120 ps pulses (Fig. 5.21a, filled squares). This crystal showed the positive nonlinear refraction, as in the case of nanoparticle suspensions. These measurements revealed some asymmetry in the shape of the z-scan, in addition to the observation of the nonlinear absorption, which causes the difference between the valley and the peak in the $T(z)$ dependence. The asymmetry in the case of the left part of z-scan probably points out the influence of crystal properties of the material.

The photorefractive effect takes approximately 1 μs to manifest itself and takes 10 ms to relax. The femtosecond and picosecond pulses were too short for the influence of the photorefractive process to be observed, and the interval between pulses (100 ms) was far greater than the photorefractive relaxation time. Therefore, the contribution of this effect to the observed nonlinear optical refraction and absorption can be neglected. One can thus assume that in the case of photorefractive media, purely local phenomena are predominant in the femtosecond and picosecond timescales of the probe pulse. Another mechanism that could contribute to the decrease in on-axis propagation through the $BaTiO_3$ crystal can be attributed to the time-resolved buildup of grating. However, a buildup time of grating in photorefractive crystals was previously measured to be 4 ns [147], which also show that this process is too slow to influence the nonlinear optical refraction and absorption in the case of shorter pulses. For comparison of the nonlinearities of the $BaTiO_3$ crystal with those of $BaTiO_3$ nanoparticles, it is preferable to carry out experiments using shortest pulses in order to exclude the influence of slower processes, which could be manifested at subnanosecond (210 ps) timescale.

The calculated values of the nonlinear optical characteristics of the $BaTiO_3$ crystal at $\lambda = 790$ nm after fitting the theoretical curve and experimental data were as follows: $\gamma = 6 \times 10^{-16}\,\text{cm}^2\,\text{W}^{-1}$ and $\beta = 1.1 \times 10^{-11}\,\text{cm}\,\text{W}^{-1}$. There no data in the literature about the nonlinear refractive index of the $BaTiO_3$ crystal, while the nonlinear absorption coefficient was almost 1 order of magnitude less than the reported value of this parameter measured at $\lambda = 596$ nm [144]. The latter difference can be attributed to both the measurements at different wavelengths and pulse durations (120 fs and 1 ps). $BaTiO_3$ has no electronic resonances between 596 and 790 nm that would cause such a large difference. Perhaps the closeness of the former wavelength to the transmission edge of the barium titanate (380 nm) caused the increase in the two-photon absorption coefficient with regard to the longer wavelength pulse.

From these studies one can compare the γ value of low-order nanoparticles and the bulk structure of the same material. The difference between them is considerable and could not be attributed to the inaccuracy in the measurements of the molar concentration of the nanoparticles in suspension due to their sedimentation. Since it was mentioned that for the 92 nm particles the role of quantum size effect could be considered insignificant, one can expect the same values of the nonlinear optical properties of these two samples. However, the discussed studies show that the nonlinear refractive indices of the $BaTiO_3$ crystal and those of the nanoparticles differ by 4 orders of magnitude (taking into account the 10^{-3} volume part of nanoparticles in the suspensions), which is a manifestation of the influence of size effect in the case of the nanoparticle-contained medium. The reason for such a difference between the nonlinearities of relatively high-sized nanoparticles and bulk material is not clear, and further studies have to be carried out using nanoparticles of different sizes. One explanation might be related to the presence of some amount of very small nanoparticles, which can cause a strong enhancement of the nonlinear optical response. However, their concentration seems insufficient, since the sizes of most particles were centered between 70 and 130 nm.

The sizes of $SrTiO_3$ nanoparticles were closer to the range where the quantum confinement could be an important factor. Although the low-order nonlinear optical properties of these nanoparticles were not compared with those of the bulk crystal, one could expect the same difference in the nonlinear optical characteristics between the 32 nm $SrTiO_3$ nanoclusters and the bulk crystal, since the parameters of γ and β for the two nanoparticle samples ($SrTiO_3$ and $BaTiO_3$) were close to each other, while the reported values of γ and β for $SrTiO_3$ crystals measured under different experimental conditions [142, 143] were 2–3 orders of magnitude less than those observed in the discussed experiments, which is a characteristic feature of the size-related enhancement of the nonlinearities of nanoparticles. One might expect that the nonlinearities of bulk materials were smaller compared to those of nanoparticles because of the disappearance of some crystal-related processes, which masked, or diminished, the nonlinear optical response of the bulk structure. At this condition when the role of size effect remains unclear, the pure electronic response from the nanoparticles could exceed the combined effect of different nonlinear optical processes in a well-ordered structure.

References

1. H. Nadjari, A.Z. Azad, Opt. Laser Technol. **44**, 1629 (2012)
2. E. Fazio, S. Patane, L. D'Urso, G. Compagnini, F. Neri, Opt. Commun. **285**, 2942 (2012)
3. W. Husinsky, A. Ajami, P. Nekvindova, B. Svecova, J. Pesicka, M. Janecek, Opt. Commun. **285**, 2729 (2012)
4. T. Cesca, P. Calvelli, G. Battaglin, P. Mazzoldi, G. Mattei, Opt. Express **20**, 4537 (2012)
5. K. Hari, S. Mathew, B. Nithyaja, S.A. Joseph, V.P.N. Nampoori, P. Radhakrishnan, Opt. Quantum Electron. **43**, 49 (2012)
6. G. Lin, F.F. Luo, H.H. Pan, M.M. Smedskjaer, Y. Teng, D.P. Chen, J.R. Qiu, Q.Z. Zhao, J. Phys. Chem. C **115**, 24598 (2011)
7. M. Hari, S.A. Joseph, N. Balan, S. Mathew, R. Kumar, G. Mishra, R.A. Yadhav, P. Radhakrishnan, V.P.N. Nampoori, J. Nonlinear Opt. Phys. Mater. **20**, 467 (2011)
8. Y.H. Wang, L. Wei, J.D. Lu, L.L. Ji, R.G. Zang, R.W. Wang, H.Q. Li, Vacuum **86**, 285 (2011)

9. F.F. Chen, S.X. Dai, T.F. Xu, X. Shen, C.G. Lin, Q.H. Nie, C. Liu, J. Heo, Chem. Phys. Lett. **514**, 79 (2011)
10. L. Tamayo-Rivera, R.C. Fernandez-Hernandez, L. Rodriguez-Fernandez, R. Rangel-Rojo, A. Oliver, J.A. Reyes-Esqueda, Opt. Mater. Express **1**, 980 (2011)
11. J.Z. Dong, X.L. Zhang, Y.A. Cao, W.S. Yang, J.G. Tian, Mater. Lett. **65**, 2665 (2011)
12. M.S.A. El-sadek, S.M. Babu, A.Y. Nooralden, P.K. Palanisamy, Nanosci. Nanotechnol. Lett. **3**, 637 (2011)
13. Z. Dehghani, S. Nazerdeylami, E. Saievar-Iranizad, M.H.M. Ara, J. Phys. Chem. Solids **72**, 1008 (2011)
14. B.H. Zhu, H.C. Zhang, J.Y. Zhang, Y.P. Cui, Z.Q. Zhou, Appl. Phys. Lett. **99**, 021908 (2011)
15. X.Y. Yang, W.D. Xiang, H.J. Zhao, H.T. Liu, X.Y. Zhang, X.J. Liang, J. Alloy. Compd. **509**, 7283 (2011)
16. X.S. Zhu, J.F. Wang, D. Nguyen, J. Thomas, R.A. Norwood, N. Peyghambarian, Opt. Mater. Express **2**, 103 (2012)
17. K. Zhang, J.F. Li, W.Z. Wang, J.H. Xiao, W.J. Yin, L. Yu, Opt. Lett. **36**, 3443 (2011)
18. M. Sheik-Bahae, A.A. Said, T.-H. Wei, D.J. Hagan, E.W. Van Stryland, IEEE J. Quantum Electron. **26**, 760 (1990)
19. R.A. Ganeev, A.I. Ryasnyansky, M.K. Kodirov, T. Usmanov, Opt. Commun. **185**, 473 (2000)
20. R.A. Ganeev, A.I. Ryasnyanskiy, U. Chakravarty, P.A. Naik, H. Srivastava, M.K. Tiwari, P.D. Gupta, Appl. Phys. B **86**, 337 (2007)
21. D. Bethune, A.J. Schmidt, Y.R. Shen, Phys. Rev. B **11**, 3867 (1975)
22. M. Sheik-Bahae, D.C. Hutchings, D.J. Hagan, E.W. Van Stryland, IEEE J. Quantum Electron. **27**, 1296 (1991)
23. A. Agnessi, G.P. Banfi, M. Ghigliazza, G.C. Reali, Opt. Commun. **92**, 300 (1992)
24. Y.H. Lee, A. Chavez-Pirson, S.W. Koch, H.M. Gibbs, S.H. Park, J. Morhange, A. Jeffery, N. Peyghambarian, L. Banyai, A.C. Gossard, W. Wiegmann, Phys. Rev. Lett. **57**, 2446 (1986)
25. M.J. Lederer, B. Luther-Davies, H.H. Tan, C. Jagadish, M. Haiml, U. Siegner, U. Keller, Appl. Phys. Lett. **74**, 1993 (1999)
26. R.A. Ganeev, Appl. Phys. B **91**, 273 (2008)
27. T. Skauli, P.S. Kuo, K.L. Vodopyanov, T.J. Pinguet, O. Levi, L.A. Eyres, J.S. Harris, M.M. Fejer, B. Gerard, L. Becouarn, E. Lallier, J. Appl. Phys. **94**, 6447 (2003)
28. B.L. Justus, R.J. Tonucci, A.D. Berry, Appl. Phys. Lett. **61**, 3151 (1992)
29. K. Akiyama, N. Tomita, Y. Nomura, T. Isu, Physica B **272**, 505 (1999)
30. M. Inoue, Jpn. J. Appl. Phys. **39**, 3971 (2000)
31. H.-C. Lee, A. Kost, M. Kawase, A. Hariz, P.D. Dapkus, E.M. Garmire, IEEE J. Quantum Electron. **24**, 1581 (1988)
32. V.L. Malevich, I.A. Utkin, Semiconductors **34**, 924 (2000)
33. S.W. Koch, S. Schmidt-Rink, H. Haug, Phys. Stat. Sol. B **106**, 135 (1981)
34. D.V. Petrov, A.S.L. Gomes, C.B. de Araujo, Appl. Phys. Lett. **65**, 1067 (1994)
35. M. Martinelli, L. Gomes, R.J. Horowicz, Appl. Opt. **39**, 6193 (2000)
36. L. Banyai, M. Lindberg, S.W. Koch, Opt. Lett. **13**, 212 (1988)
37. R.A. Ganeev, A.I. Ryasnyansky, T. Usmanov, Opt. Commun. **272**, 242 (2007)
38. R.A. Ganeev, M. Baba, A.I. Ryasnyansky, M. Suzuki, H. Kuroda, Appl. Phys. B **80**, 595 (2005)
39. A. Markano, O.H. Maillotte, D. Gindre, D. Metin, Opt. Lett. **21**, 101 (1996)
40. R.A. Ganeev, M. Baba, A.I. Ryasnyansky, M. Suzuki, M. Turu, H. Kuroda, Appl. Phys. B **78**, 433 (2004)
41. R.A. Ganeev, M. Baba, A.I. Ryasnyansky, M. Suzuki, H. Kuroda, Opt. Commun. **240**, 437 (2004)
42. F. Hache, D. Ricard, C. Flytzanis, U. Kreibig, Appl. Phys. A **47**, 347 (1988)
43. L. Yang, K. Becker, F.M. Smith, R.H. Marguder, R.F. Haglund, L. Yang, R. Dorsinville, R.R. Alfano, R.A. Zuhr, J. Opt. Soc. Am. B **11**, 457 (1994)
44. R.A. Ganeev, A.I. Ryasnyansky, Appl. Phys. B **84**, 295 (2006)
45. R.A. Ganeev, M. Suzuki, M. Baba, M. Ichihara, H. Kuroda, J. Appl. Phys. **103**, 063102 (2008)

46. R.A. Ganeev, R.I. Tugushev, T. Usmanov, Appl. Phys. B **94**, 647 (2009)
47. R.A. Ganeev, A.S. Zakirov, G.S. Boltaev, R.I. Tugushev, T. Usmanov, P.K. Khabibullaev, T.W. Kang, A.A. Saidov, Opt. Mater. **33**, 419 (2011)
48. R.A. Ganeev, A.I. Ryasnyansky, A.L. Stepanov, T. Usmanov, Phys. Stat. Sol. B **241**, R1 (2004)
49. A.I. Ryasnyansky, P. Palpant, S. Debrus, R.A. Ganeev, A.L. Stepanov, N. Gan, C. Buchal, S. Uysal, Appl. Opt. **44**, 2839 (2005)
50. A.L. Stepanov, C. Marques, E. Alves, R.C. da Silva, M.R. Silva, R.A. Ganeev, A.I. Ryasnyansky, T. Usmanov, Techn. Phys. **51**, 1474 (2006)
51. K. Uchida, S. Kaneko, S. Omi, C. Hata, H. Tanji, Y. Asahara, A.J. Ikushima, J. Opt. Soc. Am. B **11**, 1236 (1994)
52. F. Hache, D. Ricard, C. Flytzanis, J. Opt. Soc. Am. B **3**, 1647 (1986)
53. R.A. Ganeev, A.I. Ryasnyansky, A.L. Stepanov, M.K. Kodirov, T. Usmanov, Opt. Spectrosc. **95**, 967 (2003)
54. S.C. Mehendale, S.R. Mishra, K.S. Bindra, M. Laghate, T.S. Dhami, K.S. Rustagi, Opt. Commun **133**, 273 (1997)
55. R.A. Ganeev, G.S. Boltaev, R.I. Tugushev, T. Usmanov, H. Kuroda, Appl. Phys. B **100**, 571 (2010)
56. Y.L. Song, C. Zhang, X.L. Zhao, Y.X. Wang, G.Y. Fang, G.S. Jin, S.L. Qu, S.P. Wu, X.Q. Xin, H.Y. Ye, Chem. Lett. **9**, 323 (2000)
57. L. Francois, M. Mostafavi, J. Belloni, J.F. Delouis, J. Delaire, P. Feneyrou, J. Phys. Chem. B **104**, 6133 (2000)
58. H. Inouye, K. Tanaka, J. Tanahashi, K. Hirao, Phys. Rev. B **57**(11), 334 (1998)
59. H. Inouye, K. Tanaka, J Tanahashi, T. Hattori, H. Nakatsuka, Japan. J. Appl. Phys. **39**, 5132 (2000)
60. R. Sarkar, P. Kumbhakar, A.K. Mitra, R.A. Ganeev, Current. Appl. Phys. **10**, 853 (2010)
61. R. Brause, H. Moltgen, K. Kleinermanns, Appl. Phys. B **75**, 711 (2002)
62. R.A. Ganeev, A.I. Ryasnyansky, M.K. Kodirov, S.R. Kamalov, T. Usmanov, J. Phys. D **34**, 1602 (2001)
63. J.F. Reintjes, *Nonlinear Optical Parametric Processes in Liquids and Gases* (Academic Press, New York, 1984)
64. B. Ricard, P. Raussignol, C. Flytzanis, Opt. Lett. **10**, 511 (1985)
65. J. Staromlynska, T.J. McKay, P. Wilson, J. Appl. Phys. **88**, 1726 (2000)
66. J. Olivares, J. Requejo-Isidro, R. Del Coso, R. De Nalda, J. Solis, C.N. Afonso, A.L. Stepanov, D. Hole, P.D. Townsend, A. Naudon, J. Appl. Phys. **90**, 1064 (2001)
67. M. Kyoung, M. Lee, Opt. Commun. **171**, 145 (1999)
68. L. Yang, R. Dorsinville, Q.Z. Wang, R.A. Alfano, Opt. Lett. **17**, 323 (1992)
69. R.A. Ganeev, A.I. Ryasnyansky, A.L. Stepanov, T. Usmanov, Phys. Stat. Sol. B **238**, R5 (2003)
70. D.H. Osborne, R.F. Haglund, F. Gonella, F. Garrido, Appl. Phys. B **66**, 517 (1998)
71. R.A. Ganeev, A.I. Ryasnyansky, A.L. Stepanov, T. Usmanov, Phys. Stat. Sol. B **241**, 935 (2004)
72. A.L. Stepanov, D.E. Hole, Recent Res. Devel. Appl. Phys. **5**, 1 (2002)
73. A. Nadon, D. Thaiaudiere, J. Appl. Crystallogr. **30**, 822 (1997)
74. P.D. Townsend, P.J. Chandler, L. Zhang, *Optical Effects of Ion Implantation* (Cambridge University Press, London, 1994)
75. O. Centora-Gonzalez, D. Muller, C. Estournus, M. Richard-Plouet, R. Poinsot, J.J. Grob, A. Guille, Nucl. Instrum. Meth. B **178**, 144 (2001)
76. A.L. Stepanov, D.E. Hole, P.D. Townsend, J. Non-Cryst. Solids **260**, 65 (1999)
77. A.L. Stepanov, R.A. Ganeev, A.I. Ryasnyansky, T. Usmanov, Nucl. Instrum. Meth. B **206**, 624 (2003)
78. A. Owyoung, IEEE J. Quantum Electron **9**, 1064 (1993)
79. U. Kreibig, M. Vollmer, *Optical Properties of Metal Clusters* (Springer, Berlin, 1995)
80. R.H. Lehmberg, J.F. Reintjes, R.C. Eckardt, Phys. Rev. A **13**, 1095 (1976)
81. R.A. Ganeev, A.I. Ryasnyansky, A.L. Stepanov, C. Marques, R.C. Da Silva, E. Alves, Opt. Commun. **253**, 205 (2005)

82. Y. Hamanaka, H. Hayashi, A. Nakamura, S. Omi, J. Luminesc. **87–89**, 859 (2000)
83. R.A. Ganeev, A.I. Ryasnyansky, A.L. Stepanov, T. Usmanov, Opt. Quantum Electron. **36**, 949 (2004)
84. R.A. Ganeev, R.I. Tugushev, A.A. Ishchenko, N.A. Derevyanko, A.I. Ryasnyansky, T. Usmanov, Appl. Phys. B **76**, 683 (2003)
85. L.C. Hwang, S.C. Lee, T.C. Wen, Opt. Commun. **228**, 373 (2003)
86. R.A. Ganeev, A.I. Ryasnyansky, A.L. Stepanov, T. Usmanov, Solid State Phys. **46**, 351 (2004)
87. Y. Hamanaka, A. Nakamura, H. Hayashi, S. Omi, J. Opt. Soc. Am. B **20**, 1227 (2003)
88. D.D. Smith, G. Fisher, R.W. Boyd, D.A. Gregory, J. Opt. Soc. Am. B **14**, 1625 (1997)
89. S.S. Sarkisov, E. Williams, M. Curley, D. Ila, P. Venkateswarlu, D.B. Poker, D,K. Hensley, Nucl. Instrum. Meth. B **141**, 294 (1998)
90. N. Pincon, B. Palpant, D. Prot, E. Charron, S. Debrus, Eur. Phys. J. D **19**, 395 (2002)
91. H. Kurata, A. Takami, S. Koda, Appl. Phys. Lett **72**, 789 (1988)
92. S. Link, M.B. Mohamed, B. Nikoobakht, M.A. El-Sayed, J. Phys. Chem. **103**, 1165 (1999)
93. F. Mafune, J. Kohono, Y. Takeda, T. Kondow, J. Phys. Chem. B **106**, 8555 (2002)
94. N. Chandrasekharan, P.V. Kamat, J. Hu, G. Jones, J. Phys. Chem. B **104**, 11103 (2000)
95. K.P. Unnikrishnan, V.P.N. Nampoori, V. Ramakrishnan, M. Umadevi, C.P.G. Vallabhan, J. Phys. D **36**, 1242 (2003)
96. R.H. Magruder, D.H. Osborne, R.A. Zuhr, J. Non-Cryst. Solids **176**, 229 (1994)
97. R.F. Haglung, L. Yang, R.H. Magruder, J.E. Whrite, K. Becker, R.A. Zuhr, Opt. Lett. **18**, 373 (1993)
98. R. Serna, J.M. Ballesteros, J. Solis, C.N. Afonso, D.H. Osborne, R.F. Haglung, A.K. Petford-Long, Thin solid films **318**, 96 (1998)
99. L. Brzozovski, E.H. Sargent, J. Mater. Sci. Mater. Electron. **12**, 483 (2001)
100. Y.P. Sun, J.E. Riggs, H.W. Rollins, R. Guduru, J. Phys. Chem. B **103**, 77 (1999)
101. R.A. Ganeev, A.I. Ryasnyansky, M.K. Kodirov, S.R. Kamalov, V.A. Li, R.I. Tugushev, T. Usmanov, Appl. Phys. B **74**, 47 (2002)
102. K. Wundke, S. Potting, J. Auxier, Appl. Phys. Lett. **76**, 10 (2000)
103. S. Vijayalakshmi, M.A. George, H. Grebel, Appl. Phys. Lett. **70**, 708 (1997)
104. M. Falkonieri, G. Salvetti, Appl. Phys. B **69**, 133 (1999)
105. S. Couris, M. Renard, O. Faucher, B. Loverel, R. Chaux, E. Koudoumas, X. Michaut, Chem. Phys. Lett. **369**, 318 (2003)
106. H. Toda, C.M. Verber, Opt. Lett. **17**, 1379 (1992)
107. R.A. Ganeev, A.I. Ryasnyansky, M.K. Kodirov, S.R. Kamalov, T. Usmanov, Opt. Spectrosc. **90**, 568 (2001)
108. T. Tsuji, K. Iryo, N. Watanabe, M. Tsuji, Appl. Surf. Sci. **202**, 80 (2002)
109. E. Cattaruzza, G. Battaglin, P. Calvelli, F. Gonella, G. Mattei, C Maurizio, P. Mazzoldi, S. Padovani, C. Sada, B.F. Scremin, F.D. Acapito, Comp. Sci. Technol. **63**, 1203 (2003)
110. N.N. Lepeshkin, W. Kim, V.P. Safonov, J.G. Zhu, R.L. Armstrong, C.W. White, R.A. Zuhr, V.M. Shalaev, J. Nonlin. Opt. Phys. Mater. **8**, 191 (1999)
111. S.G. Rautian, V.P. Safonov, P.A. Chubakov, V.M. Shalaev, M.I. Shtokman, JETP Lett. **47**, 200 (1988)
112. W. Ji, H.J. Du, S.H. Tang, S. Shi, J. Opt. Soc. Am. B **12**, 876 (1996)
113. A.V. Butenko, V.M. Shalaev, M.I. Shtokman, JETP **94**, 107 (1988)
114. S.V. Karpov, A.L. Basko, S.V. Koshelev, A.K. Popov, V.V. Slabko, Colloidal J. **59**, 765 (1997)
115. S.V. Karpov, A.K. Popov, V.V. Slabko, JETP Lett. **66**, 97 (1997)
116. V.V. Slabko, S.V. Karpov, V.I. Zaitsev, A.K. Popov, J. Phys. Condens. Matter. **5**, 7231 (1993)
117. V.M. Shalaev, M.I. Shtokman, JETP **92**, 509 (1987)
118. V.A. Markel, L.S. Muratov, M.I. Shtokman, JETP **98**, 819 (1990)
119. J. Stone, J. Opt. Soc. Am. **62**, 327 (1972)
120. A.P. Alivisatos, Science **271**, 933 (1996)
121. R.E. Schwerzel, K.B. Spahr, J.P. Kurmer, V.E. Wood, J.A. Jenkins, J. Phys. Chem. A **102**, 5622 (1998)
122. R.A. Ganeev, A.I. Ryasnyansky, T. Usmanov, Opt. Quantum Electron **35**, 211 (2003)

123. R.A. Ganeev, A.I. Ryasnyansky, Opt. Commun. **246**, 163 (2005)
124. R.A. Ganeev, S.R. Kamalov, I.A. Kulagin, T. Usmanov, A.I. Ryasnyansky, M.K. Kodirov, N.V. Kamanina, Nonlinear Opt. **28**, 263 (2002)
125. R.A. Ganeev, A.I. Ryasnyansky, R.I. Tugushev, T. Usmanov, J. Opt. A **5**, 409 (2003)
126. H. Du, G.Q. Xu, W.C. Chin, L. Huang, W. Li, Chem. Mater. **14**, 4473 (2002)
127. H.P. Li, C.H. Kam, Y.L. Lam, W. Ji, Opt. Commun. **190**, 351 (2001)
128. V.M. Shalaev, E.Y. Polyakov, V.A. Markel, Phys. Rev. B **53**, 2437 (1996)
129. T. Tagakahara, Phys. Rev. B **36**, 9193 (1987)
130. A.E. Neeves, M.N. Birnboim, J. Opt. Soc. Am. B **6**, 789 (1989)
131. Y. Wang, W. Mahler, Opt. Commun. **61**, 233 (1987)
132. Q. Shen, P. Liang, W. Zhang, Opt. Commun. **115**, 133 (1995)
133. H. Yao, S. Takahara, H. Mizuma, T. Kozeki, T. Hayashi, Jpn. J. Appl. Phys. **35**, 4633 (1996)
134. Y.P. Rakovich, M.V. Artemyev, A.G. Rolo, M.I. Vasilevskiy, M.J.M. Gomes, Phys. Stat. Sol. B **224**, 319 (2001)
135. T.D. Krauss, F.W. Wise, Appl. Phys. Lett. **65**, 1739 (1994)
136. Y. Kayanuma, H. Momiji, Phys. Rev. B **41**, 10261 (1990)
137. R. Rossetti, J.L. Ellison, J.M. Gibson, L.E. Brus, J. Chem. Phys. **80**, 4464 (1984)
138. M.C. Baykul, A. Balcioglu, Microelectron. Eng. **51**, 703 (2000)
139. E. Hanamura, Phys. Rev. B **37**, 1273 (1988)
140. W. Wang, L. Qu, G. Yang, Z. Chen, Appl. Surf. Sci. **218**, 249 (2003)
141. D.Y. Guan, Z.H. Chen, W.T. Wang, H.B. Lu, Y.L. Zhou, K.J. Jin, G.Z. Yang, J. Opt. Soc. Am. B **22**, 1949 (2005)
142. H. Lotem, C.B. de Araujo, Phys. Rev. B **16**, 1711 (1977)
143. R. Adair, L.L. Chase, S.A. Payne, Phys. Rev. B **39**, 3337 (1989)
144. T. Boggess, J.O. White, G.C. Valley, J. Opt. Soc. Am. B **7**, 2255 (1990)
145. R.A. Ganeev, M. Suzuki, M. Baba, M. Ichihara, H. Kuroda, J. Opt. Soc. Am. B **25**, 325 (2008)
146. R.A. Ganeev, J. Phys. B: At. Mol. Opt. Phys. **40**, R213 (2007)
147. J.M.C. Jonathan, G. Roosen, P. Roussignol, Opt. Lett. **13**, 224 (1988)

Chapter 6
Applications of Nanoparticle-Containing Plasmas for High-Order Harmonic Generation of Laser Radiation

The use of nanoparticles for efficient conversion of the wavelength of ultrashort laser toward the deep UV spectral range through harmonic generation is an attractive application of cluster-containing plasmas. Note that earlier observations of HHG in nanoparticles were limited by using the exotic gas clusters formed during fast cooling of atomic flow from the gas jets [1–4]. One can assume the difficulties in definition of the structure of such clusters and the ratio between nanoparticles and atoms/ions in the gas flow. The characterization of gas phase cluster production was currently improved using the sophisticated techniques (e.g., a control of nanoparticle mass and spatial distribution, see the review [5]). In the meantime, the plasma nanoparticle HHG has demonstrated some advantages over gas cluster HHG [6]. The application of commercially available nanopowders allowed for precisely defining the sizes and structure of these clusters in the plume. The laser ablation technique made possible the predictable manipulation of plasma characteristics, which led to the creation of laser plumes containing mainly nanoparticles with known spatial structure. The latter allows the application of such plumes in nonlinear optics, X-ray emission of clusters, deposition of nanoparticles with fixed parameters on the substrates for semiconductor industry, production of nanostructured and nanocomposite films, etc.

Other nanostructures, which attract the attention during last time, are the fullerenes and carbon nanotubes (CNT). Recently, the application of laser ablation technique allowed the creation of relatively dense C_{60}-rich plasma ($\sim 5 \times 10^{16}\,cm^{-3}$). It was demonstrated the efficient broadband HHG in C_{60}-rich plasmas [7]. The changes in fundamental wave characteristics allowed the dramatic manipulation of the harmonic spectrum and intensity at well-defined conditions of C_{60}-containing plasma. It is also proposed the application of CNT-containing plasma for efficient harmonic generation. Carbon nanotubes have remarkable electronic and optical properties due to

We analyse the studies of the conditions when the plasma producing on the surface of targets contains the nanoparticles, clusters, and nanotubes. The cluster-containing plasma plumes proved to be the effective media for the high-order harmonic generation of femtosecond laser pulses. We describe the HHG in various plasmas containing nanoparticles, clusters, and nanotubes.

R. A. Ganeev, *Nonlinear Optical Properties of Materials*,
Springer Series in Optical Sciences 174, DOI: 10.1007/978-94-007-6022-6_6,

their particular structure combining one-dimensional solid-state characteristics with molecular dimensions. While their structure has extensively been studied by means of transmission electron microscopy and scanning tunneling microscopy, only few experimental studies have been reported on their nonlinear optical properties. Moreover, their high-order nonlinearities can be analyzed solely by probe experiments in the media containing sufficient amount of these species and transparent in the extreme ultraviolet range. For these purposes, one has to create the CNT-containing plasmas, where the presence of nanotubes should be proven by indirect methods.

In this Chapter, we analyse the studies of the conditions when the plasma producing on the surface of targets contains the nanoparticles, clusters, and nanotubes. The cluster-containing plasma plumes proved to be the effective media for the HHG of femtosecond laser pulses. We describe the HHG in various plasmas containing nanoparticles, clusters, and nanotubes.

6.1 Experimental Arrangements for Cluster-Containing Plasma Formation and High-Order Harmonic Generation

The details of HHG in plasmas are described in Chap. 2. Here we briefly outline most important peculiarities of experimental arrangements for HHG in cluster-contained plasmas. The studies describing below were carried out using the Ti:sapphire lasers. To create the ablation, part of uncompressed radiation from the Ti:sapphire laser ($\lambda = 800\,\text{nm}$, $t = 210 - 300\,\text{ps}$) was focused on a target placed in the vacuum chamber. The spot size of this radiation on the target surface was maintained in the range of 0.5–0.8 mm. The intensity of this radiation on the target surface was varied between 2×10^9 and $5 \times 10^{10}\,\text{W cm}^{-2}$. The laser fluence during ablation was in the range of 0.4–1 J cm^{-2}. The chamber was maintained at the vacuum pressure of 8×10^{-4} torr. The debris from nanoparticle-containing plasma plume was deposited on nearby placed semiconductor wafers, glass plates, and metal foils. The distance between the target and deposited substrates was 40–70 mm. The structure of deposited material was analyzed by the scanning electron microscope and transmission electron microscope. The absorption spectra of deposited debris were analyzed using the fiber spectrometer.

At the first set of experiments, the bulk targets, such as silver, were ablated to create the conditions for nanoparticles formation in the laser plumes. Further, the targets containing nanoparticles were ablated as well. Various commercially available nanoparticle powders purchased from different manufacturers were glued on the glass substrates by mixing with the drop of superglue. The bulk materials of the same origin as nanoparticle powders were used to compare the ablation properties of these targets. The targets containing silver, gold, platinum, ruthenium, and palladium nanoparticles were also used in these studies for creation of clustered plumes. These nanoparticles were purchased in the form of suspensions and were dried on the surfaces of glass substrates prior to laser ablation. Other ablated materials were the C_{60} and CNT powders. These powders were also glued on different substrates. More details on targets preparation are presented below in the corresponding sections.

The cluster-containing plumes were used for the HHG of femtosecond laser pulses. The harmonic generation in nanoparticle-containing laser plumes was carried out using the femtosecond pulses ($\lambda = 800\,\text{nm}$, $t_{fp} = 35$–$130\,\text{fs}$), which were focused, after some delay (6–74 ns) with regard to the beginning of laser ablation, on the plasma from the orthogonal direction. The harmonics were spectrally dispersed by an XUV spectrometer, detected by a micro-channel plate, and recorded using a CCD camera. The details of harmonic generation set-up can be found elsewhere [8], as well as in Chap. 2.

6.2 High-Order Harmonic Generation in the Plasmas Containing *In-Situ* Produced Nanoparticles and Fullerenes

The plumes containing various metal nanoparticles, fullerenes, and carbon nanotubes were used for the high-order harmonic generation of the femtosecond radiation propagating through the laser plasmas. The harmonics were generated effectively at the conditions when the presence of nanoparticles in the plumes was confirmed by morphological analysis of the debris. The high-order harmonics were observed in all of studied nanoparticle-containing laser plasmas. The focusing of femtosecond radiation in front of or after the laser plume optimized the harmonic yield. The focusing inside the plasma area led to a decrease of harmonic efficiency due to over-ionization of nanoparticles by femtosecond pulses and appearance of additional free electrons. The latter led to the phase mismatch of the HHG, which has previously been reported in the case of over-excitation of bulk targets [9].

The generation of high-order harmonics from the interaction of in-situ produced nanoparticles with intense ultrashort laser pulse has been studied in [10]. The interaction of the sub-nanosecond heating pulse with the silver target at intensity of $\sim 1 \times 10^{13}\,\text{W cm}^{-2}$ generates the nanoparticles. The mean size of silver nanoparticles was 30 nm. High-order harmonics were generated through the interaction of in-situ produced nanoparticles with ultrashort laser pulses. The spectrum of HHG from in-situ-produced nanoparticles was compared with the HHG spectrum from bulk Ag plumes and 9-nm Ag nanoparticles glued on the target (Fig. 6.1). The intensities of 9–15th harmonics were less compared to the 17th harmonic in the case of HHG from atoms/ions-containing silver plumes. In the meantime, the intensity of harmonics from the plumes created on the target coating by silver nanoparticles decreases slowly from 9th harmonic to 17th harmonic. Comparison of the HHG spectral characteristics from in-situ produced nanoparticles with that from bulk silver and glued silver nanoparticles indicates that HHG in that case is from Ag nanoparticles rather than from Ag atoms and ions. As the intensity of pump pulse on the bulk silver surface was increased from $\sim 10^{10}$ to $\sim 8 \times 10^{11}\,\text{W cm}^{-2}$, the HHG spectra gradually reduced and completely vanished. Then, at 8×10^{12}–$1 \times 10^{13}\,\text{W cm}^{-2}$, HHG suddenly reappears. The intensity of HHG radiation from in-situ produced nanoparticles

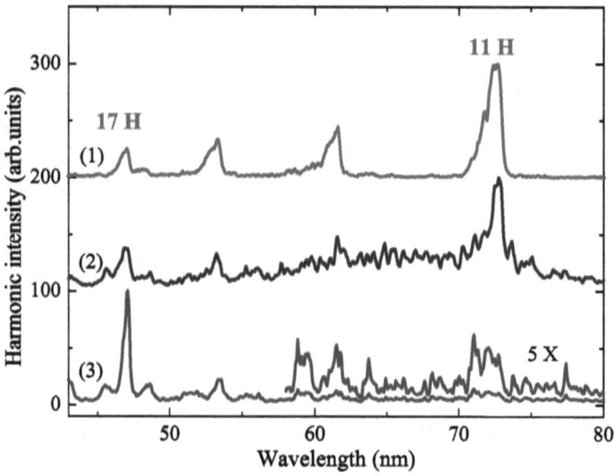

Fig. 6.1 Comparison of the HHG spectra from (*1*) silver nanoparticles glued on the target, (*2*) in-situ produced silver nanoparticles, and (*3*) silver plasma without the nanoparticles. Reproduced from [10] with permission from American Physical Society

is lower compared with that from plasma produced on the Ag nanoparticle-coated surface. However, the intensity of the HHG could be further enhanced by improving the methods for nanoparticle formation [11]. More details on the HHG in silver nanoparticles-contained plasmas will be presented in the following section.

The systematic studies of the HHG in fullerene-containing plumes have been reported in [7, 12–16]. Some new results in this area can be found in Chap. 2. Those studies have revealed the advantages of the use of the C_{60}-containing plasma as a nonlinear optical medium for harmonic generation. The examples of harmonic spectra from C_{60}-containing plasmas are presented on Fig. 6.2. Harmonics up to the 33rd order were observed in those studies [17]. Odd and even harmonics were

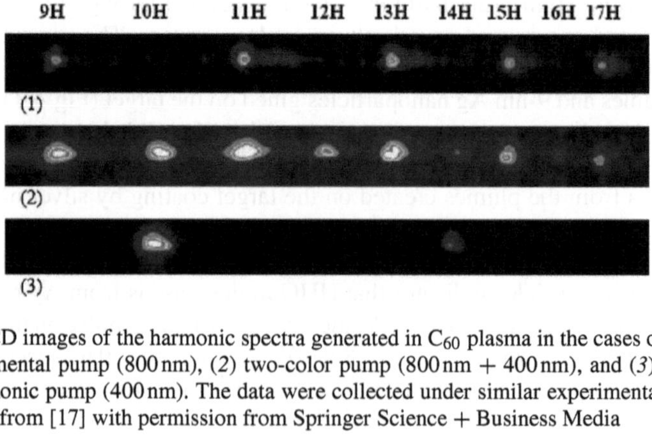

Fig. 6.2 CCD images of the harmonic spectra generated in C_{60} plasma in the cases of: (*1*) single-color fundamental pump (800 nm), (*2*) two-color pump (800 nm + 400 nm), and (*3*) single-color second harmonic pump (400 nm). The data were collected under similar experimental conditions. Reproduced from [17] with permission from Springer Science + Business Media

generated using two-color pump (800 nm + 400 nm). The comparative studies of harmonic spectra from CNTs and fullerenes have shown better HHG efficiency in the latter case.

6.3 Application of Silver Nanoparticle-Containing Laser Plasmas for HHG

Metal nanoparticles subject to intense laser pulses produce strong low-order nonlinear optical response (e.g., nonlinear refraction and nonlinear absorption), as well as can emit the coherent radiation through the low-order harmonic generation [18, 19]. These and other studies have shown that one can expect improvement of the harmonic efficiency by switching to the nanoparticle media. In the meantime, previous studies of harmonic generation in such objects were limited to exotic nanoclusters (Ar, Xe), which are formed in high-pressure gas jets due to rapid cooling by the adiabatic expansion. The physical origin of this process in the gas clusters [1–4, 20–25] is mostly attributed to standard atomic harmonic generation, modified by the fact that in clusters the atoms are disposed to each other.

Only a few high-order harmonic generation studies using 10^3–10^4 atoms/cluster gaseous media (with the cluster sizes between 2 and 8 nm) have been reported, while some theoretical simulations predicted a growth in the harmonic intensity with regard to the monoatomic media. It may be noted that a pronounced resonant enhancement for nanoparticles is found only for yields of the low-order harmonics, while the high-order harmonics completely disappear from the harmonic spectra in the simulations [26]. Clustered plasma was proposed as a nonlinear medium in which both the phase matching [21] and resonantly enhanced growth of the nonlinear susceptibility for low-order harmonics [23] could be achieved at selected cluster sizes and densities.

One of the approaches to achieve the efficient high-order harmonics is the application of commercially available nanoparticles in the pump-probe HHG experiments using the laser ablation of the targets containing Ag nanoparticles [25–27]. An important issue in the case of nanoparticles ablation is their integrity during evaporation from the target surface. One can expect the fragmentation, melting, or aggregation of nanoparticles during interaction of the strong laser radiation with the target surface. To achieve the harmonic generation from nanoparticles, one should not overexcite the nanoparticle-containing targets to produce the plasma where these species are presented in their indigenous neutral (or ionized) conditions.

6.3.1 Harmonic Generation of Laser Radiation in the Plasma Plumes Containing Large-Sized Silver Nanoparticles

Here we analyze the HHG in various commercially available nanoparticles with mean sizes of approximately 100 nm. These studies were carried out using the ablation of nanoparticles by heating subnanosecond laser pulse with further probing of nanoparticle-containing plasma by the femtosecond pulse. The presence of nanoparticles in the plumes was carefully studied by analyzing the morphology of the debris of ablated material deposited on nearby substrates. It was shown that, at the conditions of optimal ablation, the nanoparticles maintain their integrity in laser plumes, which allows for the enhancement of HHG yield in the case of nanoparticle-containing plumes with regard to the monoparticle-containing plasmas [28]. Experiments were carried out using the chirped pulse amplification Ti:sapphire laser. To create the ablation, a subnanosecond heating pulse (t_{pp} = 210 ps) was split from uncompressed radiation of Ti:sapphire laser and focused on a target placed in the vacuum chamber (see inset in Fig. 6.3). The intensity of heating picosecond pulses (I_{pp}) on the target surface was varied between 2×10^9 and $3 \times 10^{10}\,\mathrm{W\,cm^{-2}}$. After some delay (6–74 ns), the probe femtosecond pulse (t_{fp} = 35 fs, λ = 800 nm) was focused on the plasma from the orthogonal direction. The HHG experiments were carried out using the laser intensities (I_{fp}) of up to $I_{fp} = 2 \times 10^{15}\,\mathrm{W\,cm^{-2}}$, above which the HHG efficiency considerably decreased due to some impeding processes in the laser plasma. The harmonics were spectrally dispersed by an extreme ultraviolet spectrometer, detected by a micro-channel plate, and recorded using a charge-coupled device camera. In these studies, various nanosized silver powders were used. The sizes of nanoparticles were varied in the range of 30–200 nm. These powders were glued on the glass substrates by mixing with superglue. The bulk materials

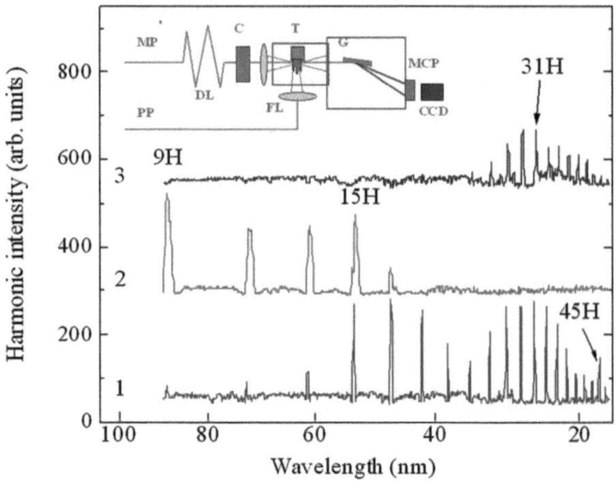

Fig. 6.3 Harmonic spectra obtained from the plasmas produced on the surfaces of (1) bulk silver, and (2,3) silver nanoparticle-containing target. The intensities of heating pulse on the target surfaces were (1) $3 \times 10^{10}\,\mathrm{W\,cm^{-2}}$, (2) $7 \times 10^9\,\mathrm{W\,cm^{-2}}$, and (3) $2 \times 10^{10}\,\mathrm{W\,cm^{-2}}$. $I_{fp} = 4 \times 10^{14}\,\mathrm{W\,cm^{-2}}$. The delay between the heating pulse and probe femtosecond pulse in these experiments was maintained at 20 ns. Inset: Experimental setup. MP: main probe pulse; PP: heating picosecond pulse; DL: delay line; C: grating compressor; FL: focusing lenses; T: target; G: gold-coated grating; MCP: micro-channel plate; CCD: charge-coupled device. Reproduced from [28] with permission from American Institute of Physics

of the same origin as nanoparticles were also used to compare the HHG from the nanoparticle-containing and atom/ion-containing plumes [28].

Those studies were concentrated on (a) comparison of the harmonic spectra generated in the plasmas containing single atomic/ionic particles and nanoparticles, and (b) morphological characterization of the plasma components. Initially, the harmonic generation in the plasmas produced on the surfaces of bulk materials was analyzed. In particular, the plasma produced on the silver solid surfaces showed the harmonics up to the 47th order. Note that the harmonic spectrum generated in silver plasma mostly consisted of the high-order components belonging to the high-energy part of plateau, while the low-wavelength harmonics of plateau were considerably suppressed due to the optimal phase matching conditions for shorter-wavelength range (Fig. 6.3, curve 1). The harmonic cutoff was shifted toward the longer wavelengths when the silver nanoparticle-containing targets were ablated and used for the harmonic generation in the produced plumes. At the same time, these harmonics were considerably stronger compared with those generated in the single particle-containing plasma created on the surface of bulk silver (Fig. 6.3, curve 2). The spectral widths of those harmonics were three to four times broader compared with the narrow lines of the harmonics generated in monoparticle-containing plasma. The enhancement factor of the harmonics generated in nanoparticle-containing plumes was varied between 5 and 12 compared with monoparticle plume, depending on the harmonic order and the conditions of excitation. Note that this enhancement of low-order harmonics was observed in the case of moderate excitation of the silver nanoparticle-containing targets ($I_{pp} = (5-7) \times 10^9\,\mathrm{W\,cm^{-2}}$). At the same time, HHG from the plasma produced on bulk surface at these intensities of heating radiation was inefficient. Further growth of heating pulse intensity on the target surface containing nanoparticles led to a dramatic change of harmonic spectra, which showed the disappearance of strong low-order harmonics and the appearance of high-order ones (Fig. 6.3, curve 3). The intensity of these harmonics was considerably weaker than the intensity of low-order harmonics in the case when low excitation was applied to the targets.

The studies of the availability of HHG during propagation of the femtosecond pulse through the glue molecules-containing plasmas were carried out as well, when no nanoparticles were mixed in this organic. No harmonics were generated in that case. The same can be said about the influence of substrates.

The growth of intensity of the femtosecond pulse did not lead to extension of the harmonics generated in the nanoparticle-containing plumes, which is a sign of saturation of the HHG in these media. Moreover, at relatively high intensities of probe femtosecond pulse, a decrease in harmonic conversion efficiency was observed due to some restricting factors (appearance of abundance of free electrons, self-defocusing, phase mismatch, etc). The same can be said about the increase in heating pulse intensity on the surface of nanoparticle-containing target over some optimal value. In that case the deterioration of harmonic generation was attributed to the increase in free electron concentration and phase mismatch.

The experiments show that the optimum delay between the heating and probe pulses was 12–50 ns (in the case of nanoparticle-contained plasma) and 40–60 ns (in the case of the plasma produced on the surface of Ag bulk target). These observations demonstrate the dynamics of harmonic generation and point out the particles responsible for this process. At moderate heating pulse intensities, the plasma contains predominantly neutral nanoparticles. In that case the harmonic cutoff is defined by interaction of the femtosecond pulse with the neutral nanoparticles. The growth of heating pulse intensity led to the increase of nanoparticle concentration and their ionization. The abundance of free electrons restricted generation of low-order harmonics, while the conditions for higher-order harmonics became more favorable.

One can assume that, in the case of small nanoparticles, the ejected electron, after returning back to the parent particle, can recombine with any of atoms in the nanoparticles due to enhanced cross section of the recombination with parent particle compared with single atom, which considerably increases the probability of harmonic emission in the former case. Thus the enhanced cross section of the recombination of accelerated electron with parent particle compared with single atom can be a reason of observed enhancement of the HHG yield from nanoparticle-containing plume compared with monoparticle-containing plasma.

The presence of nanoparticles in the plumes was confirmed by analyzing the spatial characteristics of the ablated material deposited on nearby glass and aluminum substrates. It was shown that, at optimal excitation conditions, the nanoparticles remain intact in the plasma plume until the femtosecond pulse arrives in the area of interaction. The SEM images of deposited nanostructures revealed that they remain approximately same as the initial powders. The sizes of deposited nanoparticles were in the range of 40–250 nm. These SEM studies have demonstrated that the harmonics generated at the optimal conditions of target excitation are the result of interaction of the femtosecond pulse with nanoparticles. Some additional supporting confirmation of this statement is as follows. In most cases of the HHG in the nanoparticle-containing plasma created at moderate intensities of heating pulse, no extension of high-order harmonic cut-off was seen that could be associated with the involvement of ionized particles in the process of harmonic generation, as it was observed in the case of the plasma produced on the surface of bulk materials of the same origin (Fig. 6.3). The harmonic cutoff for the nanoparticle-containing plumes should be in the range of 13–17th harmonics, taking into account the saturation intensity at which the neutral nanoparticles ionize ($\sim 1 \times 10^{14}\,\mathrm{W\,cm^{-2}}$). In most cases these harmonics decreased or even disappeared with the growth of excitation and ionization of plasma, since the phase mismatch of HHG prevented stable generation of low-order harmonics above the cutoffs defined from the three-step model for the neutral particles.

Those studies have shown that, for any nanoparticle-containing plumes, no improvements in the extension of the harmonic cutoff were observed. At the same time, an enhancement of the harmonic yield in the low-energy plateau range in the case of nanoparticle-containing plumes was achieved. These conclusions are true for the smaller-sized nanoparticles as well. The value of the enhancement factor could be attributed to rather different concentrations of nanoparticles and monoparticles in the plume at different excitation conditions, enhanced cross section of the recombination with parent nanoparticles compared with single atom, as well as to the processes related with the quantum confinement-induced growth of the nonlinear optical response of such medium.

6.4 Improvements in High-Order Harmonic Generation from Silver Nanoparticles

In this subsection, we discuss the peculiarities of the HHG from silver nanoparticles [11]. Highly efficient harmonic emission in the range of 9–19th harmonic order was generated from the plasmas containing nanoparticles. Spectral broadening of harmonic radiation through the optimization of the intensity of the femtosecond laser pulse was observed. The spectral broadening of harmonics can be attributed to the spectral broadening of laser pulse due to self-phase modulation effects inside the plasma containing silver nanoparticles. The study of HHG was mainly performed using the Ag nanoparticles having an average size of 10 nm. A comparison of HHG intensity from Ag nanoparticles with other bulk targets such as Ag and In, and various other nanoparticle targets, such as Au, $SrTiO_3$, and C_{60}, was carried out. Colloidal solutions of silver nano-platelets in water were prepared chemically by a method similar to that described in [29]. A silver salt was reduced with sodium borohydrate and hydrogen-peroxide in the presence of tri-sodium citrate. A capping agent polyvinyl pyrrolidone was used to prevent aggregation. The reaction took place at room temperature. At the end of the reaction, the solution turned blue in color. Absorption spectrum of the sample showed a peak at 638 nm. The TEM images of the samples confirmed that the particles were of triangular cross section. Some of these colloidal solutions demonstrated the SPR at 430 nm, which corresponded to the presence of spherical nanoparticles of diameter of 10 nm.

For comparison of HHG from different Ag nanoparticle targets they were prepared by three methods. In the first method, Ag nanoparticles were dissolved in polyvinyl alcohol (PVA) and then the solution was coated on a glass plate. When the solution got dried, another layer was applied over it. The process was repeated many times such that ~2 mm thick layer was formed. In the second method, the target was prepared by the same method as explained above, except that the sample was dried in oven at 60 °C. This process increased the density of Ag nanoparticles. In the third method, the nanoparticles were mixed with the fast drying glue and spread on the glass plate. This was the easiest method to create the nanoparticle targets, but surface irregularities, and density variation of the nanoparticles was large in these targets.

The HHG experimental scheme was analogous to those described in previous subsection. It was found that the HHG from multi-atomic particles started to be efficient at relatively small heating pulse intensities ($(3-7) \times 10^9\,\mathrm{W\,cm^{-2}}$), which were considerably lower compared with those required in the case of bulk graphite target ($(1-3) \times 10^{10}\,\mathrm{W\,cm^{-2}}$). HHG from silver nanoparticles is shown in Fig. 6.4 [11]. The typical harmonic emission from silver nanoparticles (solid line) is compared

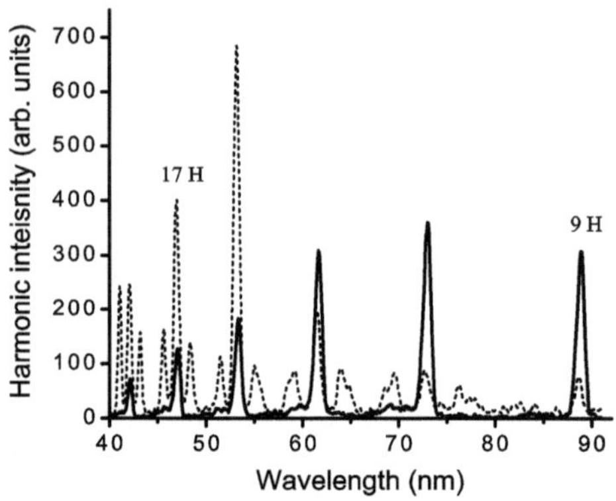

Fig. 6.4 Typical HHG spectra from silver nanoparticles (*solid line*) and bulk silver target (*dashed line*). It may be noted that the intensity of HHG spectrum from bulk Ag target is 10× multiplied for better visibility. The intensity of the 9th H from Ag nanoparticles is ~40 times higher compared to the 9th H from bulk silver. Adapted from [11] with permission from IOP Publishing

with the harmonic emission from bulk silver (dashed line) in this figure. For better visibility, the intensity of HHG from bulk silver is multiplied by 10. It can be clearly seen that lower order harmonics (9–17H) from Ag nanoparticles are very strong compared to the corresponding harmonics from the plasma produced from bulk silver. For example, the 9th H from Ag nanoparticles was ~40 times stronger than that from bulk silver. At the same time, the HHG cutoff is higher for plasma from bulk silver. The lower intensity for lower harmonics in the case of bulk Ag (9–13th H) has been observed in multiple studies. The higher harmonic intensities on the plateau range (>15th H) could be due to better phase matching conditions for the higher orders. Also, the lower intensity for the lower orders (<13th H) are more likely to be due to decrease in the spectrograph (MCP and grating) response expected for longer wavelengths. This trend has been observed in the spectra of other elements also. In these experiments, the cutoffs from bulk silver and silver nanoparticles corresponded to the 61st and 29th harmonics.

Field enhancement inside the nanoparticle through Mie resonance is especially promising explanation for the increase of harmonic conversion efficiency. Mie resonance plays an important role in cluster dynamics. Various properties of laser-cluster interaction, such as high absorption of laser energy, production of highly energetic ions from small clusters through Coulomb explosion, etc, are explained from the enhancement of laser field inside the cluster [29]. One can expect the enhancement of HHG both in the vicinity of SPR (as in the case of C_{60}, $\lambda_{SPR} \approx 60\,\text{nm}$) and far from the Mie resonances (as in the case of Ag nanoparticles, $\lambda_{SPR} \approx 430\,\text{nm}$ and $640\,\text{nm}$). Another mechanism of enhanced harmonics can be attributed to the enhanced recombination cross-section of accelerated electron with parent particle in the three-step model. In the discussed studies, a 10× to 100× enhancement in HHG for nanoparticle plasma compared to the plasma plumes produced using the same element/compound in bulk form. In the studies [27], this enhancement factor was in the range of 3–8.

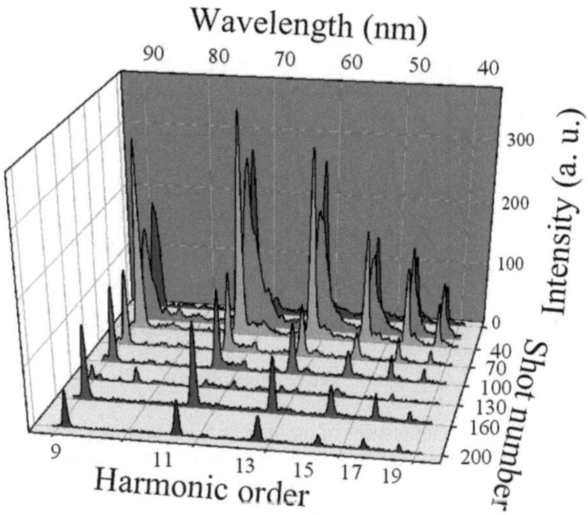

Fig. 6.5 Variation of the harmonic intensity with number of laser shots in the case of nanoparticle-containing target. The target is irradiated at same place. The HHG efficiency does not fall appreciably for ~150 shots. Reproduced from [11] with permission from IOP Publishing

The analysis of the stability of HHG from silver nanoparticles was the next set of those studies. The variation of harmonic intensity with laser shots on the same place of nanoparticle-contained target is shown in Fig. 6.5. It can be seen that there is no significant fall in the intensity of HHG up to ~150 laser shots. The stability of HHG is greatly improved in the reviewed study [11] compared to the previous reports on harmonic generation using ablated nanoparticles (for example [27, 30, 31]). Special target fabrication techniques allowed increasing the stability of the HHG from silver nanoparticle targets. The new fabrication techniques enable one to grow thicker nanoparticle targets while preserving their nanostructure. The higher thickness of targets is a dominating factor in the enhanced stability of HHG. A comparison of relative intensity of HHG in various bulk and nanoparticle targets was also made, which is shown in Fig. 6.6. It can be seen that bulk targets such as Ag and In have very small conversion efficiency for low-order harmonics compared with nanoparticles, except resonantly enhanced 13th harmonic of indium. This is an important observation, which shows that materials in their nanoparticle form give much higher harmonic intensity compared to that in bulk form. This implies that nano-size plays some crucial role in harmonic enhancement.

The harmonic yield was optimized by varying the laser and plasma parameters. During these optimizations, it was observed that the bandwidth of harmonics could be increased by optimizing the femtosecond laser intensity inside the plasma plume (Fig. 6.7). It was also reported that, as the intensity of laser pulse inside the plasma plume containing silver nanoparticles was increased from 1.8×10^{15}–$3.5 \times 10^{15}\,\mathrm{W\,cm^{-2}}$, the spectral width of HHG increased. In particular, the spectral width of 11th harmonic is increased from ~1–5 nm. It can also be seen from Fig. 6.7

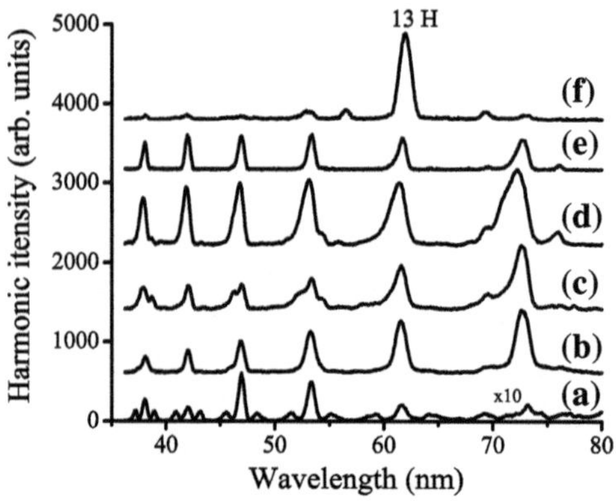

Fig. 6.6 Comparison of relative intensities of harmonics for **a** bulk Ag, **b** Ag nanoparticles, **c** C_{60} targets, **d** $SrTiO_3$ nanoparticles, **e** Au nanoparticles, and **f** bulk In. The intensity of bulk Ag harmonics is multiplied by 10 for better visibility. The intensity of harmonics from nanoparticles is of the order of the enhanced 13th harmonic from indium plasma. The spectra are sequentially shifted up for visual clarity. Reproduced from [11] with permission from IOP Publishing

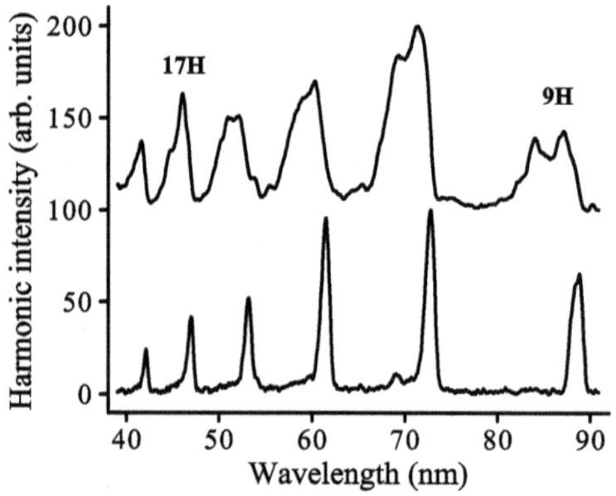

Fig. 6.7 Broadening of harmonic spectrum from nanoparticles with the increase of laser intensity. The bandwidth of 11th H changed from ~1–5 nm, when laser intensity was increased from 1.8×10^{15} to 3.5×10^{15} W cm^{-2}. The harmonic spectrum for higher laser intensity has been shifted up for visual clarity. Reproduced from [11] with permission from IOP Publishing

that the spectral broadening of harmonics occurs only towards the blue side. The spectrally broadband harmonics can be useful in spectroscopy experiments due to easier matching of harmonic wavelengths with plasma transitions. The spectral width of harmonics is related to the bandwidth of the laser. The spectral broadening in laser can be achieved through self-phase modulation effects inside the plasma. The spectral broadening in harmonics toward blue side only indicates that the blue broadening of the input laser pulse in nanoparticle-containing plasmas leads to the observed blue broadening of harmonic radiation. Since the SPM effect increases with laser intensity, the harmonic broadening is observed to increase with laser intensity.

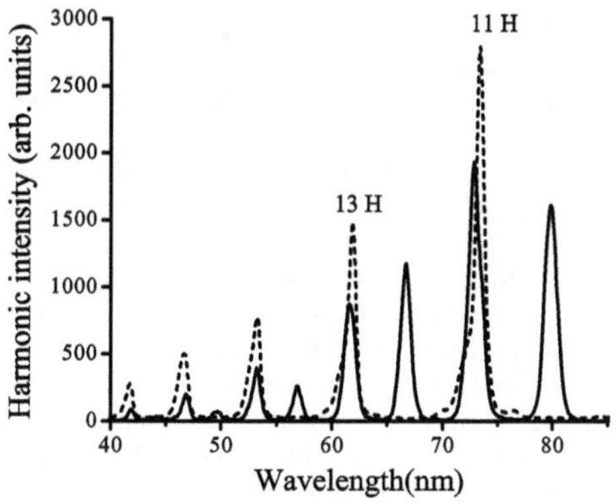

Fig. 6.8 Harmonic generation from two-color laser pulse (*solid line*) and only fundamental laser pulse (*dashed line*). Both even and odd orders of harmonics are generated from two-color laser setup. Reproduced from [11] with permission from IOP Publishing

Another objective of the reviewed study was to see if one could increase the overall conversion efficiency of HHG from Ag nanoparticles using two-color pump, as observed in gas targets. The two-color laser pulses were generated by inserting a KDP crystal in the converging laser beam between the focusing lens and plasma plume. This resulted in orthogonally polarized second harmonic pulse with a conversion efficiency of $\sim 2\,\%$. Both, even and odd harmonics were generated by the two-color pump, as seen in Fig. 6.8. The cutoff of harmonic spectrum did not increase, and no enhancement in the intensity of odd harmonics was observed with two-color laser pulses. Although the conversion efficiency of the second harmonic pump was low, the intensity of even harmonics was comparable to that of odd harmonics. The intensity of even harmonics decreased more rapidly compared to that of odd harmonics with increasing harmonic orders. Utilization of two color pump results in slight increase in overall conversion efficiency of the HHG process, since the conversion of even harmonic orders is now added with odd harmonic orders. No harmonic generation was observed by blocking the fundamental radiation and using only second harmonic pump, indicating that the intensity of second harmonic pulse by itself was not sufficient for the HHG. This study showed that even a small amount of second harmonic mixing was sufficient to break the inversion symmetry and cause both even and odd harmonic generation.

References

1. T.D. Donnelly, T. Ditmire, K. Neuman, M.D. Pery, R.W. Falcone, Phys. Rev. Lett. **76**, 2472 (1996)
2. J.W.G. Tisch, T. Ditmire, D.J. Fraser, N. Hay, M.B. Mason, E. Springate, J.P. Marangos, M.H.R. Hutchinson, J. Phys. B **30**, L709 (1997)
3. C. Vozzi, M. Nisoli, J.-P. Caumes, G. Sansone, S. Stagira, S. De Silvestri, M. Vecchiocattivi, D. Bassi, M. Pascolini, L. Poletto, P. Villoresi, G. Tondello, Appl. Phys. Lett. **86**, 111121 (2005)
4. C.-H. Pai, C.C. Kuo, M.-W. Lin, J. Wang, S.-Y. Chen, J.-Y. Lin, Opt. Lett. **31**, 984 (2006)
5. K. Wegner, P. Piseri, H.V. Tafreshi, P. Milani, J. Phys. D: Appl. Phys. **39**, R439 (2006)

6. R.A. Ganeev, Laser Phys. **18**, 1009 (2008)
7. R.A. Ganeev, L.B. Elouga Bom, J. Abdul-Hadi, M.C.H. Wong, J.P. Brichta, V.R. Bhardwaj, T. Ozaki, Phys. Rev. Lett. **102**, 013903 (2009)
8. R.A. Ganeev, Phys. Usp. **52**, 55 (2009)
9. R.A. Ganeev, J. Phys. B: At. Mol. Opt. Phys. **40**, R213 (2007)
10. H. Singhal, R.A. Ganeev, P.A. Naik, J.A. Chakera, U. Chakravarty, H.S. Vora, A.K. Srivastava, C. Mukherjee, C.P. Navathe, S.K. Deb, P.D. Gupta, Phys. Rev. A **82**, 043821 (2010)
11. H. Singhal, R.A. Ganeev, P.A. Naik, A.K. Srivastava, A. Singh, R. Chari, R.A. Khan, J.A. Chakera, P.D. Gupta, J. Phys. B: At. Mol. Opt. Phys. **43**, 025603 (2010)
12. R.A. Ganeev, L.B. Elouga Bom, M.C.H. Wong, J.-P. Brichta, V.R. Bhardwaj, P.V. Redkin, T. Ozaki, Phys. Rev. A **80**, 043808 (2009)
13. R.A. Ganeev, H. Singhal, P.A. Naik, J.A. Chakera, A.K. Srivastava, T.S. Dhami, M.P. Joshi, P.D. Gupta, J. Appl. Phys. **106**, 103103 (2009)
14. R.A. Ganeev, H. Singhal, P.A. Naik, J.A. Chakera, A.K. Srivastava, T.S. Dhami, M.P. Joshi, P.D. Gupta, Appl. Phys. B **100**, 581 (2010)
15. R.A. Ganeev, M. Baba, H. Kuroda, G.S. Boltaev, R.I. Tugushev, T. Usmanov, Eur. Phys. J. D **64**, 109 (2011)
16. R.A. Ganeev, Laser Phys. **21**, 25 (2011)
17. R.A. Ganeev, H. Singhal, P.A. Naik, J.A. Chakera, M. Tayyb, A.K. Srivastava, T.S. Dhami, M.P. Joshi, A. Singh, R. Chari, S.R. Kumbhare, R.P. Kushwaha, R.A. Khan, P.D. Gupta, Opt. Spectrosc. **108**, 787 (2010)
18. H.B. Liao, R.F. Xiao, J.S. Fu, G.K.L. Wong, Appl. Phys. B **65**, 673 (1997)
19. B. Shim, G. Hays, R. Zgadzaj, T. Ditmire, M.C. Downer, Phys. Rev. Lett. **98**, 123902 (2007)
20. S.X. Hu, Z.Z. Xu, Appl. Phys. Lett. **71**, 2605 (1997)
21. T. Tajima, Y. Kishimoto, M.C. Downer, Phys. Plasmas **6**, 3759 (1999)
22. J.W.G. Tisch, Phys. Rev. A **62**, 041802 (2000)
23. S.V. Fomichev, D.F. Zaretsky, D. Bauer, W. Becker, Phys. Rev. A **71**, 013201 (2005)
24. V. Véniard, R. Taïeb, M. Maquet, Phys. Rev. A **65**, 013202 (2001)
25. R.A. Ganeev, L.B. Elouga Bom, T. Ozaki, J. Phys. B: At. Mol. Opt. Phys. **42**, 055402 (2009)
26. M. Kundu, S. V. Popruzhenko, D. Bauer, Phys. Rev. A **76**, 033201 (2007)
27. R.A. Ganeev, M. Suzuki, M. Baba, M. Ichihara, H. Kuroda, J. Appl. Phys. **103**, 063102 (2008)
28. R.A. Ganeev, L.B. Elouga Bom, T. Ozaki, J. Appl. Phys. **106**, 023104 (2009)
29. G.S. Metraux, C.A. Mirkin, Advanced Mater. **17**, 412 (2005)
30. R.A. Ganeev, M. Suzuki, M. Baba, M. Ichihara, H. Kuroda, J. Phys. B: At. Mol. Opt. Phys. **41**, 045603 (2008)
31. R.A. Ganeev, M. Suzuki, M. Baba, M. Ichihara, H. Kuroda, J. Opt. Soc. Am. B **25**, 325 (2008)

MIX
Papier aus verantwortungsvollen Quellen
Paper from responsible sources
FSC® C105338

If you have any concerns about our products,
you can contact us on
ProductSafety@springernature.com

In case Publisher is established outside the EU,
the EU authorized representative is:
Springer Nature Customer Service Center GmbH
Europaplatz 3, 69115 Heidelberg, Germany

Printed by Libri Plureos GmbH
in Hamburg, Germany